INTRODUCTION TO

The Foundations of Mathematics

SECOND EDITION

RAYMOND L. WILDER

Dover Publications, Inc.
Mineola, New York

Bibliographical Note

This Dover edition, first published in 2012, is an unabridged republication of the 1983 printing of the work originally published in 1965 by John Wiley & Sons, Inc., New York. The First Edition was published by Wiley in 1952.

Library of Congress Cataloging-in-Publication Data

Wilder, Raymond Louis, 1896–1982.
Introduction to the foundations of mathematics : second edition / Raymond L. Wilder.
p. cm.
This Dover edition, first published in 2012, is an unabridged republication of the 1983 printing of the work originally published in 1965 by John Wiley & Sons, Inc., New York. The first edition was published by Wiley in 1952.
Includes bibliographical references and index.
ISBN-13: 978-0-486-48820-2
ISBN-10: 0-486-48820-9
1. Mathematics—Philosophy. 2. Logic, Symbolic and mathematical. I. Title.

QA9.W58 2012
510—dc23

2011052467

Manufactured in the United States by Courier Corporation
48820901
www.doverpublications.com

PREFACE
to Reprint Edition

Although this reprint is not a revision, properly speaking, typographical errors have been corrected, and additional bibliography provided. In addition, some notes have been added to familiarize the reader with events that have taken place since the 1965 edition was prepared.

I am indebted, as I assume are those who have tried in vain to obtain a copy, to Mr. Robert E. Krieger for bringing out this improved reprint. Scarcely a month goes by that I do not receive inquires asking where a copy might be obtained. Most of the material in the book is not "dated" and I presume that with new sets of problems the book could be called "revised"; but under the circumstances, I prefer to leave this aspect—provision of new problems—to the imagination and ingenuity of both teachers and students.

R. L. WILDER

Santa Barbara

Preface

In making this revision, it has been my purpose to improve the exposition, correct those errors that appear inevitable in a first edition, and augment both text and problems with material that will better achieve the original objectives of the book. Otherwise, the general plan of presentation has not been altered.

Among the changes or additions are new material on "definition" and completeness and independence of undefined terms in Chapter II; simpler proofs in Chapters III–V on the theory of sets as well as further discussion of the significance of Choice Axiom and Continuum Hypothesis in the light of the Gödel-Cohen proof of independence; greater emphasis of the genetic approach to the notion of number (cardinal, ordinal, and real); more on topology in Chapter VII; discussion of truth-table methods, predicate calculus, completeness, the Gödel and Löwenheim-Skolem theorems regarding satisfiability in the domain of natural numbers in the chapter on mathematical logic; and problems (lacking in the first edition) provided for Chapters VIII–XI.

Regarding the general intent of the book, there has been misunderstanding on the part of some readers. I wish, therefore, to emphasize that its purpose is twofold: (1) to acquaint the student, mathematical major or not, with the origin and nature of fundamental concepts of modern mathematics; and (2) to show how it became natural, and necessary, to inquire more deeply (using such tools as the axiomatic method, logical formalism, etc.) into the implications and dangers of the unrestricted use of what seem natural, necessary, or intuitively justified concepts (such as set theory, classical logic, etc.). It was by no means my intention to present a formal treatment, logically developed via theorem and proof, of axiomatic set theory or logical formalism, for example, as of the year 1965. Excellent books of the latter nature have been written during the past twelve years, and it is my hope that the student whose curiosity has been sufficiently aroused by the present book will be stimulated to read some of these more technical treatises.

For example, I have been asked if it is not an error when, after discussing the axiomatic method in Chapters I and II, I do not thereupon make use of it to develop set theory in Chapters III and IV. If this is an error, it is committed deliberately and after careful consideration. Most mathematicians now agree that geometries and algebras, for example, should be set up axiomatically. But it is by no means clear that set theory should be placed on a similar basis. The working mathematician usually wants precisely the "naive" notions regarding cardinal and ordinal numbers, Choice Axiom, etc., that are given in Chapters III–V; he is ordinarily not interested in what principles he can base them upon (in the axiomatic sense). For him, set theory is an important *method*, not a theory. And after a student becomes acquainted with the method, he may well be motivated to investigate its validity. In this respect, it is something like logic. Despite the long time that the working mathematician has used logic, he has evinced little interest in studying it for its own sake. I am still not convinced that the axiomatic framework provides the proper introduction to set theory for the average mathematician; any more than I am of the opinion that he should be introduced to logic through logical formalism.

Similar remarks can be made about the treatment of cardinal numbers in Chapter IV, ordinal numbers in Chapter V, and the real number system in Chapter VI. The natural (naive) approach is first taken, during the course of which the characterizing features (such as separability and continuity in the case of the real number continuum) are "discovered" and made available for axiomatic formulation. I suppose I might formulate my philosophy regarding this order of presentation by stating that I believe the student should get the "why" as well as the "what." I believe that it is this philosophy which has motivated much of the new curriculum for the elementary schools; traditionally the "what" was considered all that was necessary—the student was given to understand that "this is it, and you can take it or leave it." And in a tragic number of cases he was doing the latter. To present, in axiomatic form, the finished products of research into basic concepts, while providing no explanation of their genesis and evolution, comes dangerously close to providing only the "what."

For those who wanted me to go more deeply into mathematical logic, material that is relevant to the major purposes of the book has been added. To provide more than this would necessitate giving up the idea that the book is essentially designed for a one-semester course. However, in the bibliography the reader will find references to recent texts on logic, axiomatic set theory, etc., which should form entirely satisfactory sequels to the material given herein.

For encouragement and advice (not always heeded) during the process of revision, I wish to thank Professors J. Bennett, F. Harary, and K. May, as well as many students who must go unnamed. To my publisher, John Wiley and Sons, and especially Cecil Yorke, Staff Editor in the Production Division, I wish to express thanks for permitting this second edition and for expert guidance during its production.

R. L. WILDER

Ann Arbor, Michigan
March, 1965

Preface to First Edition

This book grew out of a course in Foundations of Mathematics which I have given at the University of Michigan for over twenty years. The reason for instituting the course was simply the conviction that it was not good to have teachers, actuaries, statisticians, and others who had specialized in undergraduate mathematics, and who were to base their life's work on mathematics, leave the university without some knowledge of modern mathematics and its foundations. The training of these people consisted chiefly of "classical" mathematics and its applications—that part of mathematics which is based on pre-twentieth-century and, in large part, on pre-Cantorian ideas and methods.

It seemed, too, that a course in Foundations at about the senior level might serve to unify and extend the material covered in the traditional mathematics curriculum. The "compartmentalization" of the preparatory school—arithmetic, algebra, and geometry—is usually continued in college with a further dose of algebra, followed by courses in analytic geometry and calculus in which a little unification of preceding subjects takes place, but no time is spent on the *nature* of the material or its foundation.

Also, the growing realization that mathematical logic is a new and legitimate part of mathematics made it seem advisable to institute a course which would make manifest the importance of studies in the Foundations, and the reasons for inquiring into the nature of mathematics by either the tools of logic or other methods.

To my first class in the course I owe much for inspiration and encouragement. It consisted, with one or two exceptions, of approximately thirty actuarial students, most of them first-year graduate students. Their response was surprising; for I was aware of the antagonism of many professional mathematicians to any inquiry into the nature of mathematics, especially if it leads to any questioning of the validity of time-honored principles and methods (the analogue in mathematics, perhaps, of the historical lag in the cultivation of those sciences that study man's own behavior).

Perhaps their reaction was due to the change, possibly refreshing to them, from the type of teaching which treats mathematics as a "discipline," dogmatic in character. Whatever the reason, the course seemed to "take" with them and left no question about the desirability of repeating the experiment. (Even today I occasionally meet members of that original class, many of them now insurance company executives, and find them still able to recall topics that were discussed.)

In a few years the course ceased to be an experiment and became as well established as any other course in the curriculum. This development was aided, no doubt, by the realization that the course was also fulfilling other purposes. For those students who were going on to the doctorate in mathematics, it did spadework in the ideas and methods that were going to form the principal tools in their graduate courses and research. And for mature students in other fields, such as philosophy and the social sciences, it provided an insight into basic mathematical concepts without the necessity of first wading through the traditional courses in algebra, geometry, calculus, etc.

In the belief that such a course ought, perhaps, to be offered in most universities and colleges that train mathematicians either for teaching or for any of the professional fields, I decided to incorporate the material covered in a book which would serve as a basis for such a course. Unfortunately, a book in my own special field of research took precedence and delayed my starting on the present work for at least ten years. Moreover, it seemed desirable to include in the book material that it was not possible to crowd into a one-semester course, and which has usually been suggested for collateral reading; especially material that is in languages other than English. For it continues to be true that the students in American universities are generally not prepared to read in French and German; and most of the older and basic work in Foundations was originally published in German. The material in Chapter X (Intuitionism), for instance, has heretofore been available almost entirely only in German.

In a general way, the idea of the book is similar to that which motivated J. W. Young's *Fundamental Concepts of Algebra and Geometry*, first published in 1911. In 1932 I discussed with Professor Young the desirability of a book such as this one; he agreed thoroughly that it was desirable to write it, if only to have available a book on fundamental concepts that would take into account the great strides that have been made in Foundations since the publication of his book.

As already indicated, I have given the material in the form of a course of one semester, with a calculus prerequisite. The students who have taken the course have, however, been at all levels, from undergraduate juniors to students already writing doctoral dissertations. Ideally, the course should be given at senior level or first-year graduate level. And, I have found, it is not necessary to insist on the calculus prerequisite for mature students in other fields, such as philosophy. (One of the most enthusiastic students in my experience was a medical student who was taking the course as a "cultural" subject.) No *mathematics* student, however, should take the course without having had calculus, and for the average student it is better that he have taken courses such as advanced calculus and projective geometry in order to develop the maturity requisite for abstract thinking.

Whether I have succeeded in getting down on paper a reasonable facsimile of what I have done in class only the reception of this book can tell. No false modesty prevents me from admitting the success of the course itself, as numerous past students will testify. But the enthusiasm and inspiration which come from facing a group of interested students are hard to duplicate in the seclusion of one's study, and it is difficult to recapture the many spontaneous ideas and illustrations that have revealed themselves in the classroom from time to time over the years. No doubt, now that the book is written, I shall occasionally recall some of them and regret that they are doomed to oblivion.

I have made it a general rule, incidentally, not to reveal to students my own opinion regarding a controversial point, despite frequent requests to do so. It has always seemed better to present, as emphatically as possible from the point of view of their proponents, such topics as are controversial; and frequently, in order to aggravate class discussion, it has been my custom to oppose the point of view of a student while secretly agreeing with him. There is nothing new about such methods of instruction, of course, and I bring them up here largely to afford occasion to remark that I am including my own opinions regarding the nature of mathematics in the material of Chapter XII. In the earlier chapters I have tried to follow my usual rule of acting as advocate for the view presented; and, in thus breaking my rule, I do so not only with a view to getting in my own "innings," but also to furnish additional fuel for the stirring up of controversy which, it seems to me, is the most effective stimulant to original and creative thinking. And if, after reading this book, the student is not aroused to the extent of "thinking about" mathematics, I shall have failed in one of my chief purposes in writing it.

For rapidity and exactness of reference, the decimal system of numbering sections has been employed. Cross references to items in the text are made by citing chapter and section; thus "IV 2.4" refers to Chapter IV, Section 2.4. However, for reference to a section in the chapter under consideration, only the section number is used; thus "1.2" refers to Section 1.2 of the chapter in which the citation occurs.

The Bibliography is divided into two parts, the first listing books and longer memoirs, the second listing papers and shorter articles. References to the Bibliography are enclosed in brackets, those involving capital letters such as [B], [Ha], [H_1] referring to the first part of the Bibliography, and those involving single lower-case letters such as [a], [b] referring to the second part. Page or chapter numbers will frequently be included; thus "Hilbert [H_3; 6]" will be found under Hilbert's name in the first part of the Bibliography, item [H_3] (the reference, then, is to page 6 of Hilbert's *Foundations of Geometry*). Always, "f" indicates "footnote," and "ff" indicates "and the following page or pages."

To those colleagues and students who have given me encouragement and stimulation, I wish to express sincere thanks. I am especially grateful to Professors E. T. Bell, Leon Henkin, Paul Henle, and Leo Zippin, and to Dr. C. V. Newsom for suggestions and criticisms; but the errors and shortcomings to be found herein are not their fault and are present only in spite of their wise counsel.

For aid in a material way, thanks are due to the Office of Naval Research, under whose Contract N9onr-89300 the first draft of this work was written, as well as to the California Institute of Technology, which generously afforded an office and library facilities during the academic year 1949–1950 for my writing and research.

R. L. Wilder

Ann Arbor, Michigan
September, 1952

Contents

PART ONE · FUNDAMENTAL CONCEPTS AND METHODS OF MATHEMATICS

I · The Axiomatic Method 3

1 · Evolution of the method 3
2 · Description of the method; the undefined terms and axioms 9
3 · Description of the method; the proving of theorems 13
4 · Comment on the above theorems and proofs 16
5 · Source of the axioms 18
Suggested reading 21
Problems 21

II · Analysis of the Axiomatic Method 23

1 · Consistency of an axiom system 23
2 · The proof of consistency of an axiom system 26
3 · Independence of axioms 28
4 · Completeness of an axiom system 31
5 · Independence and completeness of undefined technical terms 39
6 · Miscellaneous comment 43
7 · Axioms for simple order 45
8 · Axioms defining equivalence 48
Suggested reading 50
Problems 50

III · Theory of Sets 54

1 · Background of the theory 55
2 · The Russell contradiction 56
3 · Basic relations and operations 58

4 · Finite and infinite sets 63
5 · Relation between the ordinary infinite and the Dedekind infinite 66
6 · The Choice Axiom 72
Suggested reading 76
Problems 76

IV · Infinite Sets 80
1 · Countable sets; the number $\aleph_0$ 80
2 · Uncountable sets 87
3 · Diagonal procedures and their applications 91
4 · Cardinal numbers and their ordering 101
Suggested reading 111
Problems 111

V · Well-Ordered Sets; Ordinal Numbers 114
1 · Order types 114
2 · The order type ω 117
3 · The general well-ordered set 120
4 · The second class of ordinals 133
5 · Equivalence of Choice Axiom, Well-Ordering Theorem, and Comparability 135
Suggested reading 139
Problems 139

VI · The Linear Continuum and the Real Number System 142
1 · Analysis of the structure of the real numbers as an ordered system 144
2 · Operations in R 154
3 · The real number system as based on the Peano axioms 158
4 · The complex number system 162
Suggested reading 163
Problems 163

VII · Groups and Their Significance for the Foundations 165
1 · Groups 165
2 · Applications in algebra and to number systems 174
3 · The group notion in geometry 182
4 · Topology 188
5 · Concluding remarks 192
Suggested reading 193
Problems 193

CONTENTS

PART TWO · DEVELOPMENT OF VARIOUS VIEWPOINTS ON FOUNDATIONS

VIII · The Early Developments 199

1 · The eighteenth-century beginnings of analysis 199
2 · The nineteenth-century foundation of analysis 200
3 · The symbolizing of logic 205
4 · The reduction of mathematics to logical form 208
5 · Introduction of antinomies and paradoxes 210
6 · Zermelo's Well-Ordering Theorem 210
7 · Poincaré's views 211
8 · Zermelo's set theory 212
9 · Amendments to the Zermelo system 217
Additional bibliography 217
Problems 218

IX · The Frege-Russell Thesis: Mathematics an Extension of Logic 219

1 · The Frege-Russell thesis 219
2 · Basic symbols; propositions and propositional functions 221
3 · Calculus of propositions 225
4 · Forms of general propositions; the predicate calculus 234
5 · Classes and relations as treated in P.M. 238
6 · Concluding remarks 243
Additional bibliography 244
Problems 244

X · Intuitionism 246

1 · Basic philosophy of Intuitionism 247
2 · The natural numbers and the definition of set; spreads 248
3 · Species 251
4 · Relations between species 252
5 · Theory of cardinal numbers 252
6 · Order and ordinal numbers 256
7 · The intuitionist logic 256
8 · General remarks 260
Additional bibliography 262
Problems 262

XI · Formal Systems; Mathematical Logic 264

1 · Hilbert's "proof theory" 264
2 · Actual development of the proof theory 266

3 · Gödel's incompleteness theorem 270
4 · Consistency of a formal system 274
5 · Formal systems in general 274
6 · General significance of formal systems 277
Additional bibliography 279
Problems 279

XII · The Cultural Setting of Mathematics 281
1 · The cultural background 282
2 · The position of mathematics in the culture 283
3 · The historical position of mathematics 284
4 · The present-day position of mathematics 285
5 · What is mathematics from the cultural point of view? 286
6 · What we call "mathematics" today 290
7 · The process of mathematical change and growth 292
8 · Differences in the kind and quality of mathematics 295
9 · Mathematical existence 298

Bibliography 301

Index of Symbols 315

Index of Topics and Technical Terms 317

Index of Names 325

TEXT REVISIONS FOR 1983 REPRINT EDITION

Page	Line	Corrections
xiv	6	Replace "The Choice Axiom" by "The Axiom of Choice."
21	1	Add to Suggested Reading: Conference Board [Con].
41	-13	In the brackets, change "Ta" to T_2"
43	17	Change "which Σ' " to "which Σ "
50		Add to Suggested Reading: Addison, J.W., Henkin, L. and Tarski, A. [Ad] ; and Thompson [Th].
57	-4	Change Gr to Gr_1
72		Change the offset heading "6. the Choice Axiom" to "6. The Axiom of Choice"
		In §6, line 3, insert after the first "the" the words "Axiom of Choice"
73		Change "6.3 Choice Axiom" to "6.3 Axiom of Choice." and in line 6 of §6.3 change "Choice Axiom" to "Axiom of Choice"
		last line of second paragraph should read "principle known as the Axiom of Choice or Choice Axiom"
83	22	Change "4" to "5"
109	19	Delete ")"
110	5	Insert "h" at end of line.
125	-17	Between "in" and "B" insert "F and $b_1, b_2, \ldots, b_n$ are elements of"
126	12	$\{w'_\nu)$ should be $\{w_\nu\}$
133	2	After "Zermelo" insert "and Fraenkel"
	7	Change "Zermelo" to "Zermelo-Fraenkel"
	12	Change "Zermelo" to "Zermelo-Fraenkel"
138	-6	Change the first "y" to "m"
147	19	Change "a" to "α"
177	21	Change "8" to "28"
180	5	Change "a" to "α"
187	3	Change "notions" to "motions"
193		Add to Suggested Reading: Christie [Chr]
200	16	Place an asterisk (*) at the end of line.
		Add a new footnote as follows: *Since this book was written, there appeared a new approach to the foundation of real analysis due to Abraham Robinson [Rob]. Termed "Non-Standard Analysis" it introduces a new set-theoretic conception of real numbers and infinitesimals which differs radically from the classical foundation outlined in Chapter VI above. It permits a treatment of the calculus in which infinitesimals are used much as in the Leibnizean tradition.

Page	Line	Corrections
205	18	Place a ‡ at end of line.
		Add footnote at bottom of page as follows:
		‡ For an authoritative account of Cantor's life see Dauben [Da].
208	7	Change [St] to [a]
213		In line 2 of footnote †, delete "in minor respects."
		In same footnote, insert
		in that objects—"Urelementen"—which are not sets are not assumed.)"
217		Add to the Additional Bibliography the following:
		Dauben [Da]
		Lakatos [La]
		Monk [a]
230	21	Change (p/p)(q/q) to (p/p)/(q/q)
244		Add to the Addional Bibliography
		Cohen [Co]
		Lyndon [Ly]
259	1	Insert "valid" between "every" and "formula"
262		Add to the Additional Bibliography
		Cohen [Co]
		Lyndon [Ly]
263		In line 6 of Exercise 9, change $\neg\neg r$ to $\neg\neg p$, and in line 9 of the same exercise, change p to m and q to n.
		In line 10 of Exercise 9, change p/q to m/n
275	-16	Insert "T" at end of line (before the period).
279		Add to the Additional Bibliography
		Rosser [Ross$_1$]
		Shoenfield [Sh]
285		At end of line 23, place an asterisk (*). And at the bottom of the page insert a new footnote:
		*The English have now adopted the metric system, and it is, of course, common to the United States industrial system. Due probably to cultural lag and resistance, the United States public is slow to adopt it.
293	-2	Delete the word "not"
298	-5	Replace "Choice Axiom" by "Axiom of Choice"
299		Replace the last line of footnote ‡ by
		the author's book Evolution of Mathematical Concepts (Wilder [Wi]) and the forthcoming Mathematics as a Cultural System (Wilder [Wi$_1$])
303		To Hardy, G.H. [Ha] add: "reprinted with a Foreword by C. P. Snow, 1967."
306		Substitute the following for the "Wilder" item:
		[Wi] Evolution of Mathematical Concepts, New York, Wiley, 1968; in paper back, New York, Halsted Pr., 1975 and England, The Open University Pr., 1974.
		[Wi$_1$] Mathematics as a Cultural System, Oxford, Pergamon Pr., forthcoming.

Page	Line	Corrections
308		Insert between the "Chwistek" and "Curry" items the following: Cohen, P.J. [a] "The independence of the continuum hypothesis," Proc. Nat. Acad. Scie., vol. 50 (1963), pp . 1143-1148. [b] Ibid, II, vol. 51 (1964), pp. 105-110.
310		"Merton, R. R." should be "Merton, R. K."
312-313		The first three lines of page 313 should be inserted between lines -4 and -3 on p. 312; in other words, the entire "Weyl" reference should be between the "Weinlos" and "Wiener" references.
318		Shift "Commutative laws..." up so as to precede "Comparability"
326		"Grassman" should be "Grassmann"

PART ONE

Fundamental Concepts and Methods of Mathematics

I

The Axiomatic Method

Since the axiomatic method as it is now understood and practiced by mathematicians is the result of a long evolution in human thought, our discussion of it shall be preceded by a brief description of some older uses of the term *axiom*. The modern usage of the term represents a high degree of maturity, and a better understanding of it may be achieved by some acquaintance with the course of its evolution.

1 Evolution of the method

If the reader has at hand a copy of an elementary plane geometry, of a type frequently used in high schools, he may find two groupings of fundamental assumptions, one entitled "Axioms," the other entitled "Postulates." The intent of this grouping may be explained by such accompanying remarks as: "An *axiom* is a *self-evident truth*." "A *postulate* is a *geometrical fact* so simple and obvious that its validity may be assumed." The "axioms" themselves may contain such statements as: "The whole is greater than any of its parts." "The whole is the sum of its parts." "Things equal to the same thing are equal to one another." "Equals added to equals yield equals." It will be noted that such geometric terms as "point" or "line" do not occur in these statements; in some sense the axioms are intended to transcend geometry—to be "universal truths." In contrast, the "postulates" probably contain such statements as: "Through two distinct points one and only one straight line can be drawn." "A line can be extended indefinitely." "If L is a line and P is a point not on L, then through P there can be drawn one and only one line parallel to L." (Some so-called "definitions" of terms usually precede these statements.)

This grouping into "axioms" and "postulates" has its roots in antiquity. Thus we find in Aristotle (384–321 B.C.) the following viewpoint:†

"Every demonstrative science must start from indemonstrable principles; otherwise, the steps of demonstration would be endless. Of these

† As summarized by T. L. Heath [Hea; I, 119]; quoted by permission of Cambridge University Press. The reader is referred to this book for citations from Aristotle, Proclus, *et al.*

indemonstrable principles some are (a) common to all sciences, others are (b) particular, or peculiar to the particular science; (a) the common principles are the *axioms*, most commonly illustrated by the axiom that, if equals be subtracted from equals, the remainders are equal. In (b) we have first the *genus* or subject-matter, the *existence* of which must be assumed."

1.1 In Euclid's *Elements* (written about 300 B.C.), the two groups occur, respectively labeled "Common notions" and "Postulates." From these and a collection of definitions, Euclid deduced 465 propositions in a logical chain. Although the actual background for Euclid's work is not clear, apparently he did not originate this method of deducing logically from certain unproved propositions, given at the start, all the remaining propositions. As we have just noted, Aristotle, and probably other scholars of the period, had a well-conceived notion of the nature of a demonstrative science; and the logical deduction of mathematical propositions was common in Plato's Academy and perhaps among the Pythagoreans.

Since the origin of the axiomatic method is not known, we can only speculate regarding the reasons for its development. The early Greek philosophers developed forms of deductive reasoning, and were no doubt astute enough to realize that a deductive system must start with some kind of initial premises. Moreover, the crises in Greek mathematics attendant upon the discovery of irrationals and the paradoxes of Zeno may well have spurred a search for a secure foundation for geometry that could have culminated in the form of the axiomatic method found in Euclid.

The influence of Euclid's work has been tremendous; probably no other document has had a greater influence on scientific thought. For example, modern high school geometries have usually been modeled after Euclid's famous work, thus explaining the still common grouping into "axioms" and "postulates." Also the use in non-mathematical writings of such phrases as "It is axiomatic that . . . ," and "It is a fundamental postulate of . . . ," in the sense of something being "universal" or beyond opposition, is explained by this traditional use of the terms in mathematics.

The method featured in Euclid's work was employed by Archimedes (287–212 B.C.) in his two books which provided a foundation for the science of theoretical mechanics; in Book I of this treatise Archimedes proved 15 propositions from 7 postulates. Newton's famous *Principia*, first published in 1686, is organized as a deductive system in which the well-known laws of motion appear as unproved propositions, or postulates, given at the start. The treatment of analytic mechanics published by Lagrange in 1788 has been considered a masterpiece of logical perfection, moving from explicitly stated primary propositions to the other propositions of the system.

1.2 There exists a large literature devoted to the discussion of the *nature* of axioms and postulates and their philosophical background. Most of this is influenced by the fact that only within comparatively recent years have axioms and postulates been very generally employed in parts of mathematics other than geometry. Even though the method popularized by Euclid is acknowledged now as a fundamental part of the scientific method in every realm of human endeavor, our modern understanding of axioms and postulates, as well as our comprehension of deductive methods in general, has resulted to a great extent from studies in the field of geometry. And since geometry was conceived to be an attempt to describe the actual physical space in which we live, there arose a conviction that axioms and postulates possessed a character of *logical necessity*. For example, Euclid's fifth postulate (the "parallel postulate") was "Let the following be postulated that, if a straight line falling on two straight lines make the interior angles on the same side less than two right angles, the two straight lines, if produced indefinitely, meet on that side on which are the angles less than the two right angles."† Proclus (A.D. 410–485) described vividly in his writings the controversy that was taking place in connection with this postulate even in his time; in fact, he argued in favor of the elimination "from our body of doctrine this merely plausible and unreasoned statement."† With the renewal of interest in Greek learning during the Renaissance, controversy in regard to the fifth postulate was renewed. Attempts were made to prove the "parallel postulate," often from logical—non-geometrical—principles alone. Surely, if a statement is a "logical necessity" the assumption of its invalidity should lead to contradiction—such was the motivation of much of the work on the postulates of geometry. With the invention of non-euclidean geometries the futility of such attempts became clear.

1.3 The development of the non-euclidean geometries was evidence of a growing recognition of the independent nature of the fifth postulate; that is, this postulate cannot be demonstrated as a logical consequence of the other axioms and postulates in the euclidean system. By a suitable replacement of the fifth postulate, we may obtain the alternative and logically consistent geometry of Bolyai, Lobachevski, and Gauss in which the fifth postulate of Euclid fails to hold. In it appears, for example, the proposition that the sum of the interior angles of a triangle is less than two right angles. Riemann in 1854 developed another non-euclidean

† Quoted from T. L. Heath [Hea; I, 154–155, 203], by permission of Cambridge University Press.

geometry, likewise composed of a non-contradictory collection of propositions, in which all lines are of finite length and the sum of the interior angles of a triangle is greater than two right angles.

The invention of the non-euclidean geometries was only part of the rapidly moving developments of the nineteenth century that were to lead to the acceptance of formal geometries apart from those that might be regarded as constituting definitive sciences of space or extension. Grassmann's *Ausdehnungslehre*, published in 1844 and a critical landmark during this era of changing ideas, was described by its author in these terms: "My *Ausdehnungslehre* is the abstract foundation for the doctrine of space, i.e., it is free from all spatial intuition, and is a purely mathematical discipline whose application to space yields the science of space. This latter science, since it refers to something given in nature (i.e., space), is no branch of mathematics, but is an application of mathematics to nature."† In explanation of Grassmann's concept of a formal science, Nagel [a; 169] writes: "Formal sciences are characterized by the fact that their sole principles of procedure are the rules of logic as well as by the further fact that their theorems are not 'about' some phase of the existing world but are 'about' whatever is *postulated* by thought."†

1.4 The idea expressed by Grassmann is essentially the one held at the present time; that is, a mathematical system called "geometry" is not necessarily a description of actual space. We must distinguish, of course, between the origin of a theory and the form to which it evolves. Geometry, like arithmetic, originated in things "practical," but to assert that any particular type of geometry is a description of physical space is to make a *physical* assertion, not a mathematical statement. In short, the modern viewpoint is that we must distinguish between mathematics and *applications* of mathematics.

A natural consequence of this change in viewpoint on the significance of a mathematical system was a re-examination of the nature of the basic, unproved propositions. It became clear, for instance, that the euclidean "common notion" that "the whole is greater than the part" has no more of an absolute character than the "parallel postulate" but is contingent upon the meaning of "greater than"; in fact, the proposition may even fail to hold, as in the theory of the infinite (cf. Chapter IV). Although there was much discussion as to whether the parallel postulate should be listed as a "postulate" or as a "common notion" (axiom), it was finally realized that neither had any more universality than the other and the distinction

† As quoted by E. Nagel [a; 172, 169]; used here by permission of E. Nagel and *Osiris*.

might as well be deleted.† Accordingly we find in the classical work of Hilbert on the foundations of geometry,‡ published in 1899, that only one name, "axioms," is applied to the fundamental statements or assumptions, and that certain basic terms such as "point" and "line" are left completely undefined. To be sure, Hilbert made a grouping of his axioms—into five groups—but this pertained only to the technical character of the statements, and not to their relative status of "trueness" or "commonness."

1.5 Although this work of Hilbert has come to be regarded by many as the first to display the axiomatic method in its modern form, it should be recognized that similar ideas were appearing in works of his contemporaries.

In 1882 appeared the first edition of *Vorlesungen über Neuere Geometrie* by M. Pasch [Pa]. Pasch based his treatment of geometry on a small number of so-called "nuclear" concepts and propositions, which are introduced respectively without definition and without demonstration, but which he believes have a common basis of acceptance and understanding in man's experience. After the basic system of propositions (axioms) has been introduced, the logical deduction of the remaining propositions of the system would probably be regarded as quite "rigorous" by modern mathematical scholars. Pasch's ideas in this regard were described by him as follows:§

"Indeed, if geometry is to be deductive, the deduction must everywhere be independent of the *meaning* of geometrical concepts, just as it must be independent of the diagrams; only the *relations* specified in the propositions and definitions employed may legitimately be taken into account. During the deduction it is useful and legitimate, but in *no* way necessary, to think of the meanings of the terms; in fact, if it is necessary to do so, the inadequacy of the proof is made manifest. If, however, a theorem is rigorously derived from a set of propositions—the *basic* set—the deduction has a value which goes beyond its original purpose. For if, on replacing the geometric terms in the basic set of propositions by certain other terms, true propositions are obtained, then corresponding replacements may be made in the theorem; in this way one obtains a new theorem as a consequence of the altered basic propositions without having to repeat the proof."

It is probable that the organization and the logical perfection of the work of Pasch were influential in the thinking of Peano. However, of the

† For an excellent non-technical description of this "revolution" in thought, see E. T. Bell [B_2; XIV].

‡ Hilbert [H_1, H_3]; the work [H_3] is in English, but the seventh edition [H_2], published in 1930, incorporates some of the improvements found by various colleagues of Hilbert in the interval between the first and seventh editions.

§ [Pa; 98–99.] The translation given here is from Nagel [a; 197], with slight change of wording; used here by permission of E. Nagel and *Osiris*.

work of this imaginative Italian, Nagel (*loc. cit.*) writes: "No phase of Pasch's empiricism is reflected in Peano's version, and pure geometry in his hands became a calculus operating upon variables formally stipulated to be related to one another in certain ways."† Peano's *Principii di Geometria* [P_1], published in 1889, treated the geometric elements as mere "things," and was a demonstration of the idea, which Peano emphasized, that a relatively small number of *undefined* terms can be used in defining all the other terms that occur in geometry; he insisted that there should be as few undefined terms (mere "symbols") as possible. Peano's work employed the language of mathematical logic, of which he was one of the originators, and was based essentially upon an undefined entity called a "point" and an undefined relation of "betweenness." (The concept of betweenness was fundamental in several of the well-conceived organizations of geometry that were constructed during the latter part of the nineteenth century; it was employed by both Pasch and Hilbert.) In later work on the foundations of geometry, published in 1894, Peano [a] discussed *independence* of axioms.

1.6 Such studies as those of Pasch, Peano, Hilbert, and Pieri in euclidean geometry provided a tremendous impetus for investigations of possible formal organizations of the subject matter of this old discipline; these considerations, in turn, provided new understanding of mathematical systems in general and were partly responsible for the remarkable mathematical advances of the twentieth century. Among the other important modern organizations of euclidean geometry may be mentioned the following: Pieri, a member of Peano's school in Italy, published, in the same year (1899) in which Hilbert's book appeared, a treatment of the subject in which he based his development upon an aggregate of "points" and an undefined concept of "motion."‡ In 1904, Veblen [a] proposed an organization of euclidean geometry in which the notion of betweenness, as used by Peano and Hilbert, was replaced by an order relation.§ As a result of suggestions made by R. L. Moore, a revision of Veblen's system appeared in 1911.‖ (A rather interesting combination of the axiomatic systems of Hilbert and Veblen was employed by Robinson [Ro] in a treatment of euclidean geometry.)

It is noteworthy that these early studies in the field of geometry were revealing the great generality that was inherent in formal mathematical systems. Mathematics was evolving in a direction that was to compel the development of a method which could *encompass in a single framework* of

† From Nagel [a; 199]; quoted by permission of E. Nagel and *Osiris*.

‡ For a discussion of Pieri's system see Young [Y; Lecture XV].

§ See, however, II 5.2.

‖ See J. W. A. Young [Yo; 3–51].

undefined terms and basic statements concepts like *group* and *abstract space* that were appearing in seemingly unrelated branches of mathematics. As will be pointed out later, the economy of effort so achieved is one of the characteristic features of modern mathematics.

2 Description of the method; the undefined terms and axioms

As commonly used in mathematics today, the axiomatic method consists in setting forth certain basic statements about the concept (such as the geometry of the plane) to be studied, using certain undefined technical terms as well as the terms of classical logic. Usually no description of the meanings of the logical terms is given, and no rules are stated about how to use them or the methods allowable for proving theorems.†‡ The basic statements are called *axioms* (or, synonymously, *postulates*). It is usually assumed that in proving theorems from the axioms the rules of classical logic regarding contradictions and "excluded middle" (II 2.1) may be employed; hence the *reductio ad absurdum* type of proof is in common use. The statements of both the axioms and the theorems proved from them are said to be *implied by* or *deduced from* the axioms. An example might be instructive.

2.1 Let us consider again the subject of plane geometry. It will be unnecessary to recall many details. However, we hope the reader recalls from his high school course that *points* and *straight lines*, and such notions as that of *parallel* lines, were fundamental. Now, if we were going to set forth an axiomatic system for plane geometry in rigorous modern form, we would first select certain basic terms that we would leave undefined; perhaps "point" and "line" would be included here (the adjective "straight" can be omitted, since the *undefined* character of the term "line" enables us to choose to *mean* "straight line" in our thinking as well as in the later selection of statements for the axioms). Next, we would scan the propositions of geometry and try to select certain basic ones with an eye to both their simplicity and their adequacy for proving the ones not selected; these we would call our *primary propositions* or *axioms*, to be left unproved in our system.

† In modern textbooks of geometry for elementary school use there are of course directions to the student as to how to go about proving theorems; but such books do not generally, on the other hand (and for obvious reasons), set forth the subject in rigorous axiomatic form.

‡ We are not here describing the method as used in modern mathematical logic or the formalistic treatises of Hilbert and his followers, where the rules for operations with the basic symbols and formulas are (of necessity) set forth in the language of ordinary discourse. See Chapters IX and XI.

2.2 To be more explicit, let us proceed as though we were actually carrying out the above procedure; although we do not intend to give a complete system of axioms (cf. II 4), a miniature sample of what the axioms and *secondary propositions* or *theorems* might be like, together with sample proofs of the latter, follows:

Undefined terms: Point; line.

Axiom 1. Every line is a collection of points.

Axiom 2. There exist at least two points.

Axiom 3. If p and q are distinct points, then there exists one and only one line containing p and q.

Axiom 4. If L is a line, then there exists a point not on L.

Axiom 5. If L is a line, and p is a point not on L, then there exists one and only one line containing p that is parallel to L.

These axioms would not by any means suffice as a basis for proof of all the theorems of plane geometry, but they will be sufficient to prove a certain number of the theorems found in any organization of plane geometry. Their selection is motivated as follows: In the first place, the undefined terms "point" and "line" are to play a role like that of the variables in algebra. Thus, in the expression

$$x^2 - y^2 = (x - y)(x + y)$$

the x and y are undefined, in the sense that they may represent any individual numbers in a certain domain of numbers (as, for instance, the domain of ordinary integers). In the present instance, "point" may be any individual in a domain sufficiently delimited as to satisfy the statements set forth in the axioms. On the other hand, "line," as indicated in Axiom 1, has a range of values (= meanings) limited to certain collections of the individuals that are selected as "points." Thus Axiom 1 is designed to set up a *relationship* between the undefined entities *point* and *line*. It is *not* a *definition* of line, since (if the study is carried through) there will be other collections of points (circles, triangles, etc.) that are not lines. Furthermore, it enables us, as we shall see presently, to define certain terms needed in the statements of the later axioms. Axiom 2 is the first step toward introducing lines into our geometry, and this is actually accomplished by adding Axiom 3. Before the latter can have meaning, however, we need the following formal definition:

2.3 Definition. If a point p is an element of the collection of points which constitutes a line L (cf. Axiom 1), then we say, variously, that L *contains* p, p is *on* L, or L is a line containing p.

Having stated Axioms 2 and 3, we would have that there exists a line in our geometry; but in order to have *plane* geometry and not merely a *line* or "one-dimensional" geometry, we would have to say something to insure that not all points lie on a single line. Axiom 4 is designed to accomplish this. We would now imagine, intuitively (since we have a line L, a point p not on that line, and also a line containing p and each point q of L), that we have practically a plane; however, so far as euclidean geometry is concerned, we have not provided, in Axioms 1–4, for the *parallel* to L through p until we have stated Axiom 5. And of course Axiom 5 is not significant until we have the definition:

2.4 Definition. Two lines L_1 and L_2 are called *parallel* if there is no point which is on both L_1 and L_2. (We may also call L_1 "parallel to" L_2, or conversely.)

2.5 Let us denote the above set of five axioms, together with the undefined terms point, line, by Γ and call it the *axiom system* Γ.

(We shall also frequently use the term "axiom system" in a broader sense to include the theorems, etc., implied by the axioms.)

For future purposes we note two aspects of Γ, but we shall not go into these fully at this point: (1) In addition to the *geometrical* ("technical") undefined terms *point*, *line*, we have used certain terms such as *collection*, *there exist*, *one*, *every*, and *not*, which may be called *universal* (sometimes "logical") in that their meanings are tacitly assumed to be fixed and universally understood (compare Aristotle's "common to all sciences," in Section 1). (2) That Γ is far from being a set of axioms adequate for plane geometry may be shown as follows: Since *point* and *line* are left undefined, we are at liberty to consider possible meanings for them, subject, of course, to the restriction that we take into account the statements made in the axioms. If we have been educated in the American or English school systems, our reactions to these terms will no doubt immediately be specialized, our geometric experience in the schools having the upper hand in our response. But let us imagine that the terms are entirely unfamiliar, although the universal terms used in the axioms are not unfamiliar, so that we may consider other possible meanings for point and line. Unquestionably this will involve considerable experimentation before suitable meanings are found. For example, we might first try letting "point" mean book and "line" mean library; we know from the statement in Axiom 1 that a line is a collection of points, and libraries form one of the most familiar collections in our daily experience. We can imagine that we live in a city, C, which has two distinct libraries, and that by library we mean either one of the libraries of C, and by book any one of the books in these two libraries. Axiom 2 becomes a valid statement: "There exist at least two books." However, Axiom 3 fails, since, if p and q designate books in different libraries, there is no library that contains p

and q. However, before trying other meanings for point and line, we notice that Axioms 4 and 5 are valid, becoming, respectively, "If *L* is a library, then there exists a book not on (i.e., in) *L*"; and, "If *L* is a library and *p* a book not on (i.e., in) *L*, then there exists one and only one library containing *p* that is parallel to (has no books in common with) *L*."†

Now, impressed by our failure to satisfy Axiom 3 on our first attempt at meanings for "point" and "line," we may, with an eye on Axiom 3, try to imagine a community, which we denote by Z, of people in which everyone belongs to some club, but in such a manner that, if *p* and *q* are two persons in Z, then there is one and only one club of which *p* and *q* are both members. In other words, we may try letting "point" mean a *person in Z* and "line" mean a *club in Z*, and imagine that the club situation in Z is such that the statement just made is valid, so that Axiom 3 is satisfied. We would then have no difficulty in seeing that Axioms 1, 2, and 4 are satisfied: "A club in Z is a collection of people in Z"; "There exist at least two people in Z"; etc. However, Axiom 5 becomes (with suitable change of wording to express the new meanings): "If *L* is a club in Z, and *p* is a person in Z not in the club *L*, then there exists one and only one club in Z of which *p* is a member and which has no members in common with *L*." This is a statement which apparently makes a rather strong convention regarding the club situation in Z, and which may conceivably fail to apply; in any case, the stipulation that only one club have a given pair of persons as members can hardly be expected to suffice for Axiom 5. To clinch the matter, suppose that Z is a "ghost" community, there being only three persons, whom we shall designate by *a*, *b*, and *c*, respectively, living in Z; and that as a result of certain circumstances each pair *ab*, *bc*, and *ac* shares a secret from the third member of the community, so that we may consider this bond between each two as forming them into a club ("secret society") excluding the third member. Now, with the meaning of point and line as before, we see that Axioms 1–4 hold but Axiom 5 does not hold.

Before rejecting the latter attempt as impossible, however, let us imagine that Z has *four* citizens: *a*, *b*, *c*, and *d*. And suppose that *each pair* of these people forms a club excluding the other two members of the community; that is, there are six clubs consisting of *ab*, *ac*, *ad*, *bc*, *bd*, and *cd*. Now *all* the axioms of Γ are satisfied with the meanings *person in Z* for "point" and *club in Z* for "line." And we may then notice that we could arrive at a similar example by taking *any* collection Z of *four* things *a*, *b*, *c*, and *d*, and, by letting "point" mean a member or *element* of the collection Z, and "line" mean any pair of elements of Z, satisfy the statements embodied in the axioms of Γ.

† In parentheses we have placed the terms commonly employed in connection with libraries and books that are indicated by our definition of "on" and "parallel."

2.6 Although we may experience no particular thrill at this discovery—may, rather, begin to feel that it is a rather trivial game we are playing in toying with possible meanings for the system Γ—we might conceivably be beguiled into seeking an answer to questions such as: How many "points" must a collection have in order to serve as the basis for an example satisfying the statements in Γ? For a given collection at hand, how many "points" must a "line" have in order to satisfy Γ? (For example, a "line" above could not have consisted of *three* persons in Z in the case where Z has exactly four citizens.) Furthermore, if we have already a general knowledge of, or experience with, plane geometry, the above example shows us that Γ is far from being a sufficient basis for euclidean geometry; certainly an adequate set of axioms for plane geometry would exclude the possibility of the geometry permitting a set of only *four* points satisfying all the axioms.

Before proceeding any further with this general discussion, however, let us notice how theorems would be proved from such a system as Γ.

3 Description of the method; the proving of theorems

Having set down a system such as Γ, for instance, we then proceed to see what statements are *implied by*, or can be *proved* or *deduced from*, the system. Contrary to the manner in which we proceeded in high school, when we brought in all kinds of propositions and assumptions not included in the fundamental terms and axioms (such as "breadth"; "a line has no breadth"), and even drew diagrams and pictures embodying properties that we promptly accepted as part of our equipment,† we take care to use only points and lines, and those relations and properties of points and lines that are given in the axioms. (Of course, after we have proved a statement, we may use it in later proofs instead of going back to the axioms and proving it all over again.) There is no objection to drawing diagrams, provided they are used only to aid in the reasoning process and do not trick us into making assumptions not implied by the axioms; indeed, the professional mathematician uses them constantly. The reader might recall here the quotation from Pasch in Subsection 1.5.

3.1 Consider the following formal theorem and proof:

Theorem 1. *Every point is on at least two distinct lines.*

Proof. Consider any point p. Since by Axiom 2 there exist at least two points, there must exist a point q distinct from p. And by Axiom 3

† A classical example may be found in the well-known "proof" that all triangles are isosceles, which is based on a diagram that deceives the eye by placing a certain point *within* an angle instead of outside, where rigorous reasoning about the situation would place it. This may be found in J. W. Young [Y; 143–145].

there exists a line L containing p and q. Furthermore, by Axiom 4 there exists a point r not on L, and (again by Axiom 3) a line K containing p and r.

Now by Axiom 1 every line is a collection of points. Hence, for two lines to be distinct (i.e., different), the two collections which constitute them must be different; or, what amounts to the same thing, one of them must contain a point that is not on the other. The lines L and K are distinct, then, because K contains the point r which is not on L. As p is on both L and K, the theorem is proved.

3.2 Now it will be noticed that we have used Axioms 1–4 in the proof, but not Axiom 5. We could, then, go back to the example of the community Z, let "point" mean *person in* Z and "line" mean *pair of persons in* Z, rephrase Axioms 1–4 in these terms, and carry through the proof of Theorem 1 in these terms. That is, Theorem 1 is a "true" statement about any example, such as Z, which satisfies the statements embodied in Axioms 1–4 of Γ. In proving Theorem 1, then, *we have in one step proved many different statements about many different examples*, namely, the statements corresponding to Theorem 1 as they appear in the different examples that satisfy Axioms 1–4 of Γ. This aspect of the "economy" achieved in using the axiomatic method will be further dwelt upon later (II 4.10). If, because of some diagram or other aid to thought, we had used some property of point or line not stated in Axioms 1–4, we could not expect to make the above assertions, and the "economy" cited would be lost.

Note, too, that Theorem 1 will remain valid in any axiom system (such as Γ) that *contains* the undefined terms *point* and *line* as well as Axioms 1–4. In particular, it is valid for euclidean plane geometry, which is only one of the possible geometries embodying these four axioms, and which, as we stated before, would require many more axioms than those stated above.

3.3 Now consider the following statement, which we call a *corollary* of Theorem 1:†

Corollary. *Every line contains at least one point.*

3.4 Before considering a proof (see 3.5), we hasten to meet an objection which the "uninitiated" might make at this point; to wit, since Axiom 1 explicitly states that a line is a collection of points, *of course* every line contains at least one point, so why should this be repeated as a corollary of Theorem 1? This is not a trivial matter, and it leads directly to what is meant by *collection*—a question which causes considerable concern in modern mathematics. We said above that "collection" is an undefined

† We assume the reader recalls that this indicates that the statement so designated follows almost directly from Theorem 1.

universal term, and as such we took it for granted that its use is generally understood and employed, just as the word "the" is universally understood and used by anyone familiar with the English language. But now we find ourselves almost immediately in need of explaining the use of the term in the above corollary (see also III 1–3).

However, there is nothing so very astonishing about this if we reflect that, whenever we try to make very precise a term in ordinary use, it is usually necessary to adopt certain conventions. For example, such terms as *vegetable*, *fruit*, *animal* are commonly "understood" and used by anyone who habitually uses the English language; but, when we come to apply them to certain special objects, it is frequently necessary to agree on some convention such as, for example, that a certain type of living substance shall be called "mammal" rather than "fish" (e.g., *whale*). So, for instance, we may want to make the convention that, if person A wishes to talk about "the collection of all coins in B's pockets," he may do so even though person B is literally penniless! In other words, no matter whether there actually are coins in B's pockets or not, the collection of all such coins is to be regarded as an existing entity; we call the collection *empty* if B has no coins. (In case B is penniless, we may also talk about "the coin in B's pocket," but in this case there is no existing entity to which the phrase refers.) And this is the convention that is generally agreed on throughout mathematics and logic, namely, that a collection may "exist," as in the case of the collection of all coins in B's pockets, even though it is empty.

When we come to mention an *algebra* of sets, we shall see another reason for the convention regarding empty collections, analogous to the reason for the introduction of a number *zero* in arithmetic (cf. III 3.4).

3.5 Proof of corollary to Theorem 1. There exists a point p by Axiom 2, and by Theorem 1 there exist two distinct lines L_1 and L_2 containing p.

Now, if there exists a line L that contains no points, then both L_1 and L_2 are parallel to L (by definition). As this would stand in contradiction to Axiom 5, it follows that there cannot exist such a line L.

3.6 A statement "stronger" than the above corollary is embodied in the next theorem:

Theorem 2. *Every line contains at least two points.*

Proof. Let L be any line. We shall show that L contains at least two points. By the above corollary, L contains a point p, and by Theorem 1 there is another line K containing p. Either L or K must contain a point q distinct from p, else they are the same collection of points and hence the same line (Axiom 1). If q is on L, the proof is complete. Suppose q is on K. By Axiom 4 there is a point x which is not on K, and by Axiom 5 there is a line M containing x and parallel to K. The lines L and M must have a common point y, else L and K would be two lines containing p and parallel to M in violation of Axiom 5. Since p is not on M, the point y is distinct from p and L contains at least the two points p and y.

Now, since by Theorem 2 every line contains at least two points, and since by Axiom 3 two given points can lie simultaneously on only one line, we can state:

Corollary (to Theorem 2). *Every line is completely determined by any two of its points that are distinct.*

3.7 Theorem 3. *There exist at least four distinct points.*

Proof. By Axiom 2 there exist at least two distinct points p and q. By Axiom 3 there exists a line L containing p and q, and by Axiom 4 there exists a point x not on L. By Axiom 5 there exists a line L_1 containing x and parallel to L, and by Theorem 2 L_1 contains at least two distinct points (cf. Definition 2.4).

3.8 Theorem 4. *There exist at least six distinct lines.*

Before proving Theorem 4, we perhaps need to make sure that the meaning of another one of our "common" terms is agreed upon, namely the word "distinct." As we are using the term, two collections are distinct if they are *not the same*. Thus the lines L and K, which figure in the proof of Theorem 2, are distinct although, until shown otherwise, K might actually *contain* L, for they are *not the same* line (K contains q and L does not).

Proof of Theorem 4. We proceed, as in the proof of Theorem 3, to obtain the line L containing the points p and q, and the line L_1 parallel to L containing two distinct points (Theorem 2) x and y. By Axiom 3 there exist lines K and K_1 determined respectively by the pairs (p, x), (q, y). Now the point q is not on K, else by Axiom 3 K and L would be the same line (which is impossible since x is not on L). Also, y is not on K, else K and L_1 would be the same line. Similarly, p is not on K_1 and x is not on K_1. Now there also exist lines M and M_1 determined respectively by the pairs (p, y), (q, x); and we can show that q is not on M, x is not on M, p is not on M_1, and y is not on M_1. It follows that no two of the lines L, L_1, K, K_1, M, M_1 are the same.

4 Comment on the above theorems and proofs

If the reader has followed the proofs given above, he has probably resorted to the use of figures by this time! This would be quite natural, since in high school geometry he used figures; and they help to keep the various symbols ($L, p, q, \cdots$) and their significance in mind. However, as we stated above, no special meanings have been assigned to "point" and "line," and consequently the above proofs should, and do, hold just as well if the reader uses coins for "points" and pairs of coins for "lines." As a matter of fact, if *any* collection of four objects is employed, and "point" means any object of the collection and "line" any pair of the objects, then the reader may follow the above proofs with these meanings in mind.

Of course, the theorems we have stated in the preceding sections are not by any means all the theorems that we might state. For example, we can show that any collection of objects satisfying the axioms of the system Γ must, if not infinite as in ordinary geometry, satisfy certain conditions regarding the number of points (there cannot be just 5 points in the collection, for instance), and that there must be a relation between the number of lines and the number of points in the collection. (See Problems 10–14 at the end of the chapter.) In fact, we can continue the above study to a surprising extent; we could hardly expect to reach a point where we could confidently assert that no more theorems could be proved. It is not our intention to extend the number of theorems, however, since we believe that we have already obtained enough theorems and proofs to serve as specimens for our later purposes.

4.1 As a useful terminology in what follows, let us agree that, when we use the term "statement" in connection with an axiom system Σ, we shall mean a sentence phrased, or phrasable, in the undefined terms and universal terms of Σ; such a statement may be called a *Σ-statement*. Thus the axioms of Γ are Γ-statements (Axiom 5 contains the word "parallel," but this is "phrasable" in the undefined terms and universal terms), as are also the theorems.

4.2 In conformity with the conventions made in Section 2, we shall say that an axiom system Σ *implies* a statement S if S follows by logical argument, such as used above, from Σ. In particular, each axiom is itself implied, trivially. We shall also say that S is *logically deducible* from Σ if Σ implies S.

4.3 In the course of our work above we had to pause in two instances to explain the conventions we were making in regard to the use of two words commonly used in ordinary discourse, namely, "collection" and "distinct." These words were left undefined, to be sure, in the sense that they are supposedly universally understood non-technical terms; but, as we discovered, not so "universal" but that it was felt advisable to give some conventions we were making in regard to their use here. On the other hand, the words "point" and "line" we left strictly undefined, saying that any meaning whatsoever could be assigned to them as long as these meanings were consistent with the statements embodied in the axioms. We saw that the "collection = library," "point = book" meanings were not permissible, but that if C is any collection of four objects, then "point = object of C," "line = pair of objects of C" are permissible meanings. The terms "point," "line," "parallel," etc., we may call *technical* terms of the system, the terms "point" and "line" being the *undefined technical terms* (often called *primitive terms*). The terms "collection," "distinct" are *universal terms* of the type mentioned in 2.5.

Other examples of universal terms in Γ are "exist" (in Axiom 2), "one" (Axiom 3), "two" (Theorem 1), "four" (Theorem 3), "six" (Theorem 4), "and" (Axiom 3), "or" (Definition 2.3), "not" (Axiom 4), and "every" (Theorem 1). However, if we were setting up an axiom system for the elementary arithmetic of integers ($1 + 2 = 3$, $2 \times 2 = 4$, etc.), we might use a term like "one" as an undefined technical term. Thus the same term may have different roles in different axiom systems! The sense in which the numbers "one," "two," etc., are used above—that is, in the universal sense—will be discussed in greater detail later (IV 4). As the term "exist" is used above, it is chiefly *permissive* so far as proofs are concerned, and *stipulative* for examples; thus in the proof of Theorem 3 we were permitted to introduce the line L_1 by virtue of Axiom 5, and the example of the "ghost community" containing only three persons (2.5) failed because it could not meet the stipulation concerning the existence of a certain line parallel to another line which is made in Axiom 5. It will be necessary, as we shall see, to distinguish between this type of "existence" and what is generally called "mathematical existence." The terms "and," "or," and "not," as logical terms, will be discussed further later. We also wish to call attention to the use of the word "element" in connection with the term "collection" in Definition 2.3; more will be said about this in the sequel.

5 Source of the axioms

Let us consider more fully the *source of the statements* embodied in the axioms. We chose axioms for *geometry* in our example Γ since we felt we could assume that the reader had studied some elementary geometry in high school. That is, we were careful to pick an already familiar subject. The undefined technical terms "point" and "line" already have a meaning of some sort for us. And, as we shall see, this is the usual way in which axioms are obtained; *they are statements about some concept with which we already have some familiarity*. Thus, if we are already familiar with arithmetic, we might begin to set down axioms for arithmetic. Of course, the method is not restricted to mathematics. If we are familiar with some field such as physics, philosophy, chemistry, zoology, or economics, for instance, we might choose to set down some axioms for it, or a portion of it, and see what theorems we might logically deduce from them.† We may say, then, that an *axiom*, as used in the modern way, is *a statement which seems to hold for an underlying concept*, an *axiom system* being *a collection of such statements about the concept.*

Thus, in practice, the concept comes first, the axioms later. Theoretically this is not necessary, of course. Thus we may say, "Let us take as undefined terms *aba* and *daba*, and set down some axioms in these and universal logical terms." With no concept in mind, it is difficult to think of anything to say! That is, unless we first *give some meanings* to "aba" and "daba"—that is, introduce some *concept* to talk about—it is difficult to find anything to say at all. And if we finally do make some statements without first fitting a suitable concept to "aba" and "daba," we shall very likely make statements that contradict one another! As we shall see below, the underlying concept is not only a source of the axioms, but it also guides us to *consistency* (about which we shall speak directly).

We can take the point of view that in selecting axioms we are actually selecting, from a totality T of statements about a concept, a collection A of certain "key" or "basic" statements, which we hope will imply all statements in T. In practice we may not know all the statements in T (usually we do not), but we know many that have proved significant. Thus in axiomatizing the concept plane geometry, we do not know all the possible statements that may be valid; but we know enough of them to serve as a guide for our selection of axioms.

The process may be compared with that of making colors. Suppose T is a collection of colors, and that we are given certain rules for mixing colors to produce new colors; select a collection A of colors from T which will be sufficient, by using the given rules for mixing, to produce all colors of T. In this analogy we have substituted colors for statements and mixing of colors for implication.

A similar process governs selection of terms. As pointed out in 4.3, we differentiate between terms that are technical and terms that are universal. If T denotes the totality of all technical terms, and we have certain rules for defining terms, then we desire to select from T a collection A of terms (to be left undefined) from which we can derive definitions of all the terms of T. The defining of terms will usually involve use of the universal terms; for example, in the definition given in 2.3, we used the universal term "collection." And some definitions may depend upon previously

† As an example in genetics and embryology, see J. H. Woodger [Wo].

proven statements. In particular, we may not know that two definitions actually define the same concept unless we resort to logical implication. But if we substitute for implication the process of defining, then the situation is quite analogous to the derivation of theorems from axioms; all that has changed is what constitutes T and A, and the process of derivation.

If we employ the ideogram $\Rightarrow$ which is frequently used in mathematics to denote "implies," then the diagram

$$\mathrm{A} \Rightarrow \mathrm{T}$$

symbolizes both the process of deriving theorems and the process of deriving definitions. In neither case have we explicitly set forth the rules governing the process symbolized by "$\Rightarrow$"; in the case of implication, this would involve stating the underlying logical laws and rules for proving theorems; and in the case of definition, it would involve giving the precise modes that govern the derivation of new terms from previously given terms. We go into this more thoroughly in the next chapter.

To summarize: We select the concept; then we select the terms that are to be left undefined and the statements that are to form our axioms; and finally we prove theorems as we did above, introducing new terms as needed. This is a simplification of the process, to be sure, but in a general way it describes the method. Note how the procedure, as so formulated, differs from the classical use of the method. In the classical use the axioms were regarded as absolute truths—absolutely true statements about material space—and as having a certain character of necessity. In the past, to have stated the parallel axiom, Axiom 5 above, was to have stated something "obviously true," something taken for granted if we had thought about the character of the space in which we lived. It would have been inconceivable before the nineteenth century to state an axiom such as "If L is a line and p a point not on L, then there exist at least two distinct lines containing p and parallel to L." To have in mathematics, simultaneously, two axiom systems Γ_1 and Γ_2 with axioms in Γ_1 denying axioms in Γ_2, as is the case in mathematics today with the euclidean and non-euclidean geometries, would also have been inconceivable! But, if we take the point of view that an axiom is only a statement about some concept,† so that axioms contradicting one another in different systems only express basic differences in the concepts from which they were derived, we see that no fundamental difficulty exists. What is important is that axioms in the *same* system should not contradict one another. This brings us to the point where we should discuss consistency and other characteristics of an axiom system.

† It is only in this sense—that an axiom is a statement *true* of some concept—that the word "true" can be used of an axiom.

5.1 Remark

The derivation of an axiom system for non-euclidean geometry from axioms for euclidean geometry, using the device of replacing the parallel axiom with one of its denials, is an example of another manner in which new axiom systems may be obtained. In general, we may select a given axiom system and change one or more of the axioms therein in suitable manner to derive a new axiom system. (See II 6.1.)

SUGGESTED READING

Bell [B_1; 20–46]
Eves and Newsom [E-N; II]
Hilbert [H_3]
Kerschner and Wilcox [K-W; II, VI]
Nagel [a]
Newsom [Ne]
Northrop [N; 221–230]
Pasch and Dehn [P-D; 185–271]
Poincaré [Po; 55–91]
Richardson [R; II]
Veblen [a]
Veblen and Young [V-Y; 2–7]
Young [Y; I–IV]

PROBLEMS

1. Obtain other meanings for "point" and "line" which make the axioms of Γ true statements. Also find a collection of *nine* objects (instead of four, as in Section 2.5) "satisfying" these axioms.
2. Why is not the corollary to Theorem 2 a direct consequence of Axiom 3 of the system Γ?
3. Do the lines M and M_1 defined in the proof of Theorem 4 in Section 3.8 necessarily have a point in common?
4. Is Axiom 5 *necessary* for the validity of the corollary to Theorem 1 in the system Γ? [*Hint:* Consider a collection containing three points p, q, and r, and let each pair of these be a line; also let the empty collection be a line.]
5. May a term appear in one axiom system as an "undefined technical term" and in another axiom system as a "universal logical term"?
6. What fundamental differences distinguish the Greek conception of axioms and the modern conception?
7. Would it be feasible to use the axiomatic method in order to describe ethical, political, or other social systems?
8. Would we necessarily become involved in contradictions in an axiom system if, in formulating an axiom as a statement which "seems to hold" for some concept, the statement is actually a false statement about the concept?
9. In what respect do definitions such as those given in Definitions 2.3 and 2.4 differ from those given in Sections 4.1 and 4.2? [*Hint:* Note that 2.3 and 2.4 are especially designed for Γ.]

In Problems 10–14, theorems are stated which are to be proved as theorems of the axiom system Γ. It is assumed that the number of points is finite (although in Problems 10 and 11 this is an unnecessary assumption if familiarity with the notion of cardinal number is assumed; see Chapter IV).

10. The number of points on a line is constant; that is, any two lines have the same number of points.

11. The number of lines containing a given point is constant.

12. If m is the number of lines containing a given point, and r is the number of points on a line, then $m = r + 1$.

13. If n is the total number of points, then $n = (r - 1)m + 1$; and hence $n = r^2$.

14. The number of lines is $n + r$ and hence, in view of Problem 13, mr.

15. In the light of the theorems stated in Problems 10–14 above, show that a collection of nine things considered as "points" yields several different interpretations of Γ according to how we select the possible combinations of points to form lines.

II

Analysis of the Axiomatic Method

When the undefined terms and the primary propositions or axioms have tentatively been selected, how can it be ascertained that the resulting axiom system is suited to the purposes for which it was set up? If, for example, it was set up to serve as a base on which to prove all theorems of euclidean plane geometry, then we should like some method whereby we can show that it suffices to do so. On the contrary, we may wish an axiom system to serve as an introduction to several different kinds of geometry, sufficing to prove the theorems common to them all, and no more. Another question that might arise would concern the so-called "independence" of the axioms; are any of the axioms provable from the others, and, if so, should we not delete them from the system, relegating them to the body of theorems to be proved later?

Experience has shown, however, that a much more fundamental and critical question is: does the system imply any contradictory theorems? If it does, clearly something is wrong, and it is useless to inquire into other questions until this defect has been eliminated. We shall therefore consider this question first.

1 Consistency of an axiom system

From a logical point of view we can make the following definition:

1.1 Definition. An axiom system Σ is called *consistent* if contradictory statements are not implied by Σ.

This definition gives rise to certain questions and criticisms. In the first place, given an axiom system Σ, *how are we going to tell whether it is consistent or not*? Conceivably we might prove two theorems from Σ which contradict one another, and hence conclude that Σ is *not* consistent.

For example, if we added to the system Γ of Chapter I the new axiom, "There exist at most three points," it would become apparent, as soon as Theorem 3 of Γ was proved (cf. I 3.7), that the new system of axioms is not consistent.

But, supposing that this does not happen, are we going to conclude that Σ is consistent? How can we tell that, if we continued stating and proving theorems, we might not ultimately arrive at contradictory statements and hence *inconsistency*? We remarked in Chapter I (I 4) about the system Γ that we could hardly expect to reach a point where we could say with confidence that no more theorems could be stated. And unless we could have all possible theorems in front of our eyes, capable of being scanned for contradictions, how could we assert that the system is consistent? There are examples in mathematical literature of cases where considerable material was published concerning certain axiom systems which later were found to be inconsistent. Until someone suspected the inconsistency and set out to prove it, or (as in some cases) stumbled upon it by chance, the systems seemed quite valid and worth while. It can also happen, for example, that the theorems become so numerous and complicated that we fail to detect a pair of contradictory ones. For example, although two theorems might really be of the form "S" and "not S" respectively, because of the manner in which they are stated it might escape our attention that they contradict one another. In short, the usefulness of the above definition is limited by our ability to recognize a contradiction even when it is staring us in the face, so to speak. Is there any *procedure for proving a system of axioms consistent*? And, if so, on what basis does the proof rest, since it may not be possible to carry out the proof *within* the system as in the case of the theorems of the system?

Let us make the definition:

1.2 **Definition.** If Σ is an axiom system, then an *interpretation* of Σ is an assignment of meanings to the undefined technical terms of Σ in such a way that the axioms become simultaneously true statements for all values of the variables (such as p and q of Axiom 3, I 2.2, for instance).

This definition requires some explanation. First, as an example we can cite the system Γ (I 2.2) and let "point" mean any one of a collection of four coins and "line" mean any pair of coins in this collection. The axioms now become statements about the collection of coins and are easily seen to be true thereof. Hence, this assignment of meanings is an interpretation of Γ. As the axioms stand, with "point" and "line" having no assigned meanings, they cannot be called either true or not true. (Similarly, we cannot speak of the expression "$x^2 - y^2 = (x - y)(x + y)$" as being either true or false until *meanings*, such as "x and y are integers," are assigned.) But, with the meanings assigned above, they are true statements about a "meaningful" concept. As a rule, we shall use the word "model" to denote the concept which results from the interpretation. Thus the concept of the collection of four coins, considered a collection of points and lines according to the meanings assigned above, is a model of Γ. Generally, if an interpretation I is made of an axiom system, we shall denote the model resulting from I by $\mathfrak{M}(\text{I})$.

For some models of an axiom system Σ, certain axioms of Σ may be *vacuously satisfied.* That is, axioms of the form "If . . ., then . . .," such as Axiom 3 of Γ, which we might call "conditional axioms," may be true as interpreted only because the conditional "If . . ." part is not fulfilled by the model.

Suppose, for example, we delete Axioms 2 and 4 from Γ and denote the resulting system by Γ'. Then a collection of coins containing just one coin is a model for Γ', if we interpret "point" to mean coin and "line" to mean a collection containing just one coin. For in this model the "If . . ." parts of Axioms 3 and 5 are not fulfilled. (Note that for Axiom 3 to be false of a model $\mathfrak{M}$, there must be *two* points p and q in $\mathfrak{M}$ such that either no line of $\mathfrak{M}$ contains p and q or more than one line of $\mathfrak{M}$ contains p and q.) Indeed, whenever *all* axioms of a system are conditional, the empty collection will generally be a model.

This may be better illustrated, perhaps, by the following digression: Suppose boy A tells girl B, "If it happens that the sun shines Sunday, I will take you boating." And let us suppose that on Sunday it rains all day, the sun not once peeping out between the clouds. Then, no matter whether A takes B boating or not, it cannot be asserted that he made her a false promise. For his promise to have been false, (1) the sun must have shone Sunday, and (2) A must not have taken B boating. Thus, in general, *for a statement of the form* "If . . ., then . . ." *to be false, the* "If . . ." *condition must be fulfilled and the* "then . . ." *not be fulfilled.* For if S and T are statements, the denial of the statement "If S, then T" is "There is a fulfillment of the statement S for which T is false."

Now we did not have in mind a collection of four coins when we set down the axioms of Γ. We were thinking of something entirely different, namely, euclidean geometry as we knew it in high school. "Point" had for us then an entirely different meaning—something "without length, breadth, or thickness"; and "line" meant a "straight" line that had "length, but no breadth or thickness." Do not these meanings also yield a model of Γ—what we might perhaps call an "ideal" model? We may admit that this is so, and, as we shall see later, we resort frequently in mathematics to such ideal models; this is always the case when every collection of objects serving as a model must of necessity be infinite in number (for instance, when we have enough axioms in a geometry to insure an infinite number of lines). We return to this discussion later (see 2.3); at present, let us go on to the so-called "working definition" of consistency:

1.3 Definition. An axiom system Σ is *satisfiable* if there exists an interpretation of Σ.

What is the relation between the two definitions in 1.1 and 1.3? What we actually want of any axiom system is that it be consistent in the sense of 1.1. But we saw that 1.1 was not a practicable definition except in cases where contradictory statements are actually found to be implied by the system and inconsistency is thus recognized. Where a system is consistent, we are usually unable to tell the fact from 1.1. But, as in the case of the four-coin interpretation of Γ, we may have a model showing "satisfiability" in the sense of 1.3. Does this imply consistency in the sense of 1.1? The working mathematician takes the point of view that it does, and to explain why, we have to go into the domain of logic for a few moments.

2 The proof of consistency of an axiom system

2.1 · The Law of Contradiction and the Law of the Excluded Middle

First, let us recall two basic "laws" of classical (i.e., Aristotelian) logic, namely, the Law of Contradiction and the Law of the Excluded Middle; the latter is also called the Law of the Excluded Third ("tertium non datur"). These are frequently, and loosely, described as follows: If S is any statement, then the Law of Contradiction states that S and a contradiction (i.e., denial) of S cannot both hold. And the Law of the Excluded Middle states that either S holds or the denial of S holds. For example, let S be the statement "Today is Tuesday." The Law of Contradiction certainly holds here, for today cannot be *both* Tuesday and Wednesday, for example. And the Law of the Excluded Middle states that either today is Tuesday or it is not Tuesday.

But "things are not so simple as they seem" here. Unless we limit ourselves to a specified point on the earth (or parallel of longitude), it *can* be both Tuesday and Wednesday at the same time! Thus, unless such a geographical provision is included in S, the statement "Today is Tuesday and it is not Tuesday" can hardly be rejected. As a matter of fact, whenever such statements are made, there usually exists a tacit understanding between speaker and listener that their locale at the time is the place being referred to.

Or consider the statement "The king of the United States wears bow ties." Does the Law of the Excluded Middle hold here? Or let S be the statement "All triangles are green."

The upshot of this is merely that, although these "laws" are called "universally valid," some sort of qualifications have to be made with regard to their applicability in order for them to have validity. So far as axiomatic systems are concerned, the problem is not so great, since we can restrict our use of the term "statement" to the convention already made in Section I 4.1 ("Σ-statement"). And this will be our understanding from now on. Further qualifications regarding the applicability of these logical "laws" will be noted in the discussion of completeness below.

2.2 As soon as an interpretation of a system Σ is made, the statements of the system become statements about the resulting model. Let us assume the following, which may be considered *basic principles of applied logic*!

2.2.1 All statements implied by an axiom system Σ hold true for all models of Σ;

2.2.2 The Law of Contradiction holds for all statements about a model of an axiom system Σ, provided they are Σ-statements whose technical terms have the meanings given in the interpretation. We can make this clearer and more precise by introducing the notion of an I-Σ-statement:

2.2.3 If Σ is an axiom system and I denotes an interpretation of Σ, the result of assigning to the technical terms in a Σ-statement their meanings in I will be called an I-Σ-statement.

Then 2.2.1 and 2.2.2 become, respectively:

2.2.1 Every I-Σ-statement, such that the corresponding Σ-statement is implied by Σ, holds true for $\mathfrak{M}$(I) (cf. 1.2);

2.2.2 Contradictory I-Σ-statements cannot both hold true of $\mathfrak{M}$(I).

Under the assumption that 2.2.1 and 2.2.2 hold, satisfiability implies consistency. For if an axiom system Σ implies two contradictory Σ-statements, then by 2.2.1 these statements as I-Σ-statements hold true for the model $\mathfrak{M}$(I); but the latter is impossible by 2.2.2. Hence, we must conclude that if 2.2.1 and 2.2.2 are valid, the existence of an interpretation for an axiom system Σ guarantees the consistency of Σ in the sense of 1.1. And this is the basis for the "working definition" 1.3. For example, the existence of the "four-coin interpretation" of the system Γ guarantees the consistency of Γ if we grant 2.2.1 and 2.2.2.

The reader will have noticed that we have not proved that consistency in the sense of 1.1 implies satisfiability. To go into this question would be impractical, since it would necessitate going into detail concerning formal logical systems (cf. Chapter XI) and is too complicated to describe here.

2.3 In Section 1 we used the term "ideal" model, by way of contrast to such models as that of the four coins for Γ; the latter might be termed a "concrete" or "physically realizable" model. It was pointed out that whenever an axiom system Σ requires an infinite collection in each of its models, of necessity the models are "ideal."

This raises not only the question as to how reliable are "ideal" models, but also the question as to what constitutes an *allowable* model. What we should like, of course, is a criterion which would allow only models that satisfy assumptions 2.2.1 and 2.2.2, especially the latter. If there is any danger that an ideal model may require such a degree of abstraction that it harbors contradictions in violation of 2.2.2, then clearly the use of models is no general guarantee of consistency in spite of what we have said above.

Further light can be shed on this matter by a consideration of well-known examples. It is not an uncommon practice, for instance, to obtain a model of an axiom system Σ in another branch of mathematics—even in a branch of mathematics that is, in its turn, based on an axiom system Σ'.† How valid are such models? Do they necessarily satisfy 2.2.2? For example, to establish the consistency of a non-euclidean geometry we give a model of it in euclidean geometry. (See Richardson [R; 418–419] for instance.) But suppose that the euclidean geometry harbors contradictions; what then? Evidently all we can conclude here is that, *if* euclidean geometry is consistent, then so is the non-euclidean geometry whose model we have set up in the euclidean framework.

We are forced to admit that in such cases we have no absolute test for consistency, but only what we may call a *relative consistency proof.* The axiom system Σ' may be one in whose consistency we have great confidence and then we may feel that we achieve a high degree of plausibility for consistency, but in the final analysis we have to admit that we are not *sure* of it.‡

As we shall see later, other tests for consistency have been explored by the methods of formal logic; but until such methods are sufficiently developed, the use of models, even of an "ideal" character, will have to suffice.

3 Independence of axioms

In the first paragraph of this chapter we mentioned "independence" of axioms. By "independence" we mean essentially that we are "not saying too much" in stating our axioms. For example, if to the five axioms of

† In general, a model of an axiom system Σ in a system Σ' is the result of an interpretation of the undefined terms of Σ in the terminology of Σ' in such a way that the axioms of Σ become Σ'-statements implied by Σ'. (See Problem 29.)

‡ In one well-known case, the system Σ' is a *subsystem* of Σ; viz., the Gödel proof (Gödel [G]) of the relative consistency of the axiom of choice (III 6.3) when adjoined to the set theory axioms.

the system Γ (I 2.2) we added a sixth axiom stating, "There exist at least four points," we would provide no new information inasmuch as the axiom is already implied by Γ (see Theorem 3 of I 3.7). Of course, the addition of such an axiom would not destroy the property of consistency inherent in Γ.

3.1 In order to state a formal definition of independence, let Σ denote an axiom system and let A denote one of the axioms of Σ. Let us denote the denial of A by ∼A, and let Σ − A denote the system Σ with A deleted. If S is any Σ-statement, let Σ + S mean the axiom system containing the axioms of Σ and the statement S as a new axiom. Then we define:

3.1.1 Definition. If Σ is an axiom system and A is one of the axioms of Σ, then A is called *independent* in Σ, or an independent axiom of Σ, if both Σ and the axiom system (Σ − A) + ∼A are satisfiable.

In practice, a Σ-statement whose addition to Σ − A implies ∼A is often used instead of ∼A. For example, to show Axiom 5 independent in Γ (I 2.2), let S be the statement "There exist a line *L* and a point *p* not on *L*, such that there does not exist a line containing *p* that is parallel to *L*." Then S is not the precise denial of Axiom 5, but certainly implies it. (The denial of Axiom 5 is: "There exist a line *L* and a point *p* not on *L* such that there does not exist one and only one line containing *p* that is parallel to *L*." Cf. the italicized statement in 1.2 above.) However, denoting Axiom 5 by A and noting that (Γ − A) + S implies ∼A (since S already implies ∼A), and recalling 2.2.1 above, we see that any model of (Γ − A) + S is also a model of (Γ − A) + ∼A. Like remarks hold for axiom systems in general. Consequently, 3.1.1 may be replaced by the statement:

3.1.1′ If Σ is an axiom system and A is one of the axioms of Σ, then A is independent in Σ if (1) Σ is satisfiable, and (2) there is a Σ-statement S such that the axiom system (Σ − A) + S implies ∼A and is satisfiable.

3.2 Thus Axiom 5 is independent in Γ (I 2.2) if Γ is satisfiable and if the first four axioms of Γ together with a "non-euclidean" form of the axiom constitute a satisfiable system. For example, let S be the statement "There exist a line *L* and point *p* not on *L*, such that there does not exist a line containing *p* and parallel to *L*." To show that the system Γ with Axiom 5 replaced by S forms a satisfiable system, let us take a collection of *three* coins, let "point" mean a coin of this collection, and "line" mean any pair of coins of this collection. Then we have an interpretation of the new system, showing it to be satisfiable. We have already ascertained (2.2) that Γ is satisfiable, and so we conclude that Axiom 5 is independent in Γ.

3.3 The reader will probably gather by this time that the reason for specifying the satisfiability of Σ, in Definition 3.1.1, is to insure that $\sim$A is not a *necessary* consequence of the axioms of $\Sigma - A$; for if it were, we would not wish to call A "independent." Thus, as the definition is phrased, it insures that neither A nor the denial, $\sim$A, of A is implied by the system $\Sigma - A$, so that the addition of A to $\Sigma - A$ is really the supplying of new information.

3.4 Actually, however, we do not place the same emphasis on independence as we do on consistency. Consistency is *always* desired, but there may be cases where independence is *not* desired. Examples of this arise in the teaching of mathematics. Frequently, we develop a subject by stating axioms and proving theorems—so that the axiomatic method assumes thereby a pedagogical role. Now it sometimes happens that an early theorem is extremely difficult to prove. Then the theorem may be stated as one of the axioms. Later, when the students have gained sufficient maturity and familiarity with the subject, it may be disclosed to them that the "axiom" is not really independent and a demonstration of its proof from the other axioms of the system may be given. Generally speaking, of course, it is preferable to have all axioms independent; but if some axiom turns out not to be independent, the system is not invalidated.

As a matter of fact, some well-known and important axiom systems, when first published, contained axioms that were not independent (a fact unknown at the time to the authors, of course). An example of this is the original formulation of the set of axioms for geometry given by Hilbert [H_1] in 1899 (already referred to in I 1.4 and I 1.5). This set of axioms contained two axioms which were later discovered to be implied by the other axioms.† This in no way invalidated the system; it was only necessary to change the axioms to theorems (supplying the proofs of the latter, of course). Other cases could be cited.‡

3.5 Types of independence other than that defined in 3.1.1 have been proposed. For example, E. H. Moore [Mo_1; 82] defined what he called "complete independence":

3.5.1 A system of axioms, Σ, is called *completely independent* if, for every subsystem $A_i, A_j, \cdots, A_k$ of the axioms, both Σ and $(\Sigma - A_i - A_j - \cdots - A_k) + \sim A_i + \sim A_j + \cdots + \sim A_k$ are consistent. Here

† See E. H. Moore [a], A. Rosenthal [a, b, c], S. Weinlös [a, b], and Lindenbaum [a]. Also see the footnotes to the first chapter of Hilbert [H_2].

‡ For example, a well-known and for many years widely used (for teaching purposes) system of eight axioms for plane topology due to R. L. Moore [a] was found by the present author [a] to contain a non-independent axiom. In this case the suspicion that the axiom was not independent arose from the fact that the independence proof given for the axiom was not valid, and search for a new proof proved fruitless.

$A_i, A_j, \cdots, A_k$ are axioms of Σ and $\sim A_i, \sim A_j, \cdots, \sim A_k$ are denials of $A_i, A_j, \cdots, A_k$ respectively. If we take "consistent" to mean "satisfiable," this definition evidently yields a generalization of Definition 3.1.1. We shall not go further into this notion, however.†

3.6 Before leaving the notion of independence, it might be interesting to consider its relation to the *number* of axioms in a system. If a consistent system of axioms, Σ, contains twelve axioms, say, and one of them is discovered to be not independent, then the system is reducible to only eleven axioms. The system may then be said to be "simplified by this reduction"; the concept "described" by Σ can now be "described" in fewer words. However, this line of thought is illusory, for it does not take much reflection to convince us that mere *number* of axioms is no measure of simplicity. An axiom system may, in a valid sense, be said to be simplified if an axiom is split up into several axioms. Sometimes we decide that an axiom has said too much; so much, in fact, that it becomes difficult to digest its meaning. The difficulty may be removed by giving instead several axioms, each of a simply understood character, and *in toto* saying what the original axiom stated.

There is no general rule to guide us in these respects. It is a case of "the same thing can be said in many ways," and consequently we use our best judgment as to how we will accomplish "simplicity." We can put too much into a single axiom; as an extreme, all axioms of a system can be lumped together in one axiom—thus achieving independence neatly—or we can split an axiom to the point of triteness.

An amusing example is given by Helmer [a] which we may transfer to the system Γ (I 2.2) as follows: Consider Axiom 3 of Γ. For the sake of brevity let us denote the combination of words "The number of lines containing p and q where p and q are distinct points" by the symbols "$\varphi(p, q)$." Then Axiom 3 can be replaced by the axioms: (*a*) $\varphi(p, q)$ is odd; (*b*) $\varphi(p, q)$ is less than 8; (*c*) $\varphi(p, q)$ is not 7; (*d*) $\varphi(p, q)$ is not 5; (*e*) $\varphi(p, q)$ is not 3. Axiom 3 may be replaced by these *five* axioms if preferred! In the same article, incidentally, Helmer gives a simple axiom system that is easily seen to consist of independent axioms, and to be not completely independent.

4 Completeness of an axiom system

It was pointed out in I 2.6 that the system Γ would not yield *all* theorems of plane geometry; and, as we have seen, Γ actually has a model containing

† Harary [a] calls an axiom A *very independent* in a system Σ if the system has a model in which all the axioms of Σ other than A hold and A "never holds." For example, if Σ is a system for euclidean plane geometry and A the parallel axiom, then we can give a model of a non-euclidean geometry in which the hypothesis of A always implies the negation of its conclusion. In another paper [b] he has introduced a measure of independence.

only a finite number of points; this could not be the case if Γ were adequate as a foundation for euclidean plane geometry. In plane geometry we would wish to be able to assert, for example, that between two points on a line there is a third point (the "bisecting" point for example), which is certainly not assertable on the basis of Γ (cf. the model consisting of four coins!).

4.1 Now, *a priori*, there is no reason to believe that it is even possible to give an axiom system that would imply *all* theorems of plane geometry. Presumably we could hardly hope to set down all possible theorems of plane geometry, so that we might easily judge that it would be equally impossible to set down a system of axioms that would *imply* all theorems of plane geometry.† Assuming, however, that it *is* possible to set down such a system of axioms, how shall we phrase an explicit definition of the sort of "completeness" which we have in mind here; and, after this, how can we ascertain the fact that a given system is complete?

4.2 To fix our ideas, let us suppose that we have given an axiom system Σ which was derived from some concept C, using a collection T of undefined terms. (Thus Σ might be the system Γ (I 2.2), C the concept of plane geometry, and T the collection of terms "point," "line.") Now Σ might be inadequate as a system of axioms for C in that there are not enough axioms; more explicitly, in that the axioms do not contain enough assertions to imply all the theorems we want. On the other hand, Σ might be insufficient in the sense that T does not contain enough undefined terms. For example, for plane geometry, not only might the system Γ be lacking in axiomatic assertions, but also it may be necessary to add new terms (for instance, "congruent") to T before we could get all the theorems of plane geometry. It may also occur to the reader that in addition to inadequacies that the axioms, and the collection T, may have, there may also be defects in our *logical apparatus* and the logical processes themselves, as well as in such "universal" terms as "set." Each of these possibilities will be considered; the first two in the present section, and the others in subsequent chapters.

4.3 Definition of completeness

Let us first consider incompleteness due to inadequacy of the axioms themselves. If we revert to the idea of completeness that is based on the

† When we speak of "all theorems" we do not, of course, include "false" theorems. And we are tacitly ruling out the trivial fact that by logical manipulation (see IX 3.5.3, for instance) it is possible to prove all theorems ("true" or "false") by the simple device of introducing contradictory axioms. We therefore tacitly assume throughout the discussion of completeness that the axiom systems discussed are consistent.

notion of a system being sufficient to "imply all theorems," we might say that if we can find a theorem which can neither be proved nor "disproved" (i.e., the denial of it is not provable), then we see that if such a theorem exists it becomes a candidate for a new axiom. This leads to the following definition (recall that a Σ-statement must be phrasable in terms of T):

4.3.1 Definition. An axiom system Σ is *complete* if there is no Σ-statement A such that A is an independent axiom in the system $\Sigma + A$;† i.e., such that both $\Sigma + A$ and $\Sigma + \sim A$ are satisfiable.

Loosely speaking, according to 4.3.1 a system Σ is complete if it is impossible to add to it a new independent‡ axiom (always bearing in mind that we are keeping T fixed, of course).

For example, let Σ be the system Γ (I 2.2), and let A_6 be the statement "There exist at most four points." Both $\Gamma + A_6$ and $\Gamma + \sim A_6$ are satisfiable; the former because of the four-coin interpretation, the latter because of the usual euclidean geometry interpretation (there exist other interpretations of $\Gamma + \sim A_6$ consisting of only a finite number of points; cf. Problem 1 at the end of Chapter I). Thus Γ is not complete according to Definition 4.3.1.

How about the system $\Gamma + A_6$; is this complete? Let us denote $\Gamma + A_6$ by Γ_6. For reasons given later, we would search in vain for a Γ_6-statement, A′, such that A′ is independent in the system $\Gamma_6 + A'$. But failure to find such an A′ after any finite number of trials would hardly *prove* Γ_6 complete in the sense of 4.3.1. This fact reveals that Definition 4.3.1 is subject to the same short-coming that Definition 1.1 has, namely, the seeming impossibility of *proving* the property defined when it may actually be possessed by an axiom system. And, as in 1.1, where we gave as "working definition" the Definition 1.3, that of *satisfiability*, we shall here also give a "working definition" as an alternative to 4.3.1. And, just as in 1.1, where we had to interpolate a new notion, that of "interpretation" of an axiom system, before we could give the alternative definition, we shall here also have to introduce a new notion before stating the new definition.

4.4 One-to-one correspondence; isomorphism

Let us recall the four-coin interpretation of Γ, which we shall call interpretation I_1 for present purposes, and the "ghost town" Z containing only four persons which we described in I 2.5. In the latter case, "person"

† It must be emphasized that Σ is actually augmented by A to form $\Sigma + A$ (even though A may already be in Σ—in which case A appears twice in $\Sigma + A$). In every case, however, $[(\Sigma + A) - A] + \sim A = \Sigma + \sim A$.

‡ So far as our discussion is concerned, the importance of the notion of "independence" is due almost solely to its use in connection with completeness.

was to serve as the meaning of "point," and "pair of persons" as the meaning of "line"; assigning these meanings constitutes an interpretation I_2 of Γ. Now, in what ways do the models $\mathfrak{M}(I_1)$ and $\mathfrak{M}(I_2)$ differ? For the sake of brevity, let us denote $\mathfrak{M}(I_1)$ by $\mathfrak{M}_1$, $\mathfrak{M}(I_2)$ by $\mathfrak{M}_2$.

This may seem like a trivial question since a collection of four *coins* and a collection of four *persons* differ so obviously in so many ways. But let us consider only those properties of $\mathfrak{M}_1$ and $\mathfrak{M}_2$ that have *significance for* Γ_6. Then the obvious differences, such as the fact that a coin is metallic whereas a person is organic, are not to be taken into account. Only properties and relations having to do with *point* and *line* have any significance so far as Γ_6 is concerned. And, with regard to these, $\mathfrak{M}_1$ and $\mathfrak{M}_2$ have no differences whatsoever. To make this more precise, we first introduce the following notions:

4.4.1 **Definition.** If X and Y are two collections (not necessarily disjoint), then a *function from* X *to* Y is a collection f of ordered pairs (x, y) such that (1) x and y are elements of X and Y, respectively, and (2) each element x of X occurs as an "x" in a pair (x, y) exactly once. That f is a function from X to Y may be denoted by the expression "f: $X \to Y$." And that y is paired with x in a pair (x, y) may be denoted by the expression "$y = f(x)$." The collection X may be called the *domain* of f and the collection of all y's such that $y = f(x)$ for some x called the *range* or *image* of f. If a function f: $X \to Y$ has Y as its range, and g: $Y \to Z$ is another function having Y as its domain, then we may define a function gf: $X \to Z$ called the *composition* of f and g; for two pairs (x, y) and (y, z) of f and g, respectively, having a y in common, the pair (x, z) constitutes a pair of the function gf.

4.4.2 When it happens that X and Y are the domain and range of a function f, and each element y of Y occurs exactly once in the corresponding pairs (x, y), then f is called a (1-1)-*correspondence between the elements of* X *and the elements of* Y, or, briefly, a (1-1)-*correspondence between* X *and* Y. In this case one has also a function from Y to X, denoted by "f^{-1}," called the *inverse function* of f; its pairs are of the form (y, x) formed by reversing the order in each pair (x, y) of f. And x and $f(x)$ are called *corresponding* elements.

As we shall see later (Chapter IV), (1-1)-correspondence is the basis for ascertaining whether two collections have the same number of elements. Thus, two collections X and Y have the same number of elements if and only if there is a way of pairing off their elements, which constitutes both a function f from X to Y and a function f^{-1} from Y to X as defined above. We might also remark that even when a function f does not constitute a (1-1)-correspondence, the symbol "f^{-1}" is frequently used, $f^{-1}(y)$ being the collection of all the x's such that $y = f(x)$; $f^{-1}(y)$ is called the *counter-image* of y. A more general notion than function is that of *relation*, but we shall reserve this for later discussion.

The collections X and Y may have common elements; in fact, one of the most important cases, as we shall see later (III 4.8.1) is that in which one of these collections is a part (subcollection) of the other. Of great importance, as a matter of fact, is the *identity correspondence* in the case where X and Y are the *same* collection; this is the (1-1)-correspondence consisting of the collection of pairs (x, x), in which there is one pair (x, x) for each element x of X $(= Y)$.

4.4.3 Now $\mathfrak{M}_1$ and $\mathfrak{M}_2$ have four "points" each and, in each, every pair of "points" constitutes a "line." Considering $\mathfrak{M}_1$ and $\mathfrak{M}_2$ collections of points and lines, it is easy to see that there is a (1-1)-correspondence $\mathfrak{M}_{1,2}$ between $\mathfrak{M}_1$ and $\mathfrak{M}_2$ such that (1) if x_1 is a point in $\mathfrak{M}_1$, then the corresponding element in $\mathfrak{M}_2$ is a point; and (2) if x_1 and y_1 constitute a line in $\mathfrak{M}_1$, then the corresponding elements x_2 and y_2 constitute a line in $\mathfrak{M}_2$; and vice versa. Moreover, *any* true statement about the points and lines of $\mathfrak{M}_1$ is also a true statement about $\mathfrak{M}_2$ if the special points and lines mentioned in the statement are replaced by the corresponding points and lines of $\mathfrak{M}_2$. For example, if the statement "The lines L_1 and L_1' have the point p_1 in common" is a true statement about $\mathfrak{M}_1$, then the statement "The lines L_2 and L_2' have the point p_2 in common" is a true statement about $\mathfrak{M}_2$ if L_2, L_2', p_2 correspond respectively to L_1, L_1', p_1. Consequently, we call $\mathfrak{M}_{1,2}$ a (1-1)-correspondence between $\mathfrak{M}_1$ and $\mathfrak{M}_2$ with *preservation of* Γ_6*-statements.* And we define:

4.4.4 **Definition.** Two models $\mathfrak{M}_1$ and $\mathfrak{M}_2$ of an axiom system Σ are called *isomorphic with respect to* Σ if there exists between $\mathfrak{M}_1$ and $\mathfrak{M}_2$ a (1-1)-correspondence with preservation of Σ-statements; and such a (1-1)-correspondence is called an *isomorphism.* Thus $\mathfrak{M}_{1,2}$ constitutes an isomorphism between $\mathfrak{M}_1$ and $\mathfrak{M}_2$.

4.4.5 It should be emphasized that mere (1-1)-correspondence between two models is generally not sufficient to yield an isomorphism between the models. That any (1-1)-correspondence between the points of the model $\mathfrak{M}_1$ and the points of the model $\mathfrak{M}_2$ of Γ_6 above yields an isomorphism is an accident due to the simplicity of these models. For, since every pair of points constitutes a line in each model, the (1-1)-correspondence between points can be extended to a (1-1)-correspondence between the lines, which is sufficient in the case of these models to give an isomorphism.

But suppose that, instead of A_6, we add to Γ the axiom A_6': "There exist exactly nine points," and denote the system $\Gamma + A_6'$ by Γ_6'. As stated in Problems 13 and 14 of Chapter I, in any model of Γ_6' there must then be exactly 3 points on each line, and 12 lines in all. Since there are 84 different triples in a collection of 12 things, it is clear that the 12 triples that constitute lines must be carefully selected in defining a model of

Γ_6', because not every triple constitutes a line. Consequently, if $\mathfrak{M}_1'$ and $\mathfrak{M}_2'$ are models of Γ_6' and we wish to set up a (1-1)-correspondence between the points and lines of $\mathfrak{M}_1'$ and the points and lines of $\mathfrak{M}_2'$ in such a way that lines correspond to lines, then not every (1-1)-correspondence between the points of $\mathfrak{M}_1'$ and the points of $\mathfrak{M}_2'$ will work. (See Problem 20.)

Of course, if the numbers of elements in two models are different so that no (1-1)-correspondence exists, then isomorphism of the models is *a fortiori* impossible. But the existence of (1-1)-correspondence does not guarantee the stronger property implied by isomorphism. In short, (1-1)-correspondence is a *necessary* but not a *sufficient* condition for isomorphism. [For another example where (1-1)-correspondence exists although isomorphism does not exist, see Problem 9 at the end of this chapter.]

4.5 Categoricalness

With the notion of isomorphism thus defined, we are ready to state the "working definition" of completeness, which is embodied in the notion of *categoricalness*:†

4.5.1 Definition. An axiom system Σ is called *categorical* if *every* two models of Σ are isomorphic with respect to Σ.

4.6 Proof that categoricalness implies completeness in the sense of Definition 4.3.1‡

Suppose that an axiom system Σ is categorical. Now, if Σ were not complete in the sense of 4.3.1, there would exist a Σ-statement A such that both Σ + A and Σ + ~A are satisfiable systems. Let I_1 be an interpretation of Σ + A and let I_2 be an interpretation of Σ + ~A. Since Σ is categorical, there is a (1-1)-correspondence $\mathfrak{M}_{1,2}$ between $\mathfrak{M}(I_1)$ and $\mathfrak{M}(I_2)$, with preservation of Σ-statements. But this is impossible, since A is a Σ-statement, true of $\mathfrak{M}(I_1)$ and false of $\mathfrak{M}(I_2)$.

4.7 Completeness of Hilbert's geometry

In Γ_6 we have, then, an example of an axiom system that is complete. But Γ_6 is not, it must be admitted, a very interesting system, its chief virtue being its simplicity and consequent adaptability to purposes of exposition. The reader may still feel that, when it comes to mathematically

† The term *categorical* seems to have been introduced by O. Veblen [a], who states that it was suggested by John Dewey. As a test for completeness, however, the notion was used earlier; see Huntington [a] for instance.

‡ We recall that our definition of independence, 3.1.1, is in terms of satisfiability.

"rich" concepts, such as the usual euclidean geometry, with their seemingly unlimited number of theorems, completeness may not be so easily attained. Fortunately, this is not necessarily the case, although the *proofs* of completeness are not, as a rule, so easily attained as the above. In the case of plane euclidean geometry, a number of categorical axiom systems have been given. The system of Hilbert for the geometry of the plane and three dimensions ("solid" geometry) achieved completeness in a way that excited considerable discussion. As originally set up by Hilbert, the system included the following statement, which he called "Completeness Axiom":†

"The elements (points, straight lines, planes) form a class of objects which, under the set of all previously made assumptions, is not capable of further extension."

This meant that no new points, lines, or planes could be added to the collection of points, lines, and planes satisfying the other axioms without rendering at least one of these other axioms false.

Clearly, this is a peculiar type of axiom in that it almost seems to achieve completeness for the system by boldly *asserting* it; and its reference to "all previously made assumptions" smacks of an element of "higher caste" in a "hierarchy" of axioms. It is intimately related, moreover, to its predecessor, the "Archimedean axiom," which, roughly speaking, assures that the lines of the geometry shall be "long enough" without being "too long." Possibly it was this that led P. Bernays to the discovery that the axiom can be stated as a theorem, being replaced in the axiom system by the following "Axiom of Linear Completeness":

"The points of a line form a system which under the set of previously given axioms is not capable of further extension."‡

4.8 For other systems of axioms for geometry, the reader is referred to the "Suggested Reading" at the end of this chapter.

Axiomatic treatments of other mathematical concepts (in arithmetic, algebra, analysis, etc.) will be referred to later.

4.9 Desirability of completeness

Some comment regarding the desirability or undesirability of completeness should also be made at this point. When we wish to define a concept

† The English translation given here is derived from Young [Y]; quoted by permission of The Macmillan Co. Hilbert did not include this axiom in the first edition of his *Grundlagen* [H_1]; it did appear in the second and later editions, and was modified in the seventh edition [H_2].

‡ This axiom (the above is not a literal translation) appears in the seventh (1930) edition of Hilbert's book. See Hilbert [H_2; 30].

allowing a wide variety of interpretations, or a large range of application to different fields of mathematics or physics, then completeness will usually not be achieved or even desired. We shall see instances of this later in the axioms for *simple order*, *group*, etc. In such concepts we see one of the great advantages of the axiomatic method, what we might call the "economy aspect." For, as we have pointed out in Section 1, we are at liberty to make any interpretation we please, and the theorems which have been proved in the system of axioms become true statements about the model resulting from the interpretation. The "economy" involved here consists of the proving, once for all, of statements about a large number and variety of (seemingly unrelated) fields of study. Thus, when we come to study the notion of group, we shall notice that examples of groups (i.e., models of the group axioms) abound in mathematics (and in other fields), and the theorems proved on the basis of the group axioms are immediately available for use in all such cases. It is also a common mathematical device to prove, of a given concept, that it contains a model of some already known and developed axiom system, so that the theorems of the system may be made available for use in the study of the given concept. It is interesting to note, too (as a sort of converse), that it has not infrequently happened that mathematicians, working in different branches of mathematics, have duplicated one another's work; the only difference between their results being that of language—each speaking in terms which could be made the common terminology of some axiom system. Naturally, this is one of the things that *leads* to the invention of axiom systems, i.e., the discovery that the same "abstract notions" are operating under various guises in different branches of mathematics.

So far as a concept like that of plane euclidean geometry is concerned, the object is usually to set down, as Hilbert did, a complete set of axioms. We shall see an example of the same sort of thing in other cases, as for instance in the definition of the "mathematical continuum" which lies at the basis of analysis. It is of the very nature of such special concepts that their "definition," or *characterization* (to use the term most frequently employed), leads to complete systems of axioms. As we shall see later, we can so specialize the *order* and *group* concepts, mentioned above, as to attain completeness in describing them axiomatically, somewhat as we specialized the system Γ to obtain Γ_6.

4.10 The logical basis

In our discussion of consistency, we mentioned (2.1) the Law of Contradiction and the Law of the Excluded Third; that we place a great degree of confidence in their applicability is evidenced in 2.2.1 and 2.2.2. Independence of axioms was defined (3.1.1) in terms of satisfiability, so that the same "laws" are involved in this notion.

Definition 4.3.1 makes completeness of axioms, in turn, dependent on the notion of independence, and hence of satisfiability. The definition (4.5.1) in terms of isomorphism, as we saw, "implies" the former definition in the sense that, when a system is categorical, it is complete according to 4.3.1. Now, aside from the assumption of the "laws" of classical logic, there is the interpretation—not always clear, as we have noticed (I 3.4, for instance)—placed on the "universal" terms. The average person's only contact with geometry is in the high school course, and as a rule no mention by name is made there of the "Aristotelian logic." The logical apparatus employed (justifiably, for pedagogical reasons) is usually a "taken-for-granted" or "rule-of-thumb" logic. Its value probably lies in its naturalness, and as such it undoubtedly forms a good basis for the study of "formal logic" in college, if the student goes this far and happens to "elect" a course in the latter.

This "natural" logic tacitly assumes the Law of Contradiction and the Law of the Excluded Third; witness the "reductio ad absurdum" type of proof which is such a favorite in high school geometry. No attention at all is paid to logical or universal terminology, such as the use of the terms "collection," "all," "not" (again, justifiably).

It is not surprising, therefore, that the axiomatic method, originally derived from the axiom-postulate method of Euclid, proceeded on the same logical basis, employing a "taken-for-granted" logic. However, as we shall see, concurrently with the development of the axiomatic method during the latter part of the nineteenth and the beginning of the twentieth centuries, this "natural" logic was beginning to be attacked. The first criticism was directed not so much at the "laws" of the logic, as at the universal *terms*, particularly at the notion of "collection." And, in our return to this subject in the sequel, it will be to this notion of collection that we first direct our attention. For not only does the consistency of an axiom system depend on our assumption of the logical "laws," but also all the attributes of an axiom system discussed above, including completeness, are determined to a greater extent than we realize by the way in which we use the so-called "universal" terms.

5 Independence and completeness of undefined technical terms

In 4.2 we remarked that a system of axioms may be inadequate for plane geometry because it lacks certain technical terms (for instance, "congruent"). Indeed, how can we expect the statements incorporated in the axioms to be sufficient, if the technical language in which they are phrased is inadequate?

To make this clearer, let us look again at the manner in which an axiom system is developed. As we saw in Chapter I, theorems are deduced from the axioms; but in addition, we frequently add new technical terms by definition. That is, as the number of theorems grows, the body of technical terms also grows; but whereas the growth of theorems advances by deduction, the growth of technical terms proceeds by definition. And since definition of *all* technical terms would lead to endless regression, it is necessary to begin with some undefined technical terms. But how can we tell what is a suitable choice of undefined terms? For example, "parallel" was not included among the undefined terms of the system Γ because it turned out that it was definable in terms of "point" and "line." This raises the question, "How can we tell if a term is definable by means of the given undefined terms?" The analogy to the situation concerning the axioms themselves may be brought out by using the word "independent." But when is a new term independent of the undefined terms? This question, of course, is meaningless in the absence of a more precise meaning for the word "independent" in this context.

5.1 Such a meaning was suggested by Padoa [a] about 1900 (see Padoa [a, b, c]). He proposed that in order to show that a term t cannot be defined by means of a collection T of given undefined terms (always assuming a given axiom system Σ), it is sufficient to provide two models $\mathfrak{M}_1$ and $\mathfrak{M}_2$ of Σ in each of which the terms of T have the same meaning, but in which t has different meanings. The analogy to the test for independence in terms of models (3.1.1) is obvious, the only lack being that it may not always be clear when a term has "different meanings." Although a thorough investigation of this would necessitate a much deeper analysis than we can go into here, we shall go a little more into detail while hoping to make it more precise.

5.2 The nature of definition

We need to have a clearer idea of just what we mean by "defining" a term t by means of a collection T of terms in a given axiom system Σ. In the axiom system Γ (I 2.2), we gave two definitions: one of what is meant by "a point on a line" (I 2.3) and the other of "parallel" (I 2.4). Since "point" and "line" were undefined, Definition I 2.3 would have been meaningless unless we had known, from Axiom 1 of Γ, that a line is a collection of points. A similar remark holds for Definition I 2.4. However, if we had developed the implications of Γ further, we would probably have found it convenient to introduce new definitions which depended for their significance on theorems rather than directly upon the axioms. That is, generally a term t is defined by means of a Σ-statement involving both the undefined terms and the theorems. And *to define a term t by means of a collection T of* (defined and undefined) *terms is to give a Σ-statement which renders t fixed in any model in which the terms of T have been fixed.*

Then how are we to know that no one of a set of terms $t_1, \cdots, t_n$ is definable in terms of the others in the sense just described—i.e., is to be *independent* of the other terms? According to Padoa, if for each term t_i we can find an interpretation of all the terms $t_1, \cdots, t_n$ satisfying the axioms, and which continues to satisfy them when the interpretation of t_i is changed while the interpretations of the other terms are left unchanged, then t_i is not definable by means of the other terms; i.e., t_i is then an independent term of the system. The analogy to independence of axioms is evident. A statement defining t_i in terms of the other undefined terms should leave t_i unchanged when the other terms are left unchanged.

For example, consider the systems Γ_6 and Γ_6'. In Γ_6, "line" is not independent, since the statement "A collection of points constitutes a line if and only if it contains exactly two points" suffices to define the term.† (See Problem 5.) But in Γ_6', "line" is independent, for we can easily give two models of Γ_6' in which the same points are involved, but "line" has different meanings (i.e., without change of the points, the triples constituting lines can be specified differently for the two models).

Or consider the system of euclidean geometry given by Veblen in 1904 [a]. Only two terms, "point" and "order," were left undefined, all other geometrical terms presumably being defined in terms of these. However, as was first pointed out by Tarski [Ta, 306–307] (see also his paper [b]), the definition of "congruent" involves not just "point" and "order," but a geometric entity E which is not postulated by the system, but is arbitrarily chosen. Consequently, it is possible to find two models of Veblen's system consisting of the same points and having the same order relations between points, but in which congruence is different in the sense that two given pairs of points may be congruent in one model but not in the other.

5.3 We cannot extend the analogy to a concept of "completeness" of undefined terms in the spirit of Definition 4.3.1. It would make no sense to say that a collection T of terms is "complete" in an axiom system Σ if it is impossible to add a new undefined term t such that t is independent in the set consisting of t and the terms of T. For the axioms of Σ are stated

† Note that Axioms 1 and 3 now follow from the definition of "line."

by means of the terms of T and do not involve t at all. We can, however, consider the effect of simultaneously adding new terms and axioms involving them—a process which we shall use later. This may be possible in two ways. (1) Without enlarging the domain of individuals involved, we may introduce new relations and axioms involving them, or axioms naming specific individuals may be added. Examples of this are found in arithmetic where certain units may be specified (as "0" or "1"), or where new operations are introduced. (2) The domain of individuals may be enlarged; for instance, from the arithmetic of real numbers to that of complex numbers (involving naming a new unit "i"), or enlarging plane geometry to yield solid geometry.

In [T_2, 308], Tarski used the concept of categoricalness to define the notion of "completeness of concepts," subject to process (1). He first defined a system Σ_1 to be *essentially richer than a system* Σ_2 *relative to undefined terms* if (a) every axiom of Σ_2 is in Σ_1 (hence, all undefined terms of Σ_2 are in Σ_1); and if (b) in the axioms of Σ_1 there are undefined terms which are not among the terms of Σ_2, that cannot be defined, on the basis of Σ_1, by means of the undefined terms of Σ_2. For example, the axioms for the order type of the points on a euclidean straight line (see VI 1.4)—i.e., the "linear continuum"—given in terms of the undefined term "point" and an undefined relation between points, "$<$," form a categorical system Λ. If the relation of "congruence" is added along with suitable axioms, we obtain a new system Λ' which is both categorical and essentially richer than Λ relative to undefined terms. Then by further introducing two new undefined symbols, "0" and "1," naming certain individual points, and a new axiom stating that these symbols denote two distinct points, a system Λ'' is obtained—one which is essentially richer than Λ' relative to undefined terms. (This example is cited from Tarski, *loc. cit.*). But we can go no further by process (1), and Tarski calls system Λ'' complete relative to undefined terms. That is, an axiom system Σ is *complete relative to its undefined terms* if there does not exist a categorical system Σ' which is essentially richer than Σ relative to undefined terms.

We shall not go further in this direction except to remark that Tarski (*loc. cit.*) worked out an interesting test sufficient for this type of completeness. It may be described, roughly, as follows: Suppose Σ is an axiom system having at least one model. Also suppose that, if $\mathfrak{M}_1$ and $\mathfrak{M}_2$ are any two models of Σ, a (1-1)-correspondence constituting an isomorphism (with respect to Σ) can be set up between $\mathfrak{M}_1$ and $\mathfrak{M}_2$ in one and only one way. Then Σ is called *monotransformable.* If a system Σ is monotransformable, it is complete relative to undefined terms.

For further information concerning these notions, the reader is referred to Tarski (*loc. cit.*) and to papers cited therein, as well as to Tarski [T; IX] and E. W. Beth [a]. An interesting discussion and application of Padoa's method will be found in McKinsey [a].

6 Miscellaneous comment

It was emphasized in Chapter I (I 5) that in setting up an axiom system Σ we usually have some concept C in mind that we are actually describing. The concept C is not only the source of our axioms, but, if it is a valid model of Σ, it also *predetermines* the consistency of Σ. Moreover, if C possesses the requisite sort of uniqueness, then the axiom system Σ will be categorical if it is sufficiently descriptive of C.

6.1 Now there are cases where, instead of C preceding Σ, we have Σ preceding C. This happens whenever Σ is derived from a previously existing axiom system Σ' by alteration (such as by denying or deleting an axiom). For example, if from the concept C of euclidean geometry we derive a complete axiom system Σ' that includes the "parallel axiom" (Axiom 5 of the system Γ in I 2.2), we might derive a non-euclidean geometry Σ from Σ' by replacing the "parallel axiom" with a denial of the same—and then look for a concept which Σ' might describe. Historically, this is essentially what happened in non-euclidean geometry.† And we duplicate this process whenever we prove independence of an axiom A in an axiom system Σ, since we first set up the system $(\Sigma - A) + {\sim}A$ and then look for an interpretation of the latter.

Of course, the derivation of new axiom systems from old can degenerate into a game, but frequently the "game" turns out to be sufficiently interesting to justify itself, and is often found useful in both mathematics and its applications.‡ Many important concepts are developed in modern algebra, which makes great use of the axiomatic method, through the investigation of "altered" axiom systems.

6.2 Advantages of the axiomatic method

Certain advantages of the axiomatic method have already been alluded to. For example, we mentioned (4.9; also see I 1.5, I 3.2) the "economy" achieved when an axiom system Σ has many models in the same or different branches of mathematics. A single theorem in Σ yields a theorem in each interpretation; but the latter requires no special proof so long as the theorem was proved in the system Σ. An excellent example of this,

† Cf. Bell [B_1; 40–46], for instance.

‡ Cf. Bell [B_1; 36–40].

mentioned in 4.9, is the axiom system for a *group*, which we shall study in the sequel; this notion, defined axiomatically, is one of the most outstanding "labor-saving devices" in modern mathematics. Other examples will be given in Sections 7 and 8 which follow. And, obviously, once we have recognized, in some branch of mathematics, that a certain setup or structure constitutes a model of a previously known axiom system Σ, then we are likely to see many new possibilities opening up because of the knowledge already embodied in the study (previously made) of Σ. In this way, the axiomatic method not only achieves economy but also throws new light wherever interpretations of known axiom systems are possible.

We have also just pointed out that, by altering existing axiom systems, new mathematical concepts may be generated—not always fruitful concepts, to be sure, but nevertheless frequently of sufficient importance to justify their study.† In this manner, new branches of mathematics may be created, and the method becomes (as in modern algebra) a useful research tool.

Another distinct advantage of the method that deserves special mention is its character of implicit "definition." Although the genesis and development of a mathematical concept may proceed along several entirely different lines, once the concept has matured, so to speak, the axiomatic characterization of it may prove extremely advantageous. For example, the development of the real number system, which forms the foundation of modern analysis, was a slow evolution over many centuries. Today, as we shall see, we can give it a precise axiomatic definition and study its properties by means of the theorems based on the axioms. Many other mathematical concepts have developed in a similar manner.

An advantage, related to the above, which may or may not have occurred to the reader while we were proving theorems on the basis of the system Γ, is that of "placing the responsibility where it belongs." Frequently a mathematics student exclaims, "I could work this problem, I think, if I only knew what I could assume!" This is an oft-heard complaint. For, if he is not certain about just what assumptions he can make, how is he to proceed with a proof? However, if we present a student with an axiom system Σ and wish him to prove a certain Theorem A, he can say, "Here is a Σ-statement, A. Prove A"; we have thus given a specific direction to the student who now knows exactly what his "point of departure" is. For this reason, the axiomatic method forms a useful pedagogical device in the teaching of mature students. (See Wilder [c].)

† We recognize that some of the above commentary is probably productive of questions in the reader's mind such as, "What is a *fruitful* concept?" We prefer to delay the discussion of such matters, with the assurance that we shall discuss them later.

6.3 Disadvantages of, and objections to, the axiomatic method

"Nothing is perfect," and the axiomatic method turns out to be no exception. Most of the criticisms of the method, as it is commonly employed and as we have described it, are too complicated to permit simple exposition in the compass of a single section, and will have to be allowed to emerge as we progress through the succeeding chapters. For not the least of the disadvantages of the axiomatic method as it is usually employed are related to its utter dependence on logic—not only the Aristotelian "laws," but also the assumed "universality" of the logical terms. If it should turn out that the logical machinery itself reveals flaws, then what faith can be placed in the reliability of the theorems deduced?

Some mathematicians object to the highly formal character of the method, particularly its foundation in undefined terms, basic assumptions (axioms) in terms of these, and logical deduction capable of being carried through without any interpretation in mind whatsoever. They feel that formality is quite good and entirely justified in some instances, but that it can be carried too far, especially when it is used, for instance, to treat the ordinary integers 1, 2, 3, · · ·, as *undefined* terms. (See the Peano axioms in VI 3.1; also see the criticism by Russell [R_1; I] in that connection, and Klein [Kl; 14–15].) Clearly the method is not for the immature mind. Most mathematicians and pedagogues are agreed that in teaching geometry to high school students it is much better to allow the student to take for granted as "obvious" all those fundamental notions whose formalization would only seem a waste of time to him.

There are mathematicians, to be sure, who will maintain that all mathematics should be placed on an axiomatic basis. This does not mean that they ask that a *single* axiom system should be given for all mathematics. This would certainly be impossible, and tantamount to a *definition of mathematics*—something which, as we shall see, it would be almost a miracle to get as many as five† mathematicians to agree upon. But it is asserted by many that each branch or portion of mathematics should be axiomatized. At the other extreme are mathematicians who will have nothing to do with axioms, whose conception of mathematics is such that formalization immediately squeezes the *mathematical* substance out of the material formalized.

7 Axioms for simple order

Before leaving the study of the axiomatic method, let us describe two very simple examples of concepts which appear in various guises in mathematics, and which were axiomatized and studied for their own sake so that

† Perhaps "two" would be correct here, but we would rather allow a little margin for error.

the "economy" feature of the method (application to many different models) might be utilized. One of these is the concept of *simple order* (frequently called *linear order*). It is probably unnecessary to demonstrate to the reader that "ordered" collections abound in mathematics. The points of a line, the integers 1, 2, 3, $\cdots$ used in counting (these we shall call the *natural numbers*†), are two of the most common examples of ordered collections. And the notion is of such a fundamental character that it is abundantly exemplified in our physical surroundings; we order our streets, our books on shelves, soldiers in file, and so on.

What common element is involved in all such cases? Simply that, if p and q are distinct elements of any one of these collections, then either p precedes q or q precedes p. In the case of points on a line, "precedes" may mean "is to the left of"; in the case of the natural numbers, "2 precedes 5" means "2 is smaller than 5"; in the case of streets running "north and south," "precedes" may mean "is to the west of"; and so on. Consequently we make "precedes" the basic term which receives meaning according to the nature of the collection ordered; that is, *precedes* is adopted as an undefined technical term. And, as the objects that occur in ordered collections also vary from one collection to another, we adopt some undefined term for these; *point* is as good as any.

So far as the axioms themselves are concerned, we naturally start with the property that we have already noticed is common to all simply ordered systems, namely, "If p and q are distinct points, then either p precedes q or q precedes p." But is this all we need state? Suppose p and q are not distinct points; shall we let p precede q or vice versa? Since this is clearly not the case in any of the examples that we have scanned, we ought to rule it out by an axiom: "If p precedes q, then p and q are distinct points." And now do we have all we need by way of axioms? To see that we do not, we need only point out that, if we would avoid circular or "cyclical" order, we need to say more. Let p, q, r be the vertices of an equilateral triangle inscribed in a circle C, and let "x precedes y" mean that x and y are distinct and that the arc from x to y on C in a clockwise direction is less than a semicircle. Then if p, q, r occur in the order stated in a clockwise direction around C, p precedes q, q precedes r, and r precedes p. This example is a model of the system formed by the two axioms already stated, and, if we wish to restrict ourselves to defining *linear* rather than *cyclical* order, it should be ruled out. This may be done by stating "If p precedes q and q precedes r, then p precedes r." In all the examples previously mentioned this is a true statement, but it fails in the example of cyclical order just given.

It is not obvious, of course, that the three statements fixed upon above

† Some writers include 0 among the natural numbers.

suffice to describe exactly what we mean by linear order. This becomes apparent only when we see what the three statements imply. For certainly we are not looking for a *complete* axiom system—an ordered collection may contain any number of "points," and consequently not every two models of the axioms we decide on are going to be isomorphic. If we are interested in defining a linearly ordered *triple*, we would of course add the statement "There exist three and only three points," and we could then show, without much labor, that the collection of four statements forms a categorical axiom system; whereupon it would be apparent that we had made enough basic assumptions to accomplish our purpose (i.e., to define a linearly ordered triple).

The three statements about order can be presented succinctly in the following way: Let C be any collection whose elements we call *points* and which are denoted by small letters $x, y, z, a, b, \cdots$; and let there be given a "binary relation"† between the points of C which we denote by "$<$" and call "precedes." Then C is called *simply ordered*, or *linearly ordered*, relative to $<$, if the following axioms (cf. Huntington [Hu; 10]) hold:

(1) If x and y are distinct points of C, then $x < y$ or $y < x$.‡
(2) If $x < y$, then x and y are distinct.
(3) If $x < y$ and $y < z$, then $x < z$.

On the basis of these axioms we can now prove such theorems as "If x and y are points of C, then not both $x < y$ and $y < x$ hold"; "If n is a natural number such that C contains exactly n points, then symbols $x_1, x_2, \cdots, x_n$ can be so assigned to these points as to make the relations $x_1 < x_2, x_2 < x_3, \cdots, x_{n-1} < x_n$ all hold." Note that the latter statement goes a long way toward justifying the feeling that (1)–(3) above imply all that we wish concerning linear order, assuming that we want to keep sufficient "incompleteness" to allow many different interpretations. Let us call the above axiom system O.

(The reader may like to establish the independence of the axioms of O.)

7.1 The statement "$(C, <)$ is a simply ordered system" may be used as an abbreviation for the statement "C is a set and $<$ is a binary relation between elements of C such that C is simply ordered relative to $<$."

† Here we treat "relation" between two things as a universal logical notion (cf. 6.3 "Disadvantages . . ."). For analysis of the notion in terms of classes, see VIII 8.2.7.1; in terms of propositional functions, see IX 5.2.2.

‡ We always use "or" in the sense of "and/or." Thus, so far as Axiom (1) is concerned, both "$x < y$" and "$y < x$" might hold. However, the combined Axioms (1)–(3) will rule out this possibility.

8 Axioms defining equivalence

One of the most used (and abused) notions in mathematics is that of *equivalence relation*. Like the notion of simple order, it is also exemplified in our social and physical environment. Wherever *classification* occurs, we find what the mathematician calls *equivalence*. We merely agree to call all those things in the same class "equivalent." For example, if in a given city we classify all people according to their church membership, placing in a single class all those belonging to the same church (and in a single class all those not belonging to any church), then there is established an equivalence with respect to church membership; all Catholics would be considered "equivalent," all Baptists would be "equivalent," etc. Or we might classify them according to age, two persons being then "equivalent" if they are of the same age in years. Thus two persons, x and y, might be equivalent in the sense of church membership, and not equivalent with respect to their ages. Similarly, in a college, two students may be equivalent with respect to their status as freshmen, sophomores, etc.

8.1 Without pausing to attempt an analysis of their origin (see Peano [P_1; 9]) or to justify the particular choice of statements, we give the usual axioms for equivalence as follows: Let S be any collection whose elements we denote by small letters $x, y, \cdots$, and $\approx$ a binary relation between elements of S. (We avoid the symbol $=$ since it has too many connotations, such as "identity.") Then $\approx$ is called an equivalence relation in S if the following axioms hold:

(1) If x is an element of S, then $x \approx x$.
(2) If $x \approx y$, then $y \approx x$.
(3) If $x \approx y$ and $y \approx z$, then $x \approx z$.

Note that (3) is the same as (3) of the system O of Section 7, if the difference in symbols denoting the relations is ignored—this property of a binary relation is called *transitivity*—and that it holds for both $<$ and $\approx$ is expressed by stating that both these relations are *transitive*. Similarly, (1) is called *reflexiveness* and (2) *symmetry*. Thus the relation $\approx$ is *reflexive*, *symmetric*, and *transitive*. Let us call this axiom system E.

8.2 In the examples cited above, perhaps statements (1) and (2) appear trivial. Thus, in the classification according to age, it appears too trivial to mention that a person is the same age as himself. And, of course, if x is the same age as y, then y is the same age as x. The necessity for stating these properties as axioms becomes apparent as soon as we begin the "justification process"; i.e., showing that the axioms imply all that we wish to have implied (without implying too much). And, if x is the same age as

y, and y the same age as z, then x is the same age as z. Evidently in any decomposition of a collection into classes, where no two classes overlap, the meaning "x is in the same class as y" for "$x \approx y$" constitutes an interpretation of the axiom system E. Incidentally, the system is consistent in view of the interpretations stated above. The reader will find it amusing to establish the independence of the axioms in E. [In connection with axiom (1), note that it is never essential in a case where for each element x of the collection S there is some element y of S such that $x \approx y$, since in this case $x \approx x$ will follow from (2) and (3). In general, however, (1) is independent.]

8.3 Now the use of equivalence relations in mathematics is usually of a sort that is the direct converse of the manner in which we have presented the notion above. Instead of starting with a collection S which is already split into non-overlapping classes, and getting the corresponding equivalence relation, the mathematician notices the presence of an equivalence relation and then sets up the classes. A simple illustration of this is furnished by the collection of natural numbers, with "$x \approx y$" meaning "x differs from y by a multiple of 3" (0 is a multiple of 3—the 0th multiple). This is obviously an equivalence relation, and if we decompose the collection of natural numbers into classes, putting into the same class two numbers that differ by a multiple of 3, we get three classes of numbers: $(1, 4, 7, \cdots)$, $(2, 5, 8, \cdots)$, and $(3, 6, 9, \cdots)$.

8.4 A better example, and one of great interest in connection with the subject of this chapter, is furnished if we notice that *isomorphism* forms an equivalence relation. Specifically, if Σ is an axiom system, $\mu(\Sigma)$ is the collection of all models of Σ, and if $\mathfrak{M}$ and $\mathfrak{N}$ are models of Σ, let "$\mathfrak{M} \approx \mathfrak{N}$" mean "$\mathfrak{M}$ and $\mathfrak{N}$ are isomorphic with respect to Σ." Note that the axioms of E become true statements with this meaning for "$\approx$," so that we get an interpretation of E. In this case, if we decompose $\mu(\Sigma)$ into classes, placing two models in the same class if and only if they are isomorphic, we notice that *categoricalness of Σ corresponds exactly to the case where only one class results.* In the case of non-categoricalness there might be any number of classes. In a way, the number of such classes serves as a "measure of completeness" of an axiom system; the fewer the number of classes, the greater the amount of completeness present.

8.5 We notice then that, corresponding to any collection S and an equivalence relation $\approx$ in S, there always exists a unique decomposition of S into classes, called *the class decomposition of S corresponding to $\approx$.* We shall leave to the reader the proof of this on the basis of the axiom system E. He may accomplish this by using the following outline:

Definition. If x is an element of S, let $S(x)$ denote the collection of all elements y of S such that $x \approx y$.

Theorem 1. If x is an element of S, then $S(x)$ is not empty; in particular, x is an element of $S(x)$.

Corollary. Every element x of S is in some $S(x)$.

Theorem 2. If $x \approx y$, then the classes $S(x)$ and $S(y)$ are the same.

Corollary. A class $S(x)$ is completely determined by any one of its elements.

As a consequence of this corollary, we frequently say that any element of an "equivalence class" $S(x)$ may serve to *represent* $S(x)$, or to be a *representative element* of $S(x)$.

Theorem 3. If two classes $S(x)$ and $S(y)$ have a common element z, then they are the same class.

Theorem 4. The classes $S(x)$ constitute a unique decomposition of S into non-overlapping classes; we call this *the class decomposition of* S *corresponding to* $\approx$.

SUGGESTED READING

Birkhoff [Bi; I]
Hilbert [H_3; §§ 8–12]
Hilbert and Bernays [H-B; § 1]
Huntington [Hu; II]
Richardson [R; XVI, 448–457]
Tarski [T], [T_2]
Veblen [a]
Veblen and Young [V-Y; Intro.]
Young [Y; IV–V, VII]

PROBLEMS

1. Show that if Σ_1 and Σ_2 are axiom systems such that Σ_1 is consistent and Σ_2 has a model in Σ_1, then Σ_2 is consistent.
2. If an axiom system that is known to be consistent has a model in a system Σ, need Σ be consistent? (Illustrate by means of an example.)
3. Show that an empty collection is a model of the axiom system O of Section 7. Does the existence of this model prove O consistent?
4. Show that axiom (1) of the system E of Section 8 is independent in E.
5. Prove on the basis of Γ_6 that "line" may be defined in terms of "point." (Recall that since the axioms of Γ are included among those of Γ_6, the theorems proved in Chapter I from Γ are valid for Γ_6.)

Let us designate by O′ the axiom system O of Section 7 augmented by the following axiom:

Axiom 4. There exist two and only two distinct points in C.

6. Show that the system O′ is satisfiable.
7. Is the system O′ consistent by Definition 1.1?
8. Show that Axiom 3 is independent in the system O′.
9. Show that the axiom system obtained by deleting Axiom 3 from O′ is not categorical. (Note, incidentally, that although Axiom 3 is only vacuously satisfied in every model of O′, it does give information about the model.)

10. Show that if in axiom system O′ Axiom 3 is replaced by the axiom "Not both $x < y$ and $y < x$ hold," then the resulting axiom system is categorical.

11. Show that if the axiom system O of Section 7 is augmented by the axiom "There exist three and only three points," then the resulting system O_3 is categorical.

12. Let O_3' denote the system O_3 of Problem 11 augmented by some axiom of the form "If a, b, c, d are four distinct points, then . . ." Is this axiom independent in O_3'?

13. If $\mathfrak{S}$ is a collection whose elements are themselves collections (for example, $\mathfrak{S}$ might be a collection of libraries, each library being itself a collection of books), and A, B are elements of $\mathfrak{S}$, let A $\leftrightarrow$ B mean that there exists a (1-1)-correspondence between the elements of A and the elements of B (cf. 4.4.2). Show that $\leftrightarrow$ is an equivalence relation in the collection $\mathfrak{S}$.

14. Show that if two simply ordered sets A and B have the same number, n, of elements, then they are isomorphic relative to the simple order axioms.

15. Let Σ be an axiom system and suppose that, if $\mathfrak{M}_1$ and $\mathfrak{M}_2$ are any two models of Σ, there exists a (1-1)-correspondence between the elements of $\mathfrak{M}_1$ and the elements of $\mathfrak{M}_2$ which preserves those Σ-statements that form the axioms of Σ. Is Σ categorical? [*Hint:* Consider axiom system O′ with Axiom 3 deleted.]

16. If C is any collection and $\leqq$ a binary relation between certain pairs of elements of C, then C is called *partially ordered with respect* to $\leqq$ if the following axioms hold:

(1) For every element x of C, $x \leqq x$.

(2) If $x \leqq y$ and $y \leqq x$, then x and y denote the same element of C.

(3) If $x \leqq y$ and $y \leqq z$, then $x \leqq z$.

[Note that (1) and (3) are, respectively, the reflexiveness and transitivity of $\leqq$. The property of the binary relation stated in (2) is called *anti-symmetry*.]

Denote this axiom system by P. Prove P consistent and non-categorical; establish independence of each of its axioms.

17. (a) If M is a collection of objects of various ages, and the relation $\leqq$ is interpreted to mean "is at least as old as," is M partially ordered with respect to this relation? (b) Show that every collection of sets is partially ordered with respect to the relation "is contained in" (compare III 3.1.2.)

18. Let F be an axiom system with undefined terms *figure* and *rectangular*, the former being an undefined object and the latter an undefined property of figures. The axioms of F are:

(a) There exist at most twenty figures.

(b) If x is a figure, then x is rectangular.

(c) If x is a figure, then x is not rectangular.

Show that F is satisfiable and categorical, and that its axiom (c) is independent. How would the system be affected if in (a) "at most" were replaced by "at least"?

19. Let Σ be an axiom system and T a theorem in the proof of which an axiom A of Σ is used. Show that if we wish to demonstrate that T cannot be proved without using axiom A, it suffices to give a model of Σ – A in which T fails to hold.

20. Is the axiom system Γ_6' of 4.4.5 categorical? Is it a monotransformable system? [A determinant D of order 3 yields a model of Γ_6' if the triples used in evaluating D, as well as rows and columns, are called "lines."]

21. The definition of independence in 3.1.1 is stated in terms of satisfiability. Suppose, however, that the word "satisfiable" is replaced by "consistent" in 3.1.1. Then show that, as a result, Definition 4.3.1 becomes equivalent to: "An axiom system Σ is complete if the addition of a Σ-statement, A, not implied by Σ, results in contradiction."

22. Show (under assumptions 2.2.1 and 2.2.2) that completeness of an axiom system Σ in the new sense given in Problem 21 in quotations implies that Σ is complete in the sense of Definition 4.3.1 (where "independent" is understood in terms of satisfiability as given in 3.1.1).

23. If Σ is an axiom system, define a Σ-question to be a question of the form "Does A hold?" where A is a Σ-statement. Then define Σ to be complete if it implies an answer to every Σ-question. Show that this definition is equivalent to that given in Problem 21 (in quotations).

24. If Σ_1 and Σ_2 are axiom systems having the same set, T, of undefined terms, define $\Sigma_1 \approx \Sigma_2$ to mean that the collections $\mu(\Sigma_1)$ and $\mu(\Sigma_2)$ of models are the same. Is this relation $\approx$ an equivalence relation in the collection of all axiom systems based on T (i.e., having T as their sets of undefined terms)?

25. If Σ_1 and Σ_2 are axiom systems based on T (as in Problem 24), define $\Sigma_1 \leqq \Sigma_2$ to mean that $\mu(\Sigma_1)$ contains $\mu(\Sigma_2)$ as a subcollection. Is the collection of all axiom systems based on T partially ordered with respect to this relation $\leqq$? (See Problem 16.)

26. Let Σ_1 and Σ_2 be axiom systems as in Problem 25. Define $\Sigma_1 \Rightarrow \Sigma_2$ to mean that the axioms of Σ_2 are provable (as theorems) in the axiom system Σ_1 (we may read "$\Sigma_1 \Rightarrow \Sigma_2$" as "$\Sigma_1$ implies Σ_2"). Show that $\Sigma_1 \Rightarrow \Sigma_2$ implies that $\Sigma_2 \leqq \Sigma_1$ (the latter being defined as in Problem 25).

27. With Σ_1, Σ_2, etc., as in Problem 26, define $\Sigma_1 \Leftrightarrow \Sigma_2$ to mean that both $\Sigma_1 \Rightarrow \Sigma_2$ and $\Sigma_2 \Rightarrow \Sigma_1$. Show that $\Leftrightarrow$ is an equivalence relation in the collection of all axiom systems based on T.

28. Let P' denote the axiom system obtained when axiom (1) of the system P (Problem 16) is replaced by the axiom "If x and y are elements of C, then $x \leqq y$ or $y \leqq x$." If, in P', "$\leqq$" is replaced by "$<$ or identical with," what relation is there between the resulting axiom system and the axiom system for simple order (Section 7)?

29. Problem 28 suggests generalizations of the "$\Rightarrow$" and "$\Leftrightarrow$" of Problems 26 and 27, such as the following: If Σ_1 and Σ_2 are axiom systems based respectively on collections T_1 and T_2 of undefined technical terms, let $\Sigma_1 \Rightarrow \Sigma_2$ mean that the elements of T_2 may be so defined in terms of the elements of T_1 as to make the axioms of Σ_2 provable theorems in the axiom system Σ_1. Discuss.

30. Let us alter axiom system P of Problem 16 as in Problem 28 to obtain system P'. Show that if we define $x < y$ to mean "x is not identical with y and $x \leqq y$," then $P' \Rightarrow$ O in the sense of Problem 29. (Does O $\Leftrightarrow P'$ hold?)

Problems 31–35 will be based on the following axiom system:†

Let S be a collection, of which certain subcollections are designated as *m-classes*. By definition, two such *m*-classes are called *conjugate* if they have no element in common. The following axioms are to hold (we denote elements of S by small letters $x, y, z, \cdots$):

† See Problem 2894 in the *American Mathematical Monthly*, vol. 29 (1922), pp. 357–358, for solutions as well as a discussion by Professor O. Veblen of the origin of this axiom system in his class.

(1) If x and y are distinct, there is one and only one m-class containing x and y.
(2) For every m-class, there is one and only one conjugate m-class.
(3) There exists at least one m-class.
(4) No m-class is empty.
(5) Each m-class contains only a finite number of elements of S.

31. Show that every m-class has at least two elements.
32. Show that S contains at least four elements.
33. Show that S contains at least six m-classes.
34. Show that no m-class has more than two elements.
35. Prove that the given axiom system is categorical.

36. Show that the axiom system on which Problems 31–35 are based is equivalent (as in Problem 27) to the system Γ_6 of Section 4.3, if elements of S are called *points*, and *line* is substituted for m-class.

III

Theory of Sets

In Chapter I, Section 2, we gave an example of a simple geometric axiom system. The first axiom of this system stated: "Every line is a collection of points." The corollary to Theorem 1 (I 3.3) stated: "Every line contains at least one point"; i.e., the collection of points constituting a line is never empty. It will be recalled that in connection with this corollary some justification was given for the use of the notion of *empty collection*, and it was remarked that, although "collection" was one of the supposedly "universal" terms, we found it necessary to explain, in part at least, just how or in what sense we were using the term. In the same connection it was also remarked that we were touching upon a question that has caused considerable concern to mathematicians and logicians, namely, *What is a collection?*

Of course, we can look up the word "collection" in a dictionary, but if we do we find such definitions as "a group of collected objects or individuals," "an aggregation," "accumulation." These will hardly be of much help. And the mathematician uses "definition" in a different sense from that of the dictionary. When a mathematician gives a definition, it is intended that it will be not a mere synonym (such as "aggregation" for "collection") which the reader may happen to know the meaning of, but a criterion for identifying; a *characterization* of the thing defined.† If the "definition" is not of this kind, it is of little use to the mathematician.

During the past fifty years, there have been proposed almost as many "definitions" of the term "collection" as there have been of the term "mathematics" itself (we shall come to the latter in due time). And none has yet met the criterion that it serve satisfactorily to identify the notion. Let us look into the reasons for this state of affairs concerning a notion that must seem so trivial to anyone who has never looked into it before.

† Incidentally, this kind of definition is probably one of the most difficult notions to get over to the student of mathematics. Try as we will, it sometimes seems impossible to convince a student that, if he wants to know whether something is a certain type of mathematical entity which has previously been defined, he should *see if it satisfies the criterion given in the definition*. This sounds trivial, probably, to a person who has never *taught* mathematics, but I am sure that any teacher of mathematics will understand.

1 Background of the theory

In the first place, let us straighten out the synonyms for "collection." This chapter is labeled "theory of sets" because the American mathematician uses the word "set" more frequently than "collection"; this is probably because of its brevity, for mathematicians like abbreviations. However, our English colleagues seem to prefer "aggregate"; an Englishman would have entitled this chapter "theory of aggregates." On the other hand, our American colleagues who teach philosophy, including a course in logic, would probably label it "theory of classes." And mathematicians have frequently employed the terms "family," "system," and "group," although these words are usually reserved for more special types of collections.† It is probably safe to say that in the United States mathematicians prefer the terms *set*, *collection*, *class*, in that order.‡

Since some readers may wish to consult French or German works, we might interpolate here the remark that the French term for "collection" is "ensemble," and the German term is "Menge." (In French the title of this chapter would become "Théorie des ensembles"; of course the Germans have combined it into one word, "Mengenlehre"). In French texts the letter E (with or without an index as in E_i) is almost invariably used to denote a set, and in German texts the letter M is used.§

Of course collections have been used in mathematics since the very beginnings of the subject; by the Greeks, for instance. The modern theory of sets is usually considered to have begun with Georg Cantor (1845–1918), who devised the first numbers for infinite sets (i.e., *transfinite numbers*) during the latter part of the nineteenth century.|| It is revealing to read the "definition" of set given by Cantor. The first sentence of the first of the papers just cited is (translated): "By a 'set' we shall understand any collection into a whole, M, of definite, distinguishable¶ objects m (which will be called 'elements' of M) of our intuition or thought." This

† For example, a mathematician would be very unlikely to use the term "group" instead of "set" or "collection" if he actually meant the latter, because of the special use (see Chap. VII) for which the term "group" is reserved today.

‡ Undoubtedly a "collection of sets" sounds better, to many, than "a set of sets." We never seem to see the term "a set of collections."

§ In the early part of the present century, when so many American mathematicians went abroad for their doctoral or post-doctoral work, we could usually detect which foreign influence was dominant, on their return, by the preference for E or M to denote a collection.

|| Fortunately, Cantor's two most authoritative papers (see Cantor [C; 282ff]) are available in an English translation by Jourdain; Cantor [C_1].

¶ In his translation (Cantor [C_1]), Jourdain says "definite and separate." However, it seems likely that by "bestimmten wohlunterschiedenen Objekten," Cantor meant to imply that the objects are distinguishable from one another in some sense.

was written by a first-class mathematician who clearly did not know, at the time he wrote, that the word "set" was "loaded" (as did nobody else at the time).

1.1 Enlightenment was not long in forthcoming. As a matter of fact, it was virtually ready not long after Cantor published his ideas, as a result of the announcement† by an Italian logician, Burali-Forti, of a fundamental difficulty with one of Cantor's basic definitions. Unfortunately Burali-Forti misinterpreted the definition, so that his insight did not at first win recognition (for the objection he raised was just as valid when opposed to the correct interpretation). Soon after, Russell announced his famous "antinomy," described below, and this time there was no attempt to avoid the difficulty by subterfuge or otherwise, and the problem of "what to do?" had to be met head on. Once the dam was broken, the flood was on, and the construction of contradictions and paradoxes became almost a mathematical "indoor sport."

1.2 Notice that we said "contradictions and paradoxes." The examples constructed were generally of two kinds: those that were bona fide contradictions, and those that were only "apparent contradictions" without actually being contradictory. The former are frequently given the technical name *antinomy*; the latter we do not worry about, although we often use them for amusement or instructional purposes (for instance, in the case of the "proof" that all triangles are isosceles cited in I 3f). We shall describe both the (amended) Burali-Forti and Russell examples, since they are real contradictions.

The Burali-Forti contradiction we shall defer until a later section (V 3.5.3), however, since it requires an acquaintance with the notion of "ordinal number." The Russell antinomy involves no such technicalities, and for our future purposes we shall state it in the following form.

2 The Russell contradiction

Consider any collection S. Its elements are to be thought of as individual objects—each a "unit." However, in general the elements will themselves be sets if we care to "dissect" them. Thus a collection of books has, for each element, a single, individual book. But, if we wish, we can consider a book not as an individual, indivisible object, but as an (ordered) collection of printed words; and, in turn, each word as an (ordered) collection of letters.

Since, generally, the elements of S may be sets themselves, the possibility arises that S may happen to be an element of itself. For example, some-

† According to F. Bernstein, however, Cantor had already become aware of possible difficulties; see V 3.5.3f.

one (the name has escaped us) has suggested the set of all abstract ideas; such a set is certainly an element of itself if we grant (and who wouldn't!) that it is itself an abstract idea. If the notion seems rather far-fetched, however, there is no need to worry about it, since the Russell antinomy involves only sets that do *not* have themselves as elements. Such sets we may call *ordinary* sets. Thus, the set of natural numbers is ordinary—and most sets that we use are ordinary.

2.1 Now suppose that we assemble all ordinary sets in one collection $\mathfrak{U}$. Then the elements of $\mathfrak{U}$, by definition, are ordinary sets, and every ordinary set is an element of $\mathfrak{U}$. Using exactly the same type of reasoning that we used in high school geometry, we prove the following two theorems:

Theorem A. *The set* $\mathfrak{U}$ *is an ordinary set.*

Proof. For suppose $\mathfrak{U}$ is not an ordinary set. Then $\mathfrak{U}$ is an element of itself.

But the elements of $\mathfrak{U}$ are all ordinary sets. Hence, as $\mathfrak{U}$ is an element of $\mathfrak{U}$, it is an ordinary set.

Thus the supposition that $\mathfrak{U}$ is not ordinary leads to the conclusion that $\mathfrak{U}$ is ordinary, hence to a contradiction. We must conclude that our supposition is false, hence that $\mathfrak{U}$ is ordinary. Q.E.D.

Theorem B. *The set* $\mathfrak{U}$ *is not an ordinary set.*

Proof. For suppose $\mathfrak{U}$ is an ordinary set. Then $\mathfrak{U}$ is not an element of $\mathfrak{U}$, by definition of "ordinary."

But *all* ordinary sets are elements of $\mathfrak{U}$. Hence, if $\mathfrak{U}$ is not an element of $\mathfrak{U}$, it is because $\mathfrak{U}$ is *not* ordinary.

Thus the supposition that $\mathfrak{U}$ is ordinary leads to the conclusion that $\mathfrak{U}$ is not ordinary, hence to a contradiction. We must conclude that our supposition is false, hence that $\mathfrak{U}$ is not ordinary. Q.E.D.

2.2 Theorems A and B plainly contradict one another. And we have obtained this contradiction by purely logical deduction. Furthermore, if "set" is to be regarded as a logical notion, as many maintained, the whole affair, from the definition of the set $\mathfrak{U}$ to the end of the proof of Theorem B, is purely logical.

The set $\mathfrak{U}$ is often called "the Russell set." Actually, in view of the above, it cannot be admitted as a set if we are to continue to hold fast to the Law of Contradiction. Some have also called $\mathfrak{U}$ a "self-contradictory set."

An amusing (except to philosophers, who can substitute the word "mathematicians" for "philosophers" below) formulation of Russell's contradiction was given by Grelling [Gr; pp. 44–49] in a "proof" that "the set of all philosophers is a camel."

It is not surprising, then, that the first reaction of some mathematicians to this matter was that it concerned "logicians," not "mathematicians," so let the logicians find what was wrong and set their house in order! But this attitude was short-lived, since, if mathematics is to use logic, any defect of logic is of concern to mathematics—especially when the defect is found in a concept that is so frequently used in mathematical definition and proof. The natural result was a movement among mathematicians, slow at first but accelerating in recent years, toward the study of logic. And, whereas logic was traditionally a cut-and-dried rehash of the work of Aristotle—with great concentration on the syllogism, etc.—it has today become a live and growing field of investigation, known under the name of *symbolic logic* or *mathematical logic*.

2.3 Two comments should be made here. In the first place, symbolic logic itself had its beginnings long before the discovery of contradictions such as the one just described above. For evidence of this, the reader may glance at the bibliography of symbolic logic compiled by Church [a, b]; the works cited cover a period beginning with the year 1666. There can be little doubt, however, of the great impetus given to its development by the contradictions. Secondly, and related to the preceding sentence, the breadth of the modern interest and extent of research in symbolic logic is evidenced in the formation of the Association for Symbolic Logic, Inc., and the founding of a journal, *The Journal of Symbolic Logic*, which began publication in 1936 and has now (1963) reached its twenty-eighth volume.

It is of interest also to remark that "border-line" interests such as have been exemplified for so long among the mathematical physicists are paralleled among the "mathematical logicians." And just as the former may be found in either the mathematics or the physics department (or both) of a university, so may the latter be found in either the mathematics or the philosophy department. (Probably because it was developed by the Greek philosophers, logic has traditionally been taught in the philosophy department—a distinction of significance only in the modern university where the departmentalization due to the tremendous growth of learning has been made necessary.) That breaking down of barriers between fields in our universities, implored and exhorted by so many educators but opposed by budgetary officers, will probably be brought about by forces within the fields themselves. In the introduction to his bibliography, Church [a] was moved to state "It has been the intention to confine the bibliography to symbolic logic proper as distinguished from pure mathematics on the one hand and pure philosophy on the other. The line is, of course, difficult to draw on both sides, . . ."

3 Basic relations and operations

Deferring any attempt to explain the above contradiction—it was introduced at this point only to show the necessity for a closer analysis of the term "set"—let us consider some of the elementary parts of so-called "set theory," illustrating them with important examples of sets basic in mathematics. We shall not begin with a definition of "set," but will proceed, as before, as though we are dealing with a "universal," with the idea of becoming more familiar with its meaning and application in mathematics before attempting further analysis.

First, let us introduce a few symbols, not because we wish to become "technical," but for the sake of brevity and simplification. Probably anyone who has "taken" high school algebra thinks of symbols when the word "mathematics" is mentioned. If the course was satisfactorily (1) taught by the teacher and (2) absorbed by the pupil, then the pupil must have realized the gain accomplished in the introduction of "$x, y, z, \cdots$" and operations (addition, equating, transposition in an equation, etc.) with them. Anyone unconvinced of this should see what efforts were involved in the solution of simple problems before the introduction of modern algebraic techniques (for that matter, imagine having to add, multiply, or divide numbers in the Roman symbols: MXVI × XLVII, for instance).†

3.1 Relations between sets

If S is a set, and x is an element of S, then we express this by the symbols

(3a) $$x \varepsilon S,$$

which may be read "x is an element of S." In special instances, but rarely, we may wish to reverse (3a):

(3a′) $$S \ni x;$$

in general, however, we shall use (3a).‡ The negation of (3a) is denoted by the symbols $x \notin S$. We may abbreviate the relations $x_1 \varepsilon S, x_2 \varepsilon S, \cdots, x_n \varepsilon S$ to $x_1, x_2, \cdots, x_n \varepsilon S$.

3.1.1 Notice that in 3.1 we used different types of symbols for the element and the set: a lower-case x and capital S. Generally we shall adhere to this custom.§ It can become complicated, of course, if we start "pyramiding" sets. Thus, if x, S are already introduced, and S is in turn an element of a "still larger" set, we shall resort to German capitals: $S \varepsilon \mathfrak{S}$ for instance. Greek capitals may be used for higher types: $\mathfrak{S} \varepsilon \Sigma$, etc.

3.1.2 A set A is called a *subset* of a set B if $x \varepsilon A$ implies $x \varepsilon B$ for all x; in words, A is a subset of B if every element of A is also an element of B. The relation is expressed symbolically by the symbol $\subset$:

(3b) $$A \subset B,$$

† There is in existence a letter to an Italian of the Middle Ages advising that he send his son to the German universities because long division was taught there!

‡ The alternative form "ϵ" of the Greek epsilon is also widely used instead of "ε."

§ Exceptions may be made in the case of symbols that have become traditional; for example, the use of "f" for a function; see II 4.4.1.

which may also be reversed:

(3b′) $$B \supset A.$$

Since, by definition, $A \subset A$ for every set A, we express the fact that $A \subset B$ and A is not identical with B by the statement that A is a *proper subset* of B (sometimes the term "true subset" is used).

3.1.3 That the sets A and B are identical, i.e., have the same elements, is expressed by the equality sign:

(3c) $$A = B.$$

Evidently *a necessary and sufficient condition that $A = B$ is that both $A \subset B$ and $B \subset A$ hold.* This is the basis of one of the most frequently employed methods for establishing the identity of two sets. It should be noticed, too, that the equality relation $=$ is an equivalence relation (II 8) in any collection of sets whatsoever. In particular, in any collection of sets there can be at most one empty set, and it is usually denoted by a "zero" symbol, "0," or a modification thereof such as "$\emptyset$." We shall generally use the latter. The empty set is frequently called *the null set.*

3.1.4 *If $\mathfrak{S}$ is any collection of sets and $\emptyset \in \mathfrak{S}$, then for every $S \in \mathfrak{S}$, $\emptyset \subset S$.*

[The logical equivalent of "$x \in A$ implies $x \in B$" is "$x \notin B$ implies $x \notin A$"; the reader may prefer the latter form of the definition of "$A \subset B$" in proving that $\emptyset \subset S$.]

Negations of $\subset$, $\supset$, and $=$ are denoted by $\not\subset$, $\not\supset$, and $\neq$, respectively.

3.2 A symbol for defining special sets

An especially useful symbol for the defining of individual sets is the formula "{ | }," which is applied as follows: Suppose that we have the combination of symbols:

(3d) $$A = \{x \mid x \text{ is a human being}\}.$$

Then (3d) would be read:

(3d′) "*A* is the set of all things x such that x is a human being."

In other words, A is the set of all human beings. In (3d) the braces { } are used to denote the set of all things denoted by whatever is enclosed by the braces, and the "|" may be read "such that."

To get a better idea of the use of this symbol, suppose that we wish to define the set, S, of all points that lie within the unit square of the coordinate "xy-plane" (such as is used in the subject of "graphs" in the elementary algebra or analytic geometry courses), which is bounded by the two coordinate axes and the lines, parallel thereto, one unit to the right of the y-axis and one unit above the x-axis, respectively. This set [within the framework of the subject matter, so that the meaning of (x, y) is understood], may be defined as follows:

$$S = \{(x, y) \mid 0 < x < 1, 0 < y < 1\}.$$

Certainly this is more specific and easily understood, as well as more economical of symbols, than the definition in words which precedes it.

In cases where the "such that" is superfluous, we use the braces $\{\,\}$ within which the elements are named instead of $\{\ \mid\ \}$. Thus, $\{2, 4\}$ denotes the set whose elements are the numbers 2 and 4; $\{2\}$ denotes the set whose single element is 2 (we shall wish to distinguish this *set* from the *number* 2).

An exception which has evolved in practice often occurs when, given a set S and $x \varepsilon S$, we frequently write $S \cup x$ instead of $S \cup \{x\}$, or $S - x$ instead of $S - \{x\}$ (the operations $\cup$ and $-$ are defined in 3.3.1 and 3.3.2). However, when standing alone, the symbol $\{x\}$ *must* be used to distinguish between the *set* whose single element is x, and x itself. This is particularly important when x is itself a set, since then $\{x\}$ has only *one* element whereas x may have many elements.

3.2.1 The proposal of the symbol $\{\ \mid\ \}$ suggests that the notion of "property" might be made basic instead of the notion of "set." Just as in (3d) the set A is defined by the property of "being a human being," so in general we would expect that some property distinguishes the elements of any given set. If so desired, it would be feasible to build a theory of sets on this basis.

3.3 Operations with sets

In the classical set theory, the basic operations are *addition* and *multiplication*. As might perhaps be inferred from its connotation in ordinary use, to *add* sets is merely to combine their elements to form one set. Thus, if A is the set of all even natural numbers, and B the set of all odd natural numbers, then $A + B$ would be the set, N, of all natural numbers. If A were also to have all prime numbers—2, 3, 5, 7, etc.—as elements, $A + B$ would still give N, since the occurrence of 3, say, in both A and B, does not "double" its occurrence in $A + B$. As another example, using points in the coordinate xy-plane again, we can write:

(3e) $$\{(x, y) \mid 0 < x < 1, 0 < y < 1\} + \{(x, y) \mid 0 \leqq x < 2, 0 < y < 1\} = \{(x, y) \mid 0 \leqq x < 2, 0 < y < 1\}.$$

In relation (3e), the second term has "swallowed" the first; in general, if

$A \subset B$, then $A + B = B$. If sets A_i are given, where the index i ranges over some "index set" I, then $\Sigma_i A_i$, or $\Sigma_{i \varepsilon I} A_i$ denotes the set of all elements that belong to at least one of the sets A_i. For example, if A_1, A_2, A_3 are sets, then $A_1 + A_2 + A_3$ may be replaced by $\Sigma_{i \varepsilon I} A_i$, where I is the set of integers $\{1, 2, 3\}$; in this case, however, the symbol $\sum_{n=1}^{3} A_n$ is used more frequently, n being an almost universally employed generic symbol for a natural number (compare "n-dimensional space" for example).

3.3.1 The symbols + and Σ are today usually replaced by the symbols $\cup$ and $\bigcup$, especially where the + symbol is being used for other purposes (in algebra, for instance). The symbols $\cup$, $\bigcup$, which may be called "union," "join," or "cup," are derived from the symbol v used commonly in symbolic logic to denote "and/or." In fact, $\cup$ may be defined in terms of v by the expression

$$A \cup B = \{x \mid (x \varepsilon A) \vee (x \varepsilon B)\}; \tag{3f}$$

in words, "$A \cup B$ is the collection of all those things x such that x is an element of A and/or x is an element of B." Similarly, $\Sigma_i A_i$ is replaced by $\bigcup_i A_i$. We shall give preference to the new symbols $\cup$, $\bigcup$ in the sequel, reserving + for its traditional role in arithmetic and algebra.

3.3.2 The *difference*, $A - B$, is defined by the expression

$$A - B = \{x \mid (x \varepsilon A) \;\&\; (x \not\varepsilon B)\}, \tag{3g}$$

which can be read "$A - B$ is the collection of all those things x such that x is an element of A but not an element of B." In (3g) we use the symbol & instead of the symbol · ordinarily used in symbolic logic to denote "and," preferring to restrict · to such uses as algebraic multiplication. In (3g) it is not necessary that B be a subset of A. Thus, if $A = \{x \mid x$ is a prime number$\}$ and $B = \{x \mid x$ is an odd number$\}$, then $A - B$ is the set whose only element is the number 2; in symbols, we write $A - B = \{2\}$. When B is a subset of A, then $A - B$ may be called the *complement* of B in A.

3.3.3 To "multiply" sets A and B is to give the *product* $A \cdot B$, in the classical terminology; or, for a collection of indexed sets A_i, the product is indicated by $\prod_i A_i$ or more specifically $\prod_{i \varepsilon I} A_i$, where I is the index set. Today · and $\prod$ are usually replaced by $\cap$ and $\bigcap$ respectively, which may be called "intersection," "meet," or "cap." The definitions are:

$$A \cdot B = A \cap B = \{x \mid (x \varepsilon A) \;\&\; (x \varepsilon B)\} \tag{3h}$$

$$\prod_{i \varepsilon I} A_i = \bigcap_{i \varepsilon I} A_i = \{x \mid x \varepsilon A_i \text{ for every } i \varepsilon I\}. \tag{3h'}$$

An expression such as $\bigcap_{n=1}^{k} A_n$ indicates $\bigcap_{i \varepsilon I} A_i$, where $I = \{1, 2, \cdots, k\}$; $\bigcap_{n=1}^{\infty} A_n$ is an intersection of sets A_n where, for every natural number n, A_n is a given set. We shall prefer the symbols $\cap$, $\bigcap$.

3.4 At this point we can augment the argument given in I 3.4 for the empty set, $\emptyset$. Suppose that A and B are sets having no element whatsoever in common. Then we can express this fact, using the $\emptyset$ symbol, by the symbols

$$A \cap B = \emptyset.$$

Or, if $A \subset B$, then

$$A - B = \emptyset.$$

This use of $\emptyset$ may be compared with the use of zero in $3 - 3 = 0$. The reader familiar with the history of mathematics will recall the antagonism to the introduction of zero as a *number*;† an analogous antagonism is felt, by many who are unfamiliar with set theory, to the introduction of the null set.

3.5 Parentheses are used, as in elementary algebra, to indicate operations that are to be performed before removal of the parentheses. Thus, $S = (A - B) \cup C$ indicates that the set S is the union of (1) the set $A - B$ and (2) the set C. On the other hand, the set $A - (B \cup C)$ is the set of all elements of A that are in neither B nor C. We see that the use of parentheses in a union of sets—such as $(A_1 \cup A_2) \cup A_3$—is unnecessary, if we refer to the definition in 3.3.1. However, it is sometimes clarifying, when the union of A_1 and A_2 is of special significance, to enclose their union in parentheses.

3.6 If $A \neq \emptyset$, then A is called *non-empty* or *non-vacuous*. If A has more than one element, then it is called *non-degenerate*. If $A \cap B = \emptyset$, then A and B are called *disjoint* sets. And, more generally, if $\{A_\nu\}$ is any collection of sets A_ν, where ν runs over some set of indices or marks, such that for every two different indices ν' and ν'', $A_{\nu'} \cap A_{\nu''} = \emptyset$, then we shall again call the A_ν disjoint sets.

4 Finite and infinite sets

One of the most important distinctions in mathematics, from any standpoint (be it methodological or philosophical), is that between finite sets and infinite sets. If we wish to give them explicit definitions, we usually define one or the other and arbitrarily place all sets not satisfying the definition in the other category. Thus we may define finite set, and then infinite set as any set that is not finite; or we may define infinite set,

† The author has frequently illustrated, for his students, the difference between "0" and "nothing" by pointing out that a "0" in the grade book opposite a student's name indicates that the student actually handed in an assignment, whereas nothing opposite the name indicates no work handed in; in the former case, "good intention" is indicated on the part of the student at least.

and then finite set as any set that is not infinite. In either case, we divide any collection A of sets into two categories, each set in the collection falling definitely in one category or the other.

It would seem more natural, surely, to define finite set and then infinite set as "not finite," rather than the reverse, inasmuch as our everyday experience seems to involve encounters only with "finite" sets (sets of people, sets of books, etc.). Furthermore, it is only in comparatively recent times that there has been developed any respectable theory of infinite sets.

In defining "finite" we shall presume for the present that the reader is familiar with the sequence of natural numbers—1, 2, 3, $\cdots$.† However, there may be no serious objection to this, as it means, virtually, that we presuppose familiarity with the ordinary counting process. As we define the term, a set is finite if, roughly speaking, *its elements can be counted.* We say "roughly" because, practically speaking, we can hardly say that the set of all grains of sand on the earth allows of its elements being counted! The trouble here, however, is that we are thinking of counting as a *physical* process, hence a time-consuming process.

4.1 To avoid this difficulty, we revert to the notion of (1-1)-correspondence defined in II 4.4.2. Let us denote by N the set of all natural numbers, and by N_n the set defined by the expression

(4a) $$N_n = \{x \mid (x \,\varepsilon\, N) \,\&\, (1 \leqq x \leqq n)\}.‡$$

Then we may define "finite" as follows:

4.1.1 Definition. A set S is *finite* if either it is empty, or there exists a natural number n such that between the elements of S and the elements of the set N_n there exists a (1-1)-correspondence.

Note that, for a non-empty finite set, the pairs in the (1-1)-correspondence are of type (x, k), $x \,\varepsilon\, S$, $k \,\varepsilon\, N_n$, and, consequently, we may denote uniquely the element x in the pair (x, k) by x_k. Thus the set of all states in the United States is finite, since between its elements and the elements of N_{50} there exists a (1-1)-correspondence; the set of all (natural) moons of the earth is finite since between its elements and the elements of N_1 there exists a (1-1)-correspondence; the set of all present-day kings of France is finite, since it is empty. In the case of the set G of all grains of sand on the earth, we do not know the corresponding N_n, since we cannot count them; but, by virtue of simple physical facts and the properties of the set N (we go more fully into these later), it is not difficult to show that the required N_n exists (at a given time, because, owing to erosion, etc., there is no reason to think that G is a fixed set).

† Regarding alternative definitions of "finite," see 4.9 in this chapter.

‡ Whenever we use symbols $<$ and $=$ in connection with natural numbers, we shall, unless otherwise stated, mean $1 < 2$, $2 < 3$, and so on, and $=$ denoting "equals" in the ordinary sense of logical identity ($2 = 2$, $3 = 3$, etc.). The set N is simply ordered relative to this relation $<$ (II 7) in what we may call its *natural order*.

4.2 Having defined "finite" as above, we define *infinite* as *not finite.* Thus N itself is infinite although this has to be proved, of course; the set of all points on a euclidean line or euclidean plane is infinite; etc.

4.3 In order to differentiate between the above definitions of finite and infinite and those given later, let us agree to call them the *ordinary finite* and *ordinary infinite* (if for no other reason than that they represent what we ordinarily think of as the meanings of these terms).

4.4 The alternative definitions, which were originated by Dedekind, will seem more natural if we recall the classic example† of the (1-1)-correspondence between the set N of all natural numbers and the set N_e of all even numbers. The pairs in this case are of the type $(n, 2n)$, where $n \varepsilon N$, $2n \varepsilon N_e$; in detail, the pairs are (1, 2), (2, 4), (3, 6), (4, 8), and so on.

4.5 A more elementary example, perhaps, is the (1-1)-correspondence between the elements of N and the elements of the set $N - \{1\}$ (the latter being the set of all natural numbers except the number 1; cf. 3.3.2). The pairs representing the (1-1)-correspondence are of the type $(n, n + 1)$, where $n \varepsilon N$, $n + 1 \varepsilon N - \{1\}$. We shall make use of this example in the sequel (5.3).

4.6 The reader should note that in each of these examples the "law" or "rule" for the (1-1)-correspondence is explicitly stated. Up to now we have avoided mentioning the significance of the term "there exists" as used in Definition II 4.4.4, for instance. The "rule" $(n, n + 1)$ tells us explicitly what are the elements of the (1-1)-correspondence between the elements of N and the elements of $N - \{1\}$, and is one of the ways in which we fulfill the requirement "there *exists* a (1-1)-correspondence. . . ." We shall consider this matter more fully later. It is possible, for instance, that we may sometimes be satisfied with a logical proof of the "existence" of a (1-1)-correspondence, without knowing any such explicit definition of it such as is given above.

4.7 If we denote the set of all odd numbers by N_o, then there exists a (1-1)-correspondence between N_e and N_o whose elements are of the type $(2n, 2n - 1)$, where n is a natural number. This example differs from the two above in that here the sets involved (N_e and N_o) are disjoint, whereas in each of the above two cases one of the sets was a proper subset of the other.

4.8 Dedekind's definition of infinite is suggested by observation of such correspondences‡ as that between N and N_e, in which one of the sets (N_e)

† Galileo evidently was aware of such examples; see Bell [B_3; 272ff].

‡ As Dedekind observed ([D_2; 2nd ed.], or the *Werke*), both Cantor and Bolzano had called attention to the property of infinite sets which he (Dedekind) made the basis of his definition; however, neither elevated it to the rank of a *defining* characteristic.

is a proper subset of the other (N). Such a circumstance, Dedekind observed, must be *characteristic* of what we intuitively recognize as "infinite" sets:

4.8.1 Definition (Dedekind). A set S is *infinite* if it has a proper subset S_1 such that between the elements of S and the elements of S_1 there exists a (1-1)-correspondence.

Thus the set N is infinite, according to Definition 4.8.1. (It is interesting to notice that it is much easier to prove this than to prove N infinite by the ordinary definition; see below.)

Of course, we may now call a set finite if it is not infinite. And to distinguish this definition from that (4.1.1) of the ordinary finite, we call it the *Dedekind finite*; and, similarly, we call the "infinite" defined in 4.8.1 the *Dedekind infinite.*

4.9 In the foreword to the second edition (1893) of [D_2], Dedekind proposed a non-negative definition of "finite" as follows:

"A set S is called *finite* if there exists a mapping† of S into itself such that no proper subset of S is mapped into itself."

He predicted that it would be very difficult to develop a theory on the basis of this definition unless the sequence of natural numbers is assumed. However, in an unpublished note (see [D_3; LXIII]), he developed some of the properties of sets that are finite by the new definition; and in 1932 Cavaillès [a] continued the investigation, showing how to develop the fundamental properties of finite sets (such as the finiteness of all subsets of a finite set, the mathematical induction property, etc.). None of this necessitates any use of the Choice Axiom (see Section 6).

An extensive investigation of possible definitions of "finite" and of their interrelations was made by Tarski [a]; the reader is referred to his work for further details and citations to earlier investigations.

5 Relation between the ordinary infinite and the Dedekind infinite

When we have two definitions of the same term, here "infinite," we should investigate the relationship between them. The ordinary finite and infinite presuppose a knowledge of the set, N, of natural numbers; the Dedekind definitions of finite and infinite presuppose only the concept of (1-1)-correspondence. We have defined what we mean by a (1-1)-correspondence; but up to now we have used the set N as a universally known concept, just as we suppose that its elements, such as the number 2, are universally known in the context of such axiom systems as are set up for geometry.

† See IV 3.2.3.1.

5.1 The mathematical induction principle

In attempting to compare the two definitions of infinite given above, we shall come face to face with the fact that the set N has certain associated properties† which are necessary for a solution of the problem, and which we do not encounter in the concept of any individual number, such as 2. For example, associated with the set N is the "mathematical induction principle," which has no significance for the individual number. A knowledge of individual numbers, and how to use them in counting, by no means implies a knowledge of the induction principle, as every instructor who has tried to teach it is well aware.

To explain the principle, it is convenient to introduce the notion of an "inductive" set of natural numbers. A set, G, of natural numbers, is called *inductive* if $n \varepsilon G$ implies $n + 1 \varepsilon G$; that is, if a natural number n is in G, its "successor" (in the natural order) is in G. Then *the mathematical induction principle* states that *if G is an inductive set and* $1 \varepsilon G$, *then G is the set of all natural numbers*; i.e., $G = N$.

The intuitive notion behind the principle is simply that if $1 \varepsilon G$, then $2 \varepsilon G$; and if $2 \varepsilon G$, then $3 \varepsilon G$; and so on. We can imagine the natural numbers represented by blocks standing in juxtaposition so that a forward push on the block representing the number n forces it to knock over the block representing $n + 1$;‡ the block representing 1 is pushed over and in its fall knocks over the block representing 2; the latter, in turn, falls against the block representing 3 and knocks it over; and so on—all the blocks being knocked over as a result of the pushing over of the first block. The analogy, of course, is that G is the set of blocks that actually fall; if it contains 1, and if for each number in G the successor is in G (and hence ready to be knocked over), then *all* the blocks fall; that is, G is the whole set of blocks, the set N.

This is not a *proof* of the principle, of course. As we hope to make clear later, it is inherent in any actual definition of the "natural order" of N and is either explicitly stated therein (as in the famous axiom system of Peano; see VI 3.1), or is provable from the definition. As used in proofs which involve N (as in the proof below), it may be considered as a method which augments the classical rules of logic (which were derived from the finite sets of man's environment). However, it is not an assumption about N, but is a consequence of the type of order represented in N. Put another way, when we assume a knowledge of the complete set, N, of natural

† These associated properties are not properties of N as a *set*, any more than is the fact that $2 \times 2 = 4$ is such a property. They are associated with the fact that the elements of N have a "natural order," 1, 2, 3, and so on. Cf. 4.1f.

‡ Just as children frequently stand dominoes in this fashion, so that a push on the first domino in line is sufficient to knock over *all* the dominoes.

numbers, we have in particular assumed the mathematical induction principle.

The use made of the mathematical induction principle in practice is generally of two kinds: (1) As a mode of definition, and (2) as a method of proof. If we wish to define some entity or concept $D(n)$ for every natural number n, we may first define $D(1)$, and then show how from $D(n)$ we can define $D(n + 1)$. By the mathematical induction principle this will define $D(n)$ for all elements of N, since if we let G denote the collection of natural numbers for which we have defined $D(n)$, then $1 \varepsilon G$ since we defined $D(1)$; and if $n \varepsilon G$, then $n + 1 \varepsilon G$ since we showed that from $D(n)$ we can define $D(n + 1)$, so that G is inductive; thus $G = N$. An example of this mode of definition is given in 5.3 below. Similarly, if we wish to prove some theorem $T(n)$ about the natural number n, for every natural number n, we may prove $T(1)$, and then show how from $T(n)$ we may prove $T(n + 1)$. From the mathematical induction principle it will then follow that the set G of all numbers n for which the theorem $T(n)$ has thereby been proved consists of all the numbers of N. This type of proof is called *proof by mathematical induction*; we shall see examples of this type of proof later.

Sometimes, in practice, to prove a theorem $T(n)$ for all n we first prove $T(1)$ and then show how, from the assumption that $T(1), T(2), \cdots$, and $T(n)$ all hold we may prove $T(n + 1)$. [Compare (3a) in V.] Other variations in modes of proof and definition are also possible, all of which may be justified by the mathematical induction principle. (See Problems 25 and 26.)

In the case where $T(n)$ is a theorem concerning the existence of some entity or entities of some type T forming a set S_n of n elements, whether proved by mathematical induction or not, we must be careful not to conclude, without further justification, that there then must exist an infinite set S_∞ of entities of type T. If, however, after having proved the existence of S_n for each n, we can prove that $S_n \subset S_{n+1}$, then the conclusion that there exists an infinite set of entities of type T follows. (See Problems 23 and 27.)

5.2 Proof that a set which is Dedekind infinite is ordinary infinite

Let S be a set which is Dedekind infinite. Then S has a proper subset S' such that between the elements of S and the elements of S' there exists a (1-1)-correspondence.

Either S is ordinary finite or ordinary infinite, since we defined ordinary infinite as "not ordinary finite," and it will be sufficient to show that S is not ordinary finite. If S were ordinary finite, there would exist a set

N_k, as defined earlier (4.1.1), such that between the elements of S and the elements of N_k there exists a (1-1)-correspondence.†

Now consider the following lemma:

Lemma. *Let A and B be sets such that there exists a* (1-1)-*correspondence f between A and B, as well as a* (1-1)-*correspondence g between A and a proper subset A' of A. Then there exists a* (1-1)-*correspondence between B and a proper subset B' of B.*

Proof. Consider the following diagram:

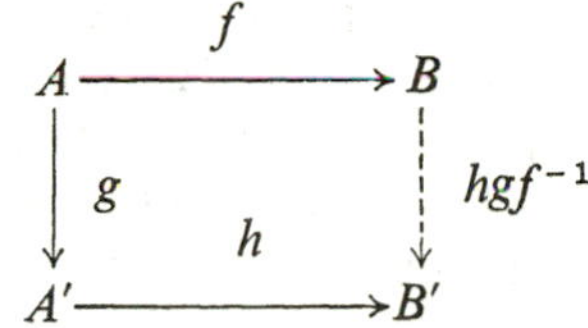

The arrow between A and B represents the function $f: A \to B$, and the vertical arrow between A and A' the function $g: A \to A'$. The function h is simply the collection of all pairs (x, y) of f whose x's are also in A'; their y's form the set B'. Then hgf^{-1}, the composition of gf^{-1} and h, is the desired (1-1)-correspondence between B and the proper subset B' of B.

It follows from this Lemma that if A, B, and A' are replaced by S, N_k, and S', respectively, we can assert that there exists a proper subset N_k' of N_k such that there exists a (1-1)-correspondence between the elements of N_k and the elements of N_k'. We shall show that this is impossible.

5.2.1 Theorem. *There exists no N_k such that there exists a* (1-1)-*correspondence between N_k and a proper subset of N_k.*

Proof. The theorem certainly holds for $k = 1$, since the only proper subset of N_1 is $\emptyset$. Hence, if G is the subset of N for which the theorem holds, $1 \varepsilon G$.

Now we shall show that if a natural number k is in G, then $k + 1 \varepsilon G$; and, hence, that G is N by the mathematical induction principle. Suppose that for some k in G, $k + 1 \not\varepsilon G$. Then N_{k+1} has a proper subset N'_{k+1} such that between N_{k+1} and N'_{k+1} there exists a (1-1)-correspondence f. Let us represent this correspondence by the following diagram:

$$\begin{array}{cccccc} N_{k+1}: & 1 & 2 & \cdots\cdots\cdots & k & k+1 \\ \Big\downarrow f & \Big\downarrow & \Big\downarrow & & \Big\downarrow & \Big\downarrow \\ N'_{k+1}: & n_1 & n_2 & \cdots\cdots\cdots & n_k & n_{k+1} \end{array}$$

$(n_1 = f(1),\ n_2 = f(2)$, etc.)

† It must be emphasized that we are here assuming a knowledge of the natural numbers just as in Section 4 (especially in terms of $<$ as in 4.1), including a knowledge of the mathematical induction principle.

There are two cases to consider: (1) If n_{k+1} is the natural number $k + 1$, then by deleting the last column of the diagram it is clear that we are left with a (1-1)-correspondence between N_k and a proper subset N_k' of N_k—which is impossible since $k \varepsilon G$. (2) If n_{k+1} is not the natural number $k + 1$, perhaps one of the numbers $n_1, n_2, \cdots, n_k$ (say, n_i) is $k + 1$; if so, we need merely exchange the position of n_i and n_{k+1} in the diagram and we are back to case (1), which we have shown impossible. But then none of the numbers $n_1, n_2, \cdots, n_k$ is $k + 1$, implying that they are all in N_k; and, moreover, that they constitute a proper subset N_k' of N_k (since n_{k+1} is missing). But, again, by deleting the last column of the diagram, we would have a (1-1)-correspondence between N_k and N_k'.

It is impossible that $k + 1 \not\varepsilon G$, then, and we conclude $G = N$; i.e., the theorem is true for all natural numbers.

Corollary. *The set N is ordinary infinite.*

The above type of argument (without using mathematical induction) can be adapted to prove:

5.2.2 Theorem. *If S is a set such that there exists no* (1-1)-*correspondence between the elements of S and the elements of any proper subset of S, and S is augmented by a new element e, then the set* $S \cup \{e\} = S_1$ *has a like property.*

[In the above terminology, *if S is not Dedekind infinite then* $S_1 = S \cup \{e\}$ *is not Dedekind infinite.*]

5.2.3 Corollary. *If a set S is Dedekind infinite, and* $x \varepsilon S$, *then* $S - \{x\}$ *is Dedekind infinite.*

5.3 Proof that a set which is ordinary infinite is Dedekind infinite

Now we consider the converse problem: Is a set which is ordinary infinite also Dedekind infinite? In the same manner as we had to introduce a new principle, mathematical induction, to prove that a Dedekind infinite set is ordinary infinite, we shall find here that we encounter another new principle in our proof.† But while the mathematical induction principle is generally accepted (as it must be, if we accept the set N), the one that we shall appeal to now is not generally accepted. First, let us consider the following argument:

Let S be a set which is ordinary infinite, i.e., not ordinary finite. Then S is not empty, and we can select an element x_1 of S. Again, $S - x_1$ is not empty; for otherwise $S = \{x_1\}$, and the pair $(x_1, 1)$ constitutes a (1-1)-correspondence with N_1 rendering S ordinary finite. So we can select an element x_2 of $S - x_1$.

† Historically, it was first noticed in another connection, which will be described later.

Continuing in this way, suppose that we have selected elements $x_1, x_2, \cdots, x_k$ of S. Then we argue that $S - x_1 - x_2 - \cdots - x_k$ is not empty; for otherwise $S = \{x_1, x_2, \cdots, x_k\}$, and the pairs $(x_1, 1), (x_2, 2), \cdots, (x_k, k)$ would constitute a (1-1)-correspondence between S and N_k, rendering S ordinary finite. So we can select an element x_{k+1} of $S - x_1 - x_2 - \cdots - x_k$.

In this manner, we have shown how to select an element x_n of S for every natural number n (by virtue of the mathematical induction principle). Let S_1 denote the collection of all such elements x_n. Now consider the collection, $\mathfrak{S}$, consisting of all pairs (x_n, x_{n+1}) as well as all pairs (x, x) such that x is not in S_1 (if such elements x exist; they may or may not exist). Then, since x_1 is never the second element in the pairs as indicated, this collection $\mathfrak{S}$ of pairs constitutes a (1-1)-correspondence between the elements of S and the elements of $S - x_1$. Hence S is Dedekind infinite.

5.3.1 Comment on the above proof

We said above that "we have shown how to select an element x_n of S for every natural number n," and no two such elements are the same. But have we? We "select" x_1, then x_2, then x_3, and so on. We don't give any *rule* for these selections—they seem to be the result of free choice. But how are we to make an infinite set of free choices, one after the other (the collection $x_1, x_2, x_3, \cdots$ is infinite in either the ordinary or Dedekind sense)?

5.3.1.1 An example may make the above criticism clearer. If we have a die with faces numbered 1 to 6, we can consider a decimal number $0.a_1$ obtained by one throw of the die; if 1 comes up, let $a_1 = 1$ and we have the number 0.1; or, if 6 comes up, we have 0.6. We can also consider a number $0.a_1a_2a_3$ obtained from three throws of the die, letting the digits a_1, a_2, a_3 be the numbers successively obtained in the three throws. To be sure, the number is not determinate from our definition; thus $0.a_1$ may be any one of six different numbers. We would not be justified in speaking of *the* number so obtained until *after* we threw the die; and then the number is quite definite. But, *before* the throw, its definition is not precise enough to fix it. In what sense, then, could we possibly speak of the number $0.a_1a_2 \cdots a_n \cdots$, where there is a digit a_n for each $n \varepsilon N$, to be obtained by a throw of a die? In this case, we cannot even complete the throws of the die and hence make the number precise.

The objection can now be raised that there is no sense in mixing numbers, which are a mathematical notion, with physical concepts such as throwing of dice, and with time (which is used up in throwing and which consequently renders the number $0.a_1a_2 \cdots a_n \cdots$ indefinable in any precise sense). This objection may be met by pointing out that it is the indeterminacy, the free choice, of a_n that is the root of the difficulty, and that this is present in the above proof which utilizes no physical concepts. For some sets S we could get around the difficulty by eliminating free choice. Thus, if S is the set of all ordinary fractions between 0 and 1 ("proper fractions"), we could *specify* $x_1 = \frac{1}{2}$, $x_2 = \frac{1}{4}$, and, in general, having x_n, let $x_{n+1} = \frac{1}{2}x_n$ [i.e., $(\frac{1}{2})^{n+1}$]. But in the above proof we don't know enough about the nature of S to make any such specifications.

Very well, we may counter-object, don't say "select"; just say, *let* x_1 *be an element of* S, then *let* x_2 *be an element of* $S - x_1$, and, in general, having obtained x_n, *let* x_{n+1} *be an element of* $S - x_1 - x_2 - \cdots - x_n$. The word "select" implies a selector, whereas the "let x_n be..." formula avoids such a connotation.

We can let the argument rest here; either let us admit that the "let x_n be..." formula solves the difficulty, or let us not admit it. There seems to be no *logical* way of settling the matter. We encounter a question of what might be called *mathematical method* or *mathematical existence* (depending on the point of view), rather than something that can be handled by appeal to the classical Aristotelian logic. Notice, however, that except for this aspect of it, the above proof that an ordinary infinite set is Dedekind infinite is evidently quite acceptable. Specifically, there is evidently no objection to the proof, contained in the above, that

5.3.2 *If a set* S *has a subset* S_1, *such that between the elements of* S_1 *and the elements of the set of natural numbers,* N, *there exists a* (1-1)*-correspondence, then* S *is Dedekind infinite.* (See Problem 21.)

6 The Choice Axiom

The difficulty that we encountered in the preceding section—the question of the existence of the set S_1 of 5.3—may be handled by appeal to what is ordinarily called the "Choice Axiom" (or the "Zermelo Postulate";† see 6.4 and 6.7). This statement must not be construed to mean that we are going to *eliminate* the difficulty; far from it. From the point of view of one who insists that generally no such set S_1 exists, we are only going to attribute the difficulty to a "principle" which will, as we shall see, serve as a haven for all sorts of such difficulties. It is convenient to have a single principle to which we can bring such matters and not be forced to leave a lot of seemingly unrelated special problems lying about unattended!

† Although the eponymous term "Zermelo Postulate" relates to E. Zermelo's use of the principle in the proof of his famous Well-Ordering Theorem in 1904 (V 3.1.2), it is asserted by E. W. Beth [Be; 376] that "As early as 1890, this axiom had been stated and applied incidentally by Peano, but only in 1902 was it seen, by Beppo Levi, to constitute an independent principle of proof."

6.1 To introduce the principle referred to, suppose that we consider the set of fifty states of the United States, where each state is considered not as a geographical entity, but as a collection of its individual citizens. And, for convenience, let us assume that no person is a citizen of more than one state. *Question*: Does there exist a set, R, consisting of one and only one citizen from each state? This must seem like a trivial question, for certainly such a set must exist. For instance, each state has two senators, already selected, and all we need do is make up a set by selecting one senator from each pair representing a state.

6.2 Now imagine that, instead of fifty states, we have an example of a "nation" whose states are in (1-1)-correspondence with the elements of N; denote the state that is paired with $n \varepsilon N$ in this correspondence by S_n. Can we still assert the existence of the set R? If so, R consists of one and only one person, p_n, from each state S_n. But how is p_n to be selected? Even if, as in the case of the fifty states, there is a senate comprised of two senators from each state, we would still have to select p_n from the pair of senators from state S_n. If in every case one senator is definitely older than the other, we could select p_n, for all n, by the rule that he is to be the older senator from S_n. If the rule does not apply, some other characteristic may serve; if not, we shall have to resort simply to asserting the existence of R, and in this case we would be making an application of the principle known as the Choice Axiom:

6.3 **Choice Axiom.** *If $\mathfrak{S}$ is a collection of disjoint non-empty sets S_ν, then there exists a set R which has as its elements exactly one element x_ν of each S_ν.*

The element x_ν is called the *representative element* of S_ν; and R may be called a *representative set* for $\mathfrak{S}$.

On the basis of the Choice Axiom we can prove:

6.4 **General Choice Axiom; or Choice Principle.** *If $\mathfrak{S}$ is a collection of non-empty sets S_ν, disjoint or not, then there exists a set $\mathfrak{R}$ whose elements are pairs (S_ν, x_ν) in which $x_\nu \varepsilon S_\nu$, and such that each S_ν occurs in one and only one pair.*

(In functional terminology, $\mathfrak{R}$ is a function whose domain is $\mathfrak{S}$, and such that $\mathfrak{R}(S_\nu) \varepsilon S_\nu$.)

6.5 Note that if in 6.4 the elements of $\mathfrak{S}$ are disjoint, then the set $R = \{x_\nu \mid x_\nu \varepsilon (S_\nu, x_\nu)\}$ is the same type of set as the set R of 6.3, each x_ν being considered the representative element of S_ν. Thus 6.4 implies 6.3. To see that 6.3 implies 6.4, we may form, for each ν, the set $\mathfrak{U}_\nu$ of all pairs $(S_\nu, x_{\nu\mu})$, where now *every* element $x_{\nu\mu}$ of S_ν occurs in a pair. Then the collection, Γ, of all sets $\mathfrak{U}_\nu$, is a collection of disjoint, non-empty sets, so that by 6.3 there exists a set $\mathfrak{R}$, which has as its elements exactly one element $(S_\nu, x_{\nu\mu})$ of each $\mathfrak{U}_\nu$.

6.6 We may continue to call the elements x_ν, which occur in pairs $(S_\nu, x_\nu) \varepsilon \mathfrak{R}$,† the *representative elements* of the sets S_ν, except that we realize that a given x_ν may be the representative element of more than one S_ν. This might be the case, for example, if an S_ν and an $S_{\nu'}$, elements of $\mathfrak{S}$, were such that $S_\nu \subset S_{\nu'}$; in such a case the element x_ν is necessarily an element of $S_{\nu'}$, and it may happen that $x_\nu = x_{\nu'}$.

Before discussing the Choice Axiom further, however, let us see how it applies to Section 5.3:

6.7 Let S be a set which is not ordinary finite. We apply 6.4 as follows: Let the sets S_ν be the non-empty subsets of S; that is, $\mathfrak{S}$ is the collection of all non-empty subsets‡ of S. In the set $\mathfrak{R}$, whose existence is asserted by 6.4, there is one and only one pair in which S (which is a subset of itself, and hence an "S_ν") occurs; denote it by (S, x_1). Also, since the set $S - x_1 = S_2$ is a subset of S, it occurs in a pair (S_2, x_2). And having assigned symbols $x_1, x_2, \cdots, x_n$ for any $n \varepsilon N$, denote by (S_{n+1}, x_{n+1}) the pair in which $S_{n+1} = S - x_1 - x_2 - \cdots - x_n$ occurs.

The set, A, of all such elements x_n is exactly of the type of the set S_1 in 5.3.2. Hence, the proof that an ordinary infinite set is Dedekind infinite may be based on the Choice Axiom.

6.8 Comment on the Choice Axiom

As we shall have occasion later to comment on the further implications of the Choice Axiom, we shall be content with a few general observations at this point.

In the first place, why call it an *axiom* or *postulate* (as in "Zermelo Postulate")? As we have presented it, it is not part of an axiom system involving undefined terms (and other axioms) subject to various interpretations. Instead, it seems to be a basic principle, of a more absolute character than an axiom system for geometry which can be applied to certain models, perhaps, but about whose truth or falsity we would hardly (at least *today*) argue. It is possible, to be sure, to devise axiom systems in which the Choice Axiom occurs as one of the axioms (see VIII 8). But the term "axiom" in this case is not due to such uses of the principle; it seems to have derived from the ancient and traditional conception of an axiom as a "universal truth," or "a self-evident proposition, requiring no

† As each S_ν occurs only once in a pair $(S_\nu, x_\nu) \varepsilon \mathfrak{R}$, we may drop the μ used in the $x_{\nu\mu}$ of the proof.

‡ The use of this collection may be avoided by using the collection of all ordinary finite subsets of S. See Sierpinski, [S_1, 116–117] or [S_3, 114].

formal demonstration to prove its truth, but received and accepted as soon as mentioned."† The Choice Axiom must seem, at first sight, to fulfill these qualifications. For what is more natural than to assume, if a collection of disjoint non-empty sets be given, that there exists a set containing one element of each set of the collection?

In particular, if we are given a single set S, we do not hesitate to think of a set $\{e\}$ consisting of a single, unique element e of S! And we probably would not hesitate to assume the existence of a representative set for the collection of fifty states, even if there were not senators all ready to be named (cf. 6.2). As a matter of fact, commencing with the premise that it is permissible to select an element from a given set, it is easy to show by mathematical induction that, for every natural number n, the Choice Axiom holds for any collection of n non-empty sets.

However, it would seem better to use the term "Choice Principle" than "Choice Axiom" or "Choice Postulate." We shall continue to use the last two terms only because they seem to be sanctioned by usage.

It is perhaps clear from the discussion given previously (5.3.1.1, for instance) that much of the objection to the Choice Axiom stems from the *lack of precise definition* of the representative set. An oft-quoted (see Sierpinski [S_1; 125] for instance) example of Russell, modified to suit our purposes, may throw further light on this aspect of the matter. Imagine a world in which there exists an infinite collection of pairs of shoes! Does there exist a set S consisting of one shoe from each pair? We may say that the answer is affirmative, since we can *define* S to consist of the *left* shoe from each pair. But suppose that this imaginary world also contains an infinite collection of pairs of socks; does a corresponding set S exist for this collection? Unfortunately, socks are manufactured in pairs of identical form, it being impossible to distinguish a left sock from a right sock, so that we have no way of defining a representative set S.‡ Despite this lack, we may still feel justified in asserting that such a set exists; but to do so is to employ the Choice Axiom, whereas in the case of the shoes no appeal to such an "axiom" is necessary, what is called an *effective* definition being possible.

This example brings out the fact that it is precisely in cases where there is insufficient information regarding the character of the sets in question that appeal must be made to the Choice Axiom. It suggests that as mathematics has grown, reaching out to concepts of a greater degree of abstraction, it has become necessary to add new principles that are not available in the classical logic (which was not devised as a tool for dealing with the infinite as we conceive of it today); and the Choice Axiom is one

† This quotation is from Huntington [a; footnote †, p. 264].

‡ Someone has suggested that if each pair were being *worn* we would!

of these new principles. (Compare the remark in II 4.2 regarding possible inadequacy of the logical apparatus.)

This may all seem like a "tempest in a teapot" at the present stage of our discussion, but we shall ask the reader to suspend judgment until we have been able to present more of the *consequences* of the Choice Axiom. After all, who pays any regard to the blaze of a match until it starts a forest fire!

SUGGESTED READING

Cantor [C_1]
Grelling [Gr; 1–15, 40–44]
Hobson [Ho; §§ 1–5]
Kamke [Ka_1; 1–5]
Peano [P_3; II § 2, pp. 1–4]
Sierpinski [S_1; 1–12, 103–117]
Suppes [Su; 98–108]
Young [Y; VI]

PROBLEMS

1. In a given university, let $\mathfrak{C}$ denote the set of all sections (mathematics sections, history sections, etc.) in session at a given time, $\mathfrak{G}$ the set of all graduate sections, $\mathfrak{M}$ the set of all sections in graduate mathematics, C a section in graduate mathematics, and p a student in C. Set up the "ε" and "$\subset$" relations between $\mathfrak{C}$, $\mathfrak{G}$, $\mathfrak{M}$, C, and p. Explain the choice of symbols (cf. 3.1.1).

2. Let $A = \{1, \emptyset, \{\emptyset\}\}$, $B = \{\emptyset\}$, $C = \{1, 2\}$. Show that $B \varepsilon A$, $B \subset A$, and that $B \cap C = \emptyset$; but that neither $A \cap B = \emptyset$ nor $A \subset B \cup C$ holds.

3. Use the $\{\ |\ \}$ symbol to define the following sets: (a) the set of all points in the coordinate plane that lie interior to the unit circle with center at the origin; (b) the set of points defined in (a) with those having negative abscissas deleted; (c) the set $(A \cap B) \cup C$ where A, B, and C are sets; (d) the set $A \cap (B \cup C)$; (e) the set $A - (B \cup C)$; (f) the set $(A - B) \cup C$.

4. Prove the following "distributive laws":

$$A \cap (B \cup C) = (A \cap B) \cup (A \cap C),$$

$$A \cup (B \cap C) = (A \cup B) \cap (A \cup C).$$

5. Let the complement of a set A relative to a set S be denoted by $\mathscr{C}_s A$, or, briefly, by $\mathscr{C}A$. Then prove:

$$\mathscr{C}(A \cap B) = \mathscr{C}A \cup \mathscr{C}B,$$

$$\mathscr{C}(A \cup B) = \mathscr{C}A \cap \mathscr{C}B.$$

6. In the notation of Problem 5, notice that $\mathscr{C}(\mathscr{C}A) = A$. Use this to derive the second formula in Problem 5 from the first.

7. Show by mathematical induction (on the number, n, of set symbols, counting repetitions) that, if M is an expression composed of set symbols $A_1, A_2, \cdots, A_n$ and the symbols $\mathscr{C}$, $\cap$, $\cup$, but containing no $\mathscr{C}(\ \ )$, then $\mathscr{C}M$ is the expression obtained by replacing, in M, each A_i, $\mathscr{C}A_i$, $\cap$, and $\cup$ by $\mathscr{C}A_i$, A_i, $\cup$, and $\cap$, respectively.

8. If $M = (A \cup \mathscr{C}B) \cap (A \cup C)$, find $\mathscr{C}M$.

9. Some have insisted that the only type of set allowable in mathematics is one such that it is determinate of everything whether it is an element of the set or not. Criticize this criterion.

10. The following problem is attributed to Tarski: Let N denote the set of all natural numbers, and for any two natural numbers m and n let "$m = n$" mean "m is identical to n." Consider any set S_1 having only one natural number as an element. Obviously, if $m,n \in S_1$, then $m = n$. Suppose it has been shown that, for any set S_n having exactly n natural numbers as elements, the relation $m,n \in S_n$ implies $m = n$. Consider any set S_{n+1} having exactly $n + 1$ natural numbers as elements; let us denote its elements by $x_1, x_2, \cdots, x_n, x_{n+1}$. As the set $S'_{n+1} = S_{n+1} - x_{n+1}$ has exactly n natural numbers as elements, the relation $m,n \in S'_{n+1}$ implies $m = n$; in particular, $x_1 = x_n$. Now consider the set $S''_{n+1} = S_{n+1} - x_n$. Since S''_{n+1} contains exactly n natural numbers, the relation $m,n \in S''_{n+1}$ implies $m = n$; in particular, $x_1 = x_{n+1}$. But, if $x_1 = x_n$ and $x_1 = x_{n+1}$, then $x_n = x_{n+1}$, and it follows that all elements of S_{n+1} are identical. Hence we have proved, in particular, that all natural numbers are identical! What is wrong with this "proof"?

11. A famous fable recounts that the barber in a certain town shaved everyone who did not shave himself, and only those. To analyze this, use the symbolism of 3.2 for defining sets. [*Hint:* Let "$x \text{ s } y$" denote that "x shaved y" and let "b" denote the barber. Then equate the set of townsmen who did not shave themselves with the set of townsmen that the barber shaved. What results when the "value" b is used for x?]

12. Analyze the Russell contradiction in the same manner as the barber fable.

13. The following is due to Grelling (1908): Divide all adjectives into two classes, calling those that describe themselves "autological" and those that do not "heterological." Thus "English" and "short" are autological, and "French" and "long" are heterological. What is the result if we ask whether the adjective "heterological" is either autological or heterological?

14. Prove that N is not ordinary finite.

15. Prove by mathematical induction that if S is ordinary infinite, then for every natural number n, S has a subset consisting of exactly n elements. Is the Choice Axiom necessary for the proof?

16. Show that if a set S has a proper subset S_1 such that there exists a (1-1)-correspondence between S and S_1, then S has subsets S_n, $n = 1, 2, 3, \cdots$, such that for each natural number n, S_{n+1} is a proper subset of S_n and there exists a (1-1)-correspondence between S_n and S_{n+1}.

17. Using the set N of natural numbers for S, and the set $N - \{1\}$ for S_1, show that the sequence $\{S_n \mid n \in N\}$ of the preceding problem is the "longest" that can be generally proved to exist (under the hypothesis of Problem 16), in that $\bigcap_{n=1}^{\infty} S_n$ may be empty.

18. Prove Theorem 5.2.2 and its corollary (5.2.3).

19. Young [Y, 63] gives the following argument for 5.2.3: S has a proper subset S' such that there exists a (1-1)-correspondence T between the elements of S and the elements of S'. Let x' be the element of S' paired with x in the correspondence T. The set $S_1' = S' - \{x'\}$ is then a proper subset of S_1 $(= S - \{x\})$ such that there exists a (1-1)-correspondence between the

elements of S_1 and the elements of S_1'. The set S_1 is, therefore, Dedekind infinite. What is missing from this argument?

20. In the same connection, Young states "an infinite class cannot be exhausted by removing its elements one at a time." But suppose that, between 12:00 NOON and 12:30 P.M., 1 is deleted from N; between 12:30 and 12:45, 2 is deleted from N; during the next $\frac{1}{8}$ hour 3 is deleted; during the next $\frac{1}{16}$ hour 4 is deleted; and so on. What is left of N at 1:00 P.M? Discuss the statement of Young.

21. Prove the converse of 5.3.2. [*Hint:* Let S' be a proper subset of S and let $\{(x, f(x))\}$ be a (1-1)-correspondence such that $x \in S$, $f(x) \in S'$. Let $x_1 \in S - S'$, and for each natural number $n > 1$ let $x_n = f(x_{n-1})$. Then $S_1 = \{x_n\}$ is the desired subset of S.]

22. In an imaginary world W, a man starts from his home at 12:00 NOON to walk to a post office. He walks at a uniform rate, reaching the post office at 1:00 P.M. He leaves home with 1 cent, intending to buy a stamp. However, between 12:00 and 12:30 a friend offers him a brighter coin than the one he has, and, intrigued by it, he trades his original coin for it. Between 12:30 and 12:45 a similar event occurs, so that at 12:45 he has a still different coin. During the next $\frac{1}{8}$ hour a similar exchange occurs; during the next $\frac{1}{16}$ hour another occurs; and so on. Assuming that W allows such unlimited possibilities, does the man have a coin at 1:00 P.M. to buy the stamp that he set out to buy? [In case the pennies are all different, a logical conclusion is possible; but, if the same two pennies are involved throughout, no conclusion is possible.]

23. If the set S of Problem 16 is the set of all pennies involved in the anecdote of Problem 22, and S_n is the set of all pennies involved from the nth interval on, show that the resulting collection of sets S_n satisfies the conditions of Problem 17 and is a case where $\bigcap_{n=1}^{\infty} S_n = \emptyset$ if the pennies are all different.

24. Prove that, if A is a non-empty subset of N (N the set of all natural numbers), then A has an element a which precedes ($<$) every other element of A in the natural order (4.1, footnote) of N.

25. Suppose we wish to define some concept $D(n)$ for every natural number n. We first define $D(1)$ and $D(2)$. Then we give a general definition of $D(n + 2)$ in terms of $D(n)$ and $D(n + 1)$. Show that we thus accomplish a definition of $D(n)$ for all n.

26. Suppose that $T(n)$ is a theorem about the natural number n such that we can prove $T(1)$, and that if we had proved $T(1), T(2), \cdots, T(n)$, then a proof for $T(n + 1)$ can be given. Show that $T(n)$ is thus proved for every natural number n.

27. Consider the accompanying figure. It consists of:

(1) Two parallel line segments L_1 and L_2 of unit length one unit distance apart. On L_1, $p_1, p_2, \cdots, p_n, \cdots$ are points such that p_n is at a distance $1/(n + 1)$ from the base of L_1; and, on L_2, $q_1, q_2, \cdots, q_n, \cdots$ are points such that q_n is at a distance $1/(n + 1)$ from the top of L_2.

(2) The line segment p_1q_1, on which $r_1, r_2, \cdots, r_n, \cdots$ are points such that r_n is at a distance $1/(n + 1)$ from q_1.

(3) A broken line $p_2s_1r_1t_1q_2$ as shown.

(4) In general, a broken line $p_{n+1}s_nr_nt_nq_{n+1}$ as shown. Let a *path* from L_1 to L_2 consist of *any* broken line—$p_2s_1r_1r_6t_6q_7$ for instance—made up of segments of the given broken lines, but with only its endpoints, p_2 and q_7

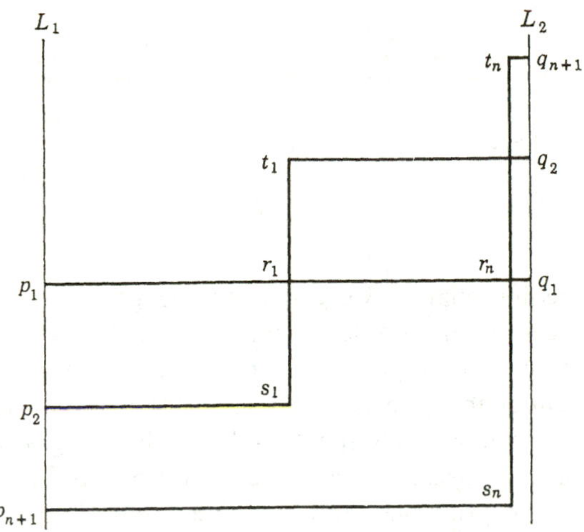

in the instance cited, on L_1 and L_2 respectively. Let two paths be called *disjoint* if they have no point in common.

Show that, for every natural number n, there exist n disjoint paths from L_1 to L_2. Show also that there does not exist an *infinite* number of disjoint paths from L_1 to L_2 (even though there do exist infinitely many different but intersecting paths). What general "moral" can be drawn from this example in regard to existence theorems provable for every natural number n?

28. Carry out the mathematical induction argument to prove the Choice Axiom for finite sets, as suggested in the third paragraph of 6.8.

IV

Infinite Sets

In the preceding chapter we discussed the general notion of set and noticed how certain debatable principles (Choice Axiom) and contradictions (Russell antinomy) may be encountered in the theory of sets.

In the present chapter we set forth sufficient material regarding infinite sets—especially with regard to the "numbers" that may be used in handling them—to give us a basis for further definition of basic notions, as well as to uncover other principles and possible contradictions. Some basic mathematical entities, for example, the "real number continuum," cannot be fully appreciated without some knowledge of this sort. And we shall encounter mathematical problems which, like that of showing that the ordinary infinite implies the Dedekind infinite, cannot be solved without appeal to principles that have never been encountered in the classical logic.

As the reader may know, until modern times the infinite was simply "the infinite"; something represented by a symbol like a figure eight lying on its side: ∞. The "number" of natural numbers was "∞," the number of real numbers "∞," the number of points in a coordinate xy-plane, ∞^2 [since ∞ x's together with ∞ y's can give ∞^2 *pairs* (x, y)—although the *doubly ordered* character of the pairs (x, y) was undoubtedly the inspiration for the symbol ∞^2]. Curious results followed from the use of this symbol (such as attempts to include it in arithmetic—$2 \times \infty = \infty$, for instance—although any competent mathematician would warn of the dire consequences of such attempts!).

With Cantor's invention of the theory of transfinite numbers, a new perspective opened up. Actually infinite numbers and an arithmetic for them were developed. What might at first have seemed to be adventures in fantasy turned out to have important applications in mathematics; as a matter of fact, it was certain applications that led Cantor to his studies in the infinite. Some of these we shall see in the sequel.†

1 Countable sets; the number $\aleph_0$

We saw in III 5.2 that N is ordinary infinite. As we shall presently see, it represents the simplest type of infinite set—as a consequence of which,

† For an excellent discussion of Cantor's work and of set theory in general, see Fraenkel [F_3]. Attention is called to the comprehensive bibliography in that book.

we might remark, we used its subsets N_k as criteria for defining the ordinary finite and infinite.

1.1 Let us consider another infinite set, i.e., the set F of all real rational numbers, which we term briefly, *rational numbers*; that is, any number which can be symbolized by a form p/q, where p is any integer (positive or negative) and q any natural number. (Thus $\frac{1}{2}$, $\frac{17}{12}$, $\frac{32}{8}$, $-\frac{8}{7}$, $\dfrac{\frac{2}{3}+\frac{6}{7}}{4}$, etc., represent rational numbers; and every natural number is a rational number, since, for example, $2 = \frac{2}{1}$.)

As in the case of the natural numbers, we do not at present try to make any rigorous distinction between a *symbol* and the *concept* which it denotes. We assume the reader has studied "fractions" and their applications in grade school, and that the question of the meaning of symbols like "$\frac{1}{2}$," "$\frac{3}{4}$," etc., may be deferred for the present. To prevent confusion, we shall usually observe the convention that by "fraction" is meant the *symbol*, and by "rational number" is meant the *concept denoted* by the symbol. Thus, "$\frac{1}{2}$," "$\frac{2}{4}$," "$\frac{3}{6}$," $\cdots$ are different fractions denoting the same rational number. When we say "$\frac{1}{2}$ is a rational number," as we do in ordinary discourse, what we really mean is that "$\frac{1}{2}$ is a fraction denoting a rational number." Analogous remarks hold for the use of the term "algebraic number" in 1.3 in this chapter.

Obviously, there are "more" elements in F than there are in N—although a mathematician would at first be startled by such a statement; it is all a matter of the meaning to be attributed to "more than." And anyone who has never heard of a "transfinite number" will probably admit that if to any set (such as N) we *add* new elements to obtain a new set (such as F), the latter has *more* elements than the former. And this is quite true, unless by "more than" we mean "a greater cardinal number† than," in which case it is quite false. Again we see the necessity of making the conventions regarding the meanings of words clear, since to one group of people they mean one thing and to another they mean something else (as any traveler knows).

The reason for the above "misunderstanding" is that if A and B are sets such that between the elements of A and the elements of B there exists a (1-1)-correspondence, the mathematician thinks of A and B as having the *same number* of elements; and that he is inclined to interpret the statement "A has more elements than B" as meaning (in a sense to be made precise below) that "the number of elements in A is greater than the number of elements in B." In the next breath, however, he will admit that if B is a *proper* subset of A, then A has more elements than B, since the new connotation of "more" is implied by "B is a *proper* subset of A." The latter is exactly what was intended by the use of "more than" in the preceding paragraph. However, the mathematician should not be taken to

† To be defined below.

task for this inconsistency, since he never uses "more than" as a technical term, and in his professional work would never be found guilty of this inconsistency. He will be the first to admit, moreover, that the above is simply an example of the necessity for precise definition of terms.

In the case of F and N, F does have more elements than N in the sense that N is a proper subset of F. But, in the "number" sense, F does not have "more" elements than N, since there does exist a (1-1)-correspondence between the elements of N and the elements of F.

1.2 One-to-one correspondence between N and F

The easiest way to see that such a (1-1)-correspondence exists (although perhaps not the most preferable) is to notice that the elements of F can be arranged in an order like that $(1, 2, 3, \cdots)$ of the natural numbers when these are placed in their "natural" order. To do this, we first split the elements of F into finite sets, and then we arrange these finite sets in the desired order.

For the first step, that of splitting F into finite sets, recall the definition of the elements of F; a number is an element of F if it can be represented by a form p/q, where p is any integer and q is a natural number. Let us imagine each element of F, then, as already represented by such a form and, moreover, in its "lowest terms" (i.e., so that p and q have no common factor greater than 1; "two-thirds" is then written as $\frac{2}{3}$, not $\frac{4}{6}$, for example); the number zero will be represented by $\frac{0}{1}$ only. Define the *index* of a number represented by p/q as $|p| + q$ (where, as usual, $|p|$ is the absolute or numerical value of p); we call this the "index of p/q" for short. Thus the index of $\frac{0}{1}$ is 1; of $-2\left(=\frac{-2}{1}\right)$ is 3; of $3\left(=\frac{3}{1}\right)$ is 4; etc. Then we place in the same set all elements of F that have the same index.

Now the set of numbers with index 1 has only 0 as its element; the set of numbers of index 2 has -1 and 1 as its elements; and so on. Let us agree to order the elements of a set of given index in pairs according to absolute value, each pair consisting of a negative and positive number of equal absolute value, the relative order between pairs being determined by numerical magnitude; for example, according to this rule, the set of index 5 will be ordered as follows:

$$-\tfrac{1}{4}, \tfrac{1}{4}, -\tfrac{2}{3}, \tfrac{2}{3}, -\tfrac{3}{2}, \tfrac{3}{2}, -4, 4.$$

Next we arrange these finite sets according to the size of their indices. The ordering of elements of F then turns out like this:

$$(F) \qquad 0, -1, 1, -\tfrac{1}{2}, \tfrac{1}{2}, -2, 2, -\tfrac{1}{3}, \tfrac{1}{3}, -3, 3, -\tfrac{1}{4}, \tfrac{1}{4}, -\tfrac{2}{3}, \tfrac{2}{3}, \cdots.$$

We call this "sequence" (F). Since every element of F has a definite index and a uniquely determined position in the set of that index, it will have a definite position in (F). And to get a (1-1)-correspondence between N and F, let the nth number in (F), starting from the left, be denoted by f_n, and make the pairing (n, f_n), $n \varepsilon N, f_n \varepsilon F$.

1.2.1 Many ways have been given for defining a (1-1)-correspondence between N and F. The one just given is not the best from the standpoint of effectiveness. For example, if we ask, "What rational number corresponds to 312,500?" the only apparent way of finding out is to run through the first 312,500 terms of the sequence (F) to find out. However, it is possible to set up correspondences in which for each number n the element of F corresponding to N is directly calculable. Consider the following:†

Let f be a (1-1)-correspondence between N and the set of all non-zero integers such that $f(2n) = -n$ and $f(2n - 1) = n$. Now each natural number greater than 1 is a unique product of powers of primes of the form $p_1^{r_1} \cdot p_2^{r_2} \cdots p_k^{r_k}$, where each $r_i \geqq 1$. If for each such number we define $g(p_1^{r_1} \cdot p_2^{r_2} \cdots p_k^{r_k}) = p_1^{f(r_1)} \cdot p_2^{f(r_2)} \cdots p_k^{f(r_k)}$, and $g(1) = 1$, then g is a (1-1)-correspondence between N and the set F^+ of positive rational numbers. In particular, $g(312{,}500) = g(2^2 \cdot 5^7) = 2^{-1} \cdot 5^4 = 625/2$. To extend this method to a (1-1)-correspondence between N and all rational numbers, positive, negative, and zero, is a simple matter. (See Problem 18.)

Other effective (See III 6.8, paragraph 4) enumerations may be found in Faber [a] and Niven [a].

1.3 One-to-one correspondence between N and A

We can, as Cantor showed, go even further with the above process. Specifically, let A be the class of all real roots of ordinary algebraic equations of the type

(1a) $$a_0x^n + a_1x^{n-1} + \cdots + a_{n-1}x + a_n = 0,$$

where n is a natural number, a_i an integer, $i = 0, 1, \cdots, n$, and $a_0 > 0$. The elements of A are called *algebraic numbers*. The elements of F all belong to A—i.e., $F \subset A$—since p/q is a root of the equation

(1b) $$qx - p = 0.$$

But F is a *proper* subset of A since the equation

$$x^2 - 2 = 0$$

has roots $\pm\sqrt{2}$, and it is not difficult to show that $\sqrt{2}$ cannot be represented by the p/q form which elements of F can assume. So A is a

† This was communicated to me by Professor Philip Obreanu.

"larger" set than F. Yet, surprisingly enough, there still exists a (1-1)-correspondence between the elements of N and the elements of A.

To show this, we proceed as we did in Section 1.2 by first splitting the elements of A into finite ordered sets, and then ordering these sets. To get the finite sets, let us define the index of an equation such as (1a) as the number

$$a_0 + |a_1| + \cdots + |a_{n-1}| + |a_n| + n.$$

As the a's and n are integers, the index is always a natural number. There exists no equation of index 1, since both n and a_0 are at least 1. The only equation of index 2 is $x = 0$, and the only root of equations of index 2 is the number zero. The only equations of index 3 are $2x = 0$, $x \pm 1 = 0$, $x^2 = 0$, yielding roots 0, -1, 1. Thus for each index there exists only a finite number of equations and, hence, a finite number of roots corresponding; of the latter, we reject the non-real roots (thus $x^2 + 1 = 0$ affords no real numbers for us) and order the remainder by the same rule that we used in the finite sets for F, except that we reject any number that was a root of an equation of lower index.

As in the case of F, we may finally arrange the sets according to the size of their indices. The sequence A then starts off like this:

$$(A)\quad 0, -1, 1, -\tfrac{1}{2}, \tfrac{1}{2}, -2, 2, -\tfrac{1}{3}, \tfrac{1}{3}, -\tfrac{1}{2}\sqrt{2}, \tfrac{1}{2}\sqrt{2}, -\sqrt{2}, \sqrt{2}, -3, 3, \cdots.$$

1.4 Non-existence of (1-1)-correspondence between N and R

Now A is itself a subset of a still larger class, namely, the class of all *real numbers*, which we shall denote by R. We have to be specific about what we mean by a real number here, just as we were in regard to the elements of F and A. For present purposes, then, a *real number*† is any number which can be represented by an unending decimal, in the ordinary decimal system, of the form

$$\text{(1c)}\qquad \pm k_1k_2\cdots k_m.a_1a_2\cdots a_n\cdots$$

where each k and each a are one of the digits $0, 1, \cdots, 9$; $k_1k_2\cdots k_m$ is the "integral" part of the number; $a_1a_2\cdots a_m\cdots$ is the "decimal part," also called a *decimal fraction*; and no n exists such that all digits $a_n, a_{n+1}, \cdots$ are zero (with the single exception that the number zero is represented by $0.a_1a_2\cdots a_n\cdots$, where *all* a_n are zero). For instance, the number $122\frac{1}{3}$ is an element of R, since it is represented by $k_1 = 1$, $k_2 = 2$, $k_3 = 2$ (m being 3 here), and all a's are 3: thus, $122.33\cdots3\cdots$. Similarly, $\sqrt{2}$ is real, since here $k_1 = 1$ (m being 1 in this case), $a_1 = 4$, $a_2 = 1$,

† As in the case of rational and algebraic numbers, we defer comment regarding the concept "real number." A more complete discussion of the concept will be given in Chapter VI.

$a_3 = 4$, etc.; and it is impossible that all $a_n, a_{n+1}, \cdots$ be zero for some n, since this would imply that $\sqrt{2}$ is rational. Also, the natural numbers are real numbers, since, for example, 2 can be represented by letting k_1 be 1 (m being 1), and letting all a's be 9; thus, $1.99 \cdots 9 \cdots$ (see Problem 8 at end of this chapter).

Now Cantor discovered that there does not exist any (1-1)-correspondence between the elements of N and the elements of R. For suppose such a correspondence did exist. Let the real number paired with each natural number n in this correspondence be denoted by "r_n." Also, denote the digit in the nth decimal place of r_n by "a_n^n." Then we can define a real number $r = 0.a_1a_2 \cdots a_n \cdots$ such that, for each n, $a_n = 1$ if $a_n^n \neq 1$, and $a_n = 2$ if $a_n^n = 1$. Since $r \varepsilon R$, it must be an r_n itself, say r_i. But r_i has a_i^i as the ith decimal digit, whereas r has a different digit a_i! Because of this contradiction we must conclude that no (1-1)-correspondence of the supposed type exists.†

1.4.1 The reader, if entirely unfamiliar with the above type of argument, may wish to explore some of the details more closely. For this purpose, it may help to imagine the digits of the numbers r_n arranged in a square array, obtained by setting the numbers down in a vertical sequence (omitting the integral parts since they do not influence the result) thus:

$$\begin{array}{ll} r_1: & .a_1^1a_2^1 \cdots a_n^1 \cdots \\ r_2: & .a_1^2a_2^2 \cdots a_n^2 \cdots \\ \cdot\ \cdot & \cdot\ \cdot\ \cdot\ \cdot\ \cdot\ \cdot \\ r_n: & .a_1^na_2^n \cdots a_n^n \cdots \\ \cdot\ \cdot & \cdot\ \cdot\ \cdot\ \cdot\ \cdot\ \cdot \end{array}$$

To get the number r, follow the main diagonal of the array beginning at the upper left-hand corner, and write down a digit $a_1 \neq a_1^1$, $a_2 \neq a_2^2, \cdots$, $a_n \neq a_n^n, \cdots$ according to the rule given above. For example, if the array started like this:

$$\begin{array}{ll} r_1: & .4146 \cdots \\ r_2: & .9999 \cdots \\ r_3: & .1019 \cdots \\ r_4: & .2682 \cdots \\ \cdot & \cdot\ \cdot\ \cdot\ \cdot \end{array}$$

then r would look like $.1121 \cdots$; and we already see that r will not be one of the numbers r_1, r_2, r_3, r_4.

† Cantor's proof is in *Jahresb. der Deut. Math. Ver.*, 1 (1892), p. 75.

1.5 The countable sets

It appears, then, in view of 1.4, that, by using the notion of (1-1)-correspondence, we can make a distinction between certain infinite sets. Thus we already see that N and R can be distinguished from one another in this way. And, indeed, if $\mathfrak{S}$ is any collection of sets whatsoever, we can decompose $\mathfrak{S}$ into disjoint classes, using the fact that (1-1)-correspondence between elements of two sets is an equivalence relation (see II 8.5). If $\mathfrak{S}$ has as its elements just the sets N, F, A, and R, for example, then the class decomposition of $\mathfrak{S}$ corresponding to (1-1)-correspondence consists of a class $\mathfrak{S}_1$ whose elements are N, F, A, and a class $\mathfrak{S}_2$ whose *single* element is R.

If $\mathfrak{S}$ is any collection of finite sets and $\mathfrak{S}$ contains a finite set, S, having, say, exactly 15 elements, then the class containing S would contain *all* the elements of $\mathfrak{S}$ each of which has exactly 15 elements. And if there were a set consisting of exactly 3 elements, the class containing this set would also contain *all* elements of $\mathfrak{S}$ each of which has exactly 3 elements, and only these. That is, (1-1)-correspondence as an equivalence relation induces a class decomposition of $\mathfrak{S}$ such that each class consists of all sets having a certain fixed number of elements. It is natural, therefore, to think of any two sets whose elements can be put in (1-1)-correspondence as "having the same number" of elements. Accordingly, Cantor made the definition:

1.5.1 **Definition.** If A and B are two sets such that there exists a (1-1)-correspondence between the elements of A and the elements of B, then we shall say that A and B *have the same cardinal number.*

Cantor used the term "have the same power (*Mächtigkeit*)." Notice, incidentally, that we have *not* defined "number" or "cardinal number," but only the term "have the same cardinal number." A definition of "cardinal number" follows later.

If two (ordinary) finite sets have the same cardinal number, they have the same "number" of elements in the usual sense; if one has 15 elements, then the other has 15 elements, etc. (cf. III 5.2.1). As we showed above, the sets N, F, and A all have the same cardinal number, but N and R do not; we say that N and R have "different cardinal numbers." And, although we have not yet defined "cardinal number," we can introduce, by analogy with the natural numbers, the numbers $\aleph_0$ ("aleph-null") and c.

When we say that "*a set S has $\aleph_0$ elements*," or, alternatively, that "*its cardinal number is $\aleph_0$*," we mean that *between its elements and the elements of N there exists a* (1-1)-*correspondence.* In particular, N, F, and A each has $\aleph_0$ elements.

If there exists a (1-1)-*correspondence between the elements of a set M*

and the elements of R, then we say "M has c elements," or alternatively, that "*its cardinal number is c.*"

The symbols "$\aleph_0$" and "c" would come, with practice, to have the same significance for us as the number 15, for example; for we never have defined the latter and we never *define* any of the natural numbers in the primary schools, because by the time a child enters school he has usually been introduced to the counting process and he "*knows*" what we mean if we say, "You may have only *two* pieces of candy."

The following terms will be useful in the sequel:

1.5.2 **Definition.** If a set has $\aleph_0$ elements, we call it *denumerable.* If a set is either (ordinary) finite or denumerable, we call it *countable.*

The reader should be warned that, although these *terms* are standard, their *meanings* vary in use. Thus "denumerable" is sometimes used to mean "countable," as we have defined the latter (English authors commonly use the term "enumerable," incidentally). The only safe way, when these terms are encountered in a mathematical work, is to ascertain explicitly the author's use of the terms. The usual German and French terms are "abzählbar" and "dénombrable." Bourbaki, for example, uses the latter in the sense of "countable" as defined above, and "infini dénombrable" in the sense of "denumerable" as defined above; see Bourbaki [B_1; 40]. On the other hand, Sierpinski [S_1; 41] (see also [S_3; 34]) uses "dénombrable" in the sense of denumerable as defined above.

1.5.3 A set which is not countable is called *uncountable.* [Many authors use the term "non-denumerable." For example, Sierpinski (*loc. cit.*) uses the French "non-dénombrable."] Thus R is an uncountable set. And we may, *a priori*, divide sets generally into two categories, the countable and the uncountable. The proof (III 5.3) that an ordinary infinite set is also Dedekind infinite devolved to showing that every ordinary infinite set contains a denumerable subset; II 5.3.2 may be restated: "If a set has a denumerable subset, then it is Dedekind infinite." And if we grant the use of the Choice Axiom, we can state (as a result of III 6.7):

1.5.4 *Every infinite*† *set contains a denumerable set.*

2 Uncountable sets

Thus far the only example of an uncountable set that we have introduced is the set R. In this section we shall give other examples, although we shall reserve some standard examples for later purposes. In particular, we give first some examples of sets that have the same cardinal number as R.

† Since we make this statement on the basis of the Choice Axiom, we need not specify "ordinary" or "Dedekind" here, the two being equivalent in this case (III 5).

2.1 The simplest of these is *the set, E, of all points on a euclidean straight line.* Anyone who has studied *analytic geometry* will recall that the basic frame of reference was the "coordinate axes." For *plane* analytic geometry, these were the "x-axis" and "y-axis": two straight lines usually perpendicular to one another, on each of which were "marked off" numbers starting from zero at the intersection of the axes (the "origin"), with positive numbers in one direction, negative in the opposite. It was *assumed* that to each point on the x-axis (and similarly on the y-axis) there corresponded a "number" and, conversely, that to every "number" there corresponded a point on the axis. In the terminology introduced above, it was assumed that there exists a (1-1)-correspondence between the elements of R and the points of the axis;† that is, the set of points on each axis has the cardinal number c. No proof is given in the analytic geometry, as a rule, that this assumption is valid. However, on the basis of a complete set of axioms for euclidean geometry the assertion is provable‡ in the sense that the set of all points on any straight line whatsoever has the cardinal number c. Moreover, if L is such a line, and S a *segment* of L, the set of points on S again has the cardinal number c (see Problem 14 at the end of the chapter).

2.2 That two line segments (considered as collections of points) have the same cardinal number may be demonstrated by construction: Let abc be any plane triangle, and d a point between b and c. To show that the set of points on bc has the same cardinal number as the set of points on dc, let $b'd$ be a segment parallel to ac with its endpoint b' on ab. Then let $b'c'$ be a segment parallel to bc with endpoint c' between a and c. Consider any point x of bc. The line ax meets $b'c'$ in a point x'; and in turn a line through x' parallel to $b'd$ meets dc in a point x''. The collection $\{(x, x'') \mid x \varepsilon bc\}$ constitutes a (1-1)-correspondence between the points of bc and the points of dc. Since bc and dc can have arbitrary lengths, so long as dc is not longer than bc, this construction is readily adapted to proving the desired theorem.

2.3 That the set of points on a euclidean straight line segment is *uncountable* is frequently proved as follows: Let ab be a segment of a straight line L; we may take the length of ab to be 1 unit. Then let us assume that the points of ab form a countable—hence denumerable (since they do not form a finite set)—collection. From the (1-1)-correspondence that exists between the points and the natural numbers, which follows from the assumption, we derive symbols x_n just as in III 4.1.1, so that each point of ab can be "named" x_n, $n = 1, 2, 3, \cdots$, no two points being assigned the same "name." Now, for each n, let I_n be an interval of length $(\frac{1}{2})^{n+1}$ with center x_n. Then the intervals I_n cover the segment ab; that is, every point of the interval is in at least one I_n. Consequently, their lengths should add up to 1, at least, since ab has unit length. But the sum of the series whose general term

† In addition, it is assumed that this correspondence is "order-preserving," but this does not concern us here. The assertion of this correspondence is sometimes called the "Cantor axiom."

‡ We refer here to such a system as that of Hilbert [H$_2$] (see H; 36–37). The Birkhoff [a] system makes the assertion an axiom.

is $(\frac{1}{2})^{n+1}$ is only $\frac{1}{2}$. Thus the assumption that the points of ab form a countable collection leads to contradiction.

[If we are familiar with the "Borel theorem," according to which a *finite* number of the segments I_n must cover ab, then the justification (omitted above) for using the *infinite* series $\frac{1}{4} + \cdots$ is avoided. Note that the above type of argument shows that any countable set of points on a line L can be covered by a set of intervals whose total length is as small as we please; we merely use intervals I_n of length $c/2^{n+1}$, where c is whatever (fixed) length we please.]

2.4 *The set of all points in the euclidean plane has the cardinal number c.* In the general remarks at the beginning of this chapter we recalled that the symbol "∞^2" was frequently used to designate the "doubly ordered" array of points in a plane, with the unfortunate connotation that somehow there is a "larger infinity" of points in the plane than on the line. It is not surprising, then, that it was anticipated by some mathematicians that the Cantor theory of cardinal numbers would provide a means of distinguishing *two*-dimensional space from *one*-dimensional space. Actually, however, the set E^2 of all points in a euclidean plane has the cardinal number c; i.e., the sets R and E^2 have the same cardinal number.†

The proof is very simple.‡ It will be sufficient to show that the set E^1 of real numbers x such that $0 < x \leqq 1$ and the set E^2 of all points in the coordinate plane defined by

$$E^2 = \{(x, y) \mid (0 < x \leqq 1) \ \& \ (0 < y \leqq 1)\},$$

have the same cardinal number (see Problem 14 at the end of the chapter). As $E^1 \subset R$, each element of E^1 is representable in the form $0.a_1a_2 \cdots a_n \cdots$ as in 1.4. We divide the array $a_1a_2 \cdots a_n \cdots$ into "blocks"; the following example illustrates what we mean. Consider a number such as $0.32046008 \cdots$. The successive "blocks" are 3, then 2, then 04, then 6, then 008, and so on; i.e., each "block" contains one digit different from 0, and this its last digit. We then construct an ordered pair $(0.a_1{}^1a_2{}^1 \cdots, 0.a_1{}^2a_2{}^2 \cdots)$ such that $a_1{}^1 = 3$, $a_1{}^2 = 2$, $a_2{}^1 = 04$, $a_2{}^2 = 6$, $a_3{}^1 = 008$, and so on—$(0.304008 \cdots, 0.26 \cdots)$. That is, we assign the blocks alternately to the two coordinates of a point of E^2. Since this process can be reversed, it provides a (1-1)-correspondence between the elements of E^1 and the elements of E^2. (See Problems 9–11 at the end of the chapter.)

2.5 *The set, $I = R - F$, of all irrational real numbers has the cardinal number c.* To prove this, we first prove:

† The problem of distinguishing between different dimensions is satisfactorily solved in that branch of mathematics known as Topology (cf. VII 3.5.2 and VII 4).

‡ The proof given in Young [Y; 167ff] is not valid. The construction which Young gives is that of Hilbert's "space-filling continuous curve," which sets up a many-one correspondence (see 3.2.3.1 below), not a (1-1)-correspondence (the latter being *impossible* in the case where the elements of E^2 are to lie on a continuous curve given by the correspondence).

2.5.1 Lemma.† *If A and B are countable sets, then $A \cup B$ is countable; and if either is denumerable, then $A \cup B$ is denumerable.*

Proof. We first consider the case where A and B are disjoint. If A and B are both finite, then $A \cup B$ is finite (see III 5, Theorem 5.2.2 for example). If A is denumerable, and B is finite, then, as in III 4.1.1, let the elements of A be assigned symbols x_n. If B is empty, then $A \cup B = A$; otherwise its elements are representable by symbols $y_1, y_2, \cdots, y_k$, where $k \varepsilon N$. Change the symbol x_n to y_{n+k}; then all elements of $A \cup B$ are assigned symbols y_m, $m \varepsilon N$. If *both* A and B are denumerable, let the symbols for elements of A be x_n as before, and for B let them be y_n, $n \varepsilon N$. Change x_n to z_{2n-1} and y_n to z_{2n}.

The case where A and B are not disjoint may be handled by considering the sets A and $B - A$; we leave this to the reader.

2.5.2 Theorem. *If S is a Dedekind infinite set and S' is a countable set, then $S \cup S'$ has the same cardinal number as S.*

Proof. We may assume S and S' disjoint; otherwise, we may use sets S and $S' - S$ (cf. Problem 12). As S is Dedekind infinite, it has a denumerable subset S_1 (cf. Problem III 21). Let $S - S_1 = M$; then $S = M \cup S_1$, where M and S_1 are disjoint sets.

Now, $S \cup S' = (M \cup S_1) \cup S' = M \cup (S_1 \cup S')$, and, by Lemma 2.5.1, $S_1 \cup S'$ is denumerable. As S_1 is also denumerable, there is a (1-1)-correspondence T_1 between the elements of $S_1 \cup S'$ and the elements of S_1; let T_2 be the identity correspondence (II 4.4.2) between the elements of M and the elements of M. The union $T_1 \cup T_2$ gives a (1-1)-correspondence between the elements of $S \cup S'$ and the elements of S.

2.5.3 Corollary. *The set I of all irrational real numbers has the cardinal number c.*

[For $R = I \cup F$, and I is Dedekind infinite (Problem 5); and by 1.2, F is denumerable.]

2.5.4 The elements of $R - A$ are called *transcendental numbers.* And since, in 1.3, we showed A to be denumerable, we also have

2.5.5 Corollary. *The set of all transcendental numbers has the cardinal c.*

2.5.6 Remark. Lemma 2.5.1 is usually considered the basis of a proof of the *existence of transcendental numbers.* For, if $R - A$ were countable, then, in view of 1.3, $(R - A) \cup A = R$ would be countable, contra-

† As we use the term "lemma," it is a statement which will be an aid in proving theorems but is not of itself of sufficient importance to be called "theorem" (because of its being a very special case of a later theorem, for instance, or not being of great import for the general theory under development). Not infrequently, however, a "lemma" proves to be more important than the theorem for whose proof it was designed.

dicting 1.4. It is to be noted that we thus have a proof of the existence of an *uncountable* set of numbers, *not one of which have we exhibited or constructed in any way*! (See 3.1.9, however.) We can, of course, exhibit such numbers—we give an example below—but the above argument has the feature of revealing the *uncountable* character of their totality. It was remarked by Sierpinski [S_1; 64] that, in the same year (1873) in which Cantor announced this result, Hermite demonstrated the transcendence of e, the base of natural logarithms, and ten years later Lindemann demonstrated the transcendence of π (from which follows the impossibility of "squaring the circle" with ruler and compass).

3 Diagonal procedures and their applications

The proof of the uncountability of R, given in 1.4 (see particularly 1.4.1), introduces a method commonly called "Cantor's diagonal method" or "diagonal procedure." As we shall see later, some mathematicians do not admit the existence of an uncountable set of real numbers as a legitimate consequence of the argument given in 1.4. However, it is generally admitted that the diagonal procedure itself, as it was used to construct the number r from the numbers r_n in 1.4, is a valid method for constructing numbers. (See 3.1.1.) Now how many such numbers can be constructed by using the diagonal method?

3.1 Construction of irrational and transcendental numbers

We showed in 1.2 how to arrange the set F of all *rational numbers* in a sequence:

$$(F) \qquad 0, -1, 1, -\tfrac{1}{2}, \tfrac{1}{2}, -2, 2, -\tfrac{1}{3}, \tfrac{1}{3}, -3, 3, -\tfrac{1}{4}, \tfrac{1}{4}, -\tfrac{2}{3}, \tfrac{2}{3}, \cdots.$$

From this sequence let us select the numbers which lie in order of magnitude between 0 and 1:

$$(F_1) \qquad \tfrac{1}{2}, \tfrac{1}{3}, \tfrac{1}{4}, \tfrac{2}{3}, \cdots.$$

If we now express these numbers in the decimal form prescribed for the elements of R in 1.4, and imagine them forming a square array as in 1.4.1, we get

$$(F_1) \qquad \begin{array}{l} 0.4999\cdots \\ 0.3333\cdots \\ 0.2499\cdots \\ 0.6666\cdots \\ \quad\cdot\quad\cdot\quad\cdot \end{array}$$

The application of the rule for obtaining r in 1.4 gives a number $0.1111 \cdots$; let us denote this number by r_ω. Evidently r_ω is a well-defined number since the order of the terms in (F), and hence in (F_1), was explicitly given and the rule of 1.4 is quite explicit; we can calculate r_ω to as many decimal places as we wish.

3.1.1 It is sometimes protested that r_ω cannot be well defined, since it is defined in terms of an infinite set of numbers [the elements of (F_1)], each of which is itself an infinite decimal. This ignores the fact, however, that, so far as determining r_ω is concerned, not a single element of (F_1) is needed in its *infinite* decimal form; all that is needed of each number in (F_1) is a knowledge of a *single one* of its digits, namely, the digit on the main diagonal of the above array. The digits constituting r_ω are as well defined, for instance, as are those of $\sqrt{2}$, whose digits are obtainable only by means of *algorithms*, such as the process taught in the grade schools for extracting square roots. As a matter of fact, the digits of r_ω are *better known* than those of $\sqrt{2}$, if we use, instead of the rule given in 1.2, a rule (Section 1.2.1) which enables us to find *directly* (without first setting down the preceding terms) the 100th term, for instance, of the sequence, hence the 100th digit of r_ω *without prior computation of the preceding* 99 *digits of* r_ω—something we cannot do for $\sqrt{2}$!

3.1.2 Let us now form a new array by placing r_ω at the head of (F_1):

$$(F_2) \qquad \begin{array}{c} 0.1111\cdots \\ 0.4999\cdots \\ 0.3333\cdots \\ \cdot\ \ \cdot\ \ \cdot \end{array}$$

and again apply the diagonal procedure. This time we get a number (whose first five significant digits we already know): $0.21111 \cdots$ [since the terms on the main diagonal of (F_2) are $1, 9, 3, 9, 6, \cdots$]. Let us denote this number by $r_{\omega+1}$.†

We may repeat the above procedure for finding $r_{\omega+1}$ by placing $r_{\omega+1}$ at the head of (F_2), thus forming a new array (F_3); application of the diagonal procedure to (F_3) yields a number $0.121111 \cdots$, which we denote by $r_{\omega+2}$.

3.1.3 For *every* natural number n we can now *define by induction* a number $r_{\omega+n}$ as follows: Having defined $r_{\omega+n-1}$ from an array (F_n), we place $r_{\omega+(n-1)}$ at the head of (F_n) to form a new array (F_{n+1}). Application of the diagonal procedure to (F_{n+1}) using the rule of 1.4, gives a new number which we denote by $r_{\omega+n}$. The number $r_{\omega+n}$ is different from all numbers in the array (F_{n+1}), since it differs in at least one digit from each

† The origin of the subscripts that we are using will be clear when we introduce the ordinal numbers; part of our present purpose is to introduce these (ordinal) numbers in a natural fashion.

of the numbers forming this array. In particular, then, $r_{\omega+n}$ is not an element of F and is therefore an irrational number.

With the definition of r_ω ($= r_{\omega+0}$) and the general rule for finding $r_{\omega+n}$, for any natural number n, as given in the preceding paragraph, we have given *by induction* a definition of an infinite sequence $r_\omega, r_{\omega+1}, \cdots, r_{\omega+n}, \cdots$ of irrational numbers, all different from one another (3.1.2 is actually superfluous, then, except as explanatory material introducing the general definition of 3.1.3).

3.1.4 It is possible to go further, however. If we follow the device used in the proof of Lemma 2.5.1 for the case where A and B are denumerable, we may form a *single* array (F_ω) from the elements of (F_1) and the elements of the collection $\{r_{\omega+n}\}$; the order from the top of the array down will correspond to $\frac{1}{2}, r_\omega, \frac{1}{3}, r_{\omega+1}, \frac{1}{4}, r_{\omega+2}, \cdots$. And by the diagonal procedure we obtain from (F_ω) a number that we denote by "$r_{\omega\cdot 2}$." Then, from arrays $(F_{\omega+1})$, $(F_{\omega+2})$, etc., obtained by placing $r_{\omega\cdot 2}$ at the head of (F_ω), etc., we can again set down a definition by induction of a set of numbers $\{r_{\omega\cdot 2+n}\}$, all different from those in (F_ω) and from one another.

3.1.5 It is not difficult to set down a definition by induction of a general array, $(F_{\omega\cdot k+n})$, for every natural number pair k, n, and hence a number $r_{\omega\cdot k+n+1}$ obtained from $(F_{\omega\cdot k+n})$ by the diagonal procedure (using the rule of 1.4). Note that this defines a *denumerable set of denumerable sets* of numbers: $r_\omega, \cdots, r_{\omega+n}, \cdots; r_{\omega\cdot 2}, \cdots, r_{\omega\cdot 2+n}, \cdots; \cdots; r_{\omega\cdot k}, \cdots, r_{\omega\cdot k+n}, \cdots; \cdots$. But we can go further! First, we prove:

3.1.6 Theorem.† *If $\{S_n\}$ is a denumerable‡ collection of denumerable sets S_n, then the set $\{x \mid x \,\varepsilon\, S_n$ for some $n\}$ is denumerable.*

Proof. For each n, denote the elements of S_n by symbols x_{nk}, $k = 1, 2, 3, \cdots$. Assign the index $j = n + k$ to each x_{nk}. Then for a given natural number j there exists only a finite number of elements x_{nk} (allowing both n and k to vary) of index j. (Thus, for $j = 3$, only x_{12} and x_{21} qualify.) We may now arrange the symbols x_{nk} in groups according to index, and order them within the group according to the first subscript n:

$$x_{11}, x_{12}, x_{21}, x_{13}, x_{22}, x_{31}, \cdots.$$

Finally, we pair each x_{nk} with the natural number that corresponds to its

† Compare with Lemma 2.5.1. Unless a (1-1)-correspondence *is given* for each n, between the elements of S_n and N, as is the case to which we shall apply this theorem below, then we can maintain that the Choice Axiom is needed in order to *assign* the correspondence to each S_n. Cf. Sierpinski [S_1; 124ff], [S_3; 117ff].

‡ We may say "countable" here if we wish, but we prefer to treat only one case; the more general case will follow immediately as soon as we study the "order" relations of cardinal numbers.

position in the above linear order, thus: $(x_{11}, 1), (x_{12}, 2), (x_{21}, 3), \cdots$. The desired (1-1)-correspondence follows immediately.

It will be noticed that if we arranged the sets S_n in a vertical array, the elements corresponding to S_n being in the nth row of the array, then the group of a given index lies on a "cross-diagonal"; thus,

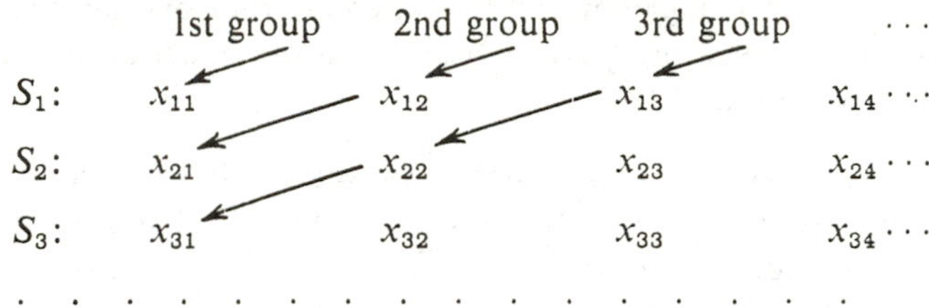

3.1.7 With the aid of Theorem 3.1.6, we may set up a (1-1)-correspondence between N and the set formed by the union of (F_1) and all irrational numbers defined by induction above. Denoting the elements of (F_1) by $r_1, r_2, r_3, \cdots$, and using the symbols of 3.1.5, we may (as suggested above for the proof of Theorem 3.1.6) arrange the sets (F_n) in a vertical array:

$$\begin{array}{llll} r_1 & r_2 & r_3 & \cdots \\ r_\omega & r_{\omega+1} & r_{\omega+2} & \cdots \\ r_{\omega\cdot 2} & r_{\omega\cdot 2+1} & r_{\omega\cdot 2+2} & \cdots \\ \cdot & \cdot & \cdot & \cdot \end{array}$$

By following the respective cross-diagonals we get the ordering $r_1, r_2, r_\omega, r_3, r_{\omega+1}, r_{\omega\cdot 2}, \cdots$. From this follows the desired (1-1)-correspondence with the elements of N, and *we may begin all over again*, as in 3.1, to construct entirely new irrational numbers. (The first of these we could denote by r_{ω^2}, the second by r_{ω^2+1}, etc.)

3.1.8 Remarks. By this time the reader should have become rather fatigued by the above procedures, and perhaps be inclined to ask: "How long can this go on?" Analysis of Sections 3.1.1 to 3.1.7 suggests the answer "Just so long as the numbers defined form a denumerable set."

But here we need to use extreme care. In 3.1.1 we elaborated on the "well-defined" character of the number r_ω, pointing out that we possess an *algorithm*, i.e., a well-defined finite procedure, which enables us to state exactly what any particular digit of r_ω is—whether it is 0, or 1, or 2, $\cdots$, or 9. In this sense, we are able to *exhibit* r_ω. And the same remark holds for each and every one of the numbers defined above (even $r_{\omega^2} = 0.112121\cdots$ is quickly calculated to 6 decimal places from the information given above). We have *effectively* or *constructively* defined, or given an *effective* definition of, each of these numbers. And the definition involved *effective* definitions of the way in which the numbers already defined at any stage can be put in (1-1)-correspondence with the natural numbers; in the latter case, we say that the sets of these numbers (previously defined) were *effectively countable* or *effectively denumerable.* It is preferable, then, that we say in answer to the above question, "Just so long as the set of numbers already defined forms an effectively denumerable collection." For the very essence of what we have done above is *to give a definite process* for obtaining the numbers. It will aid in understanding the significance of this if we interpolate the following:

3.1.8.1 Fallacious theorem. *Any collection, H, of numbers definable by continuing as far as possible the process commenced above in* 3.1.1–3.1.7 *is uncountable.*

Proof. For suppose that such a collection H is countable. Then its elements form a denumerable collection to which the diagonal procedure may be applied to define a new number, r, which is different from all numbers in H. But this implies that the process can be continued, and the assumption that H is countable leads to a contradiction.

3.1.8.2 What is wrong here? The proof is entirely analogous to that of 1.4; however, in that case we were discussing a set, R, of supposedly *already existing* numbers. Very well, one may counter, why is not H a set of already existing numbers? For one thing, we do not have a satisfactory description of H. What do we mean by "continuing as far as possible the process commenced in 3.1.1–3.1.7?" In 3.1.7 we had to *redefine* the (1-1)-correspondence with the natural numbers before we could carry on with the diagonal procedure; we had accumulated so many numbers at this stage ("a denumerable set of denumerable sets") that this became necessary. Clearly, then, "continuing" the process "commenced in 3.1.1–3.1.7" involves *redefinition* at successive critical points in the process. And until the necessary definitions have been specifically given, H is not itself really defined.

Thus the process may be continued as long as *we* are able to define new ways (at the critical points of the process) of setting up (1-1)-correspondences between the numbers already defined and the natural numbers; and these definitions have to be, by the very nature of the process, *effective definitions* which enable us to exhibit or calculate the number thereafter defined. What we *can* say, then, is that, as long as the set of numbers already defined at any point in the process is *effectively denumerable*, the process may be continued.

3.1.8.3 The "we" at the beginning of the previous paragraph was italicized in order to emphasize the contrast in the points of view which may be taken here. *Effective* definition depends on someone *effecting*, i.e., giving, the definition. However, someone may take the point of view that a definition may exist without anyone actually *giving* it. The "hard-boiled realist" would ask "Where?" In justice to the idealist, however, we may say that he could take the position that, given a dictionary D, he may consider the set $\mathfrak{D}$, of all possible finite arrangements, repetitions allowed, of words from that dictionary. Some of these will form definitions which, if "discovered" by someone, will become effective definitions adaptable to continuing the (possibly heretofore stopped for lack of effective definition) above process. For the sake of argument, we might call these *potential definitions*. And then we could ask, what is the character of a set H of potentially definable numbers obtained by continuing the above process?

Unfortunately the question is still inexact, since at no stage of the process is the manner of continuing *unique*. The diagonal procedure is itself precisely defined, but the ways of setting up the (1-1)-correspondence with the natural numbers at each stage are "unlimited." [Even in 3.1.2, we could begin by placing r_ω *after* the first element of (F_1) instead of at the head of (F_1).] To get rid of this difficulty, let us set up a (1-1)-correspondence between the elements of N and the elements of $\mathfrak{D}$.

3.1.8.4 *Ordering of all "sentences" using a dictionary D.* This may be done as follows: The elements (words and symbols) of D are already (alphabetically) ordered. If each such element is considered an arrangement of words, they form an ordered subset $\mathfrak{D}_1$ of $\mathfrak{D}$. Then the arrangements of elements of D, two at a time, can be ordered in obvious ways to form an ordered subset $\mathfrak{D}_2$ of $\mathfrak{D}$. In general, the arrangements of elements of D, n at a time, can be ordered to form an ordered subset $\mathfrak{D}_n$ of $\mathfrak{D}$. And, since each $\mathfrak{D}_n$ is finite, ordering according to index (each element of $\mathfrak{D}_n$ being of index n) induces an ordering in one array of the elements of $\mathfrak{D}$ (as in the treatment of F and A in 1.2 and 1.3 respectively).

With such a (1-1)-correspondence, we may stipulate that at any point in the above process the new definition of ordering the numbers previously defined shall be the first to occur in the ordered arrangement of $\mathfrak{D}$. In this way, we may conceive of H as a definite class of numbers. And we may then state:

3.1.8.5 Theorem. *The collection H as redefined is not effectively countable.*

The proof is like that of 3.1.8.1, except that we now have that, if H is effectively countable, then there exists an element of $\mathfrak{D}$ constituting a rule for establishing a (1-1)-correspondence between the elements of H and the elements of N and hence a new number definable by the diagonal process.

3.1.8.6 The preceding remarks obviously suffer from inexactness, being largely descriptive. We should be more precise about what dictionary, D, we are using; for instance, it should contain the symbol "ω" used so frequently in 3.1.1–3.1.7, although the word "omega" would do. Also, it should contain other symbols, such as (,), $\cdots$, ., 0, etc.

Having specified D accurately, we could then define an *effective definition* as a potential definition which has been observed and *recorded* so that it has become available for use by anyone. [The "realist" would omit the "potential definition" and say that "an effective definition is one that is expressed (already recorded) in terms of D."] And a collection would be effectively countable if there existed an effective definition of a (1-1)-correspondence between its elements and those of N.

As we shall see later, we come near, here, to the notion of a *formal system*, and for the present we leave the matter where it stands.

3.1.9 We remarked in 2.5.6 that we had obtained, in Corollary 2.5.5, a proof of the existence of an uncountable set of transcendental numbers without even exhibiting a single such number! If, in 3.1, we commence with the set A instead of F, then r_ω and all numbers later defined are transcendental. Moreover, these numbers are effectively defined, since in Sturm's functions† we have an algorithm for isolating and calculating (as closely as we wish) the real roots of a given algebraic equation. And, if we accept the existence of a set H as subsequently defined, one then has a non-effectively countable class of transcendental numbers, for some of which we have effective definitions, and for the remainder only potential definitions (relative to D).

3.2 The general diagonal procedure

We saw in 3.1 how, given an effectively denumerable set M of real numbers, we may effectively define, in terms of M, a real number that is not in M. From the standpoint of the complete totality, R, of real numbers, this is (at least theoretically) possible because R is uncountable.

3.2.1 We are now going to show how, given (1) a set S, and (2) a collection $\mathfrak{S}$ of subsets S_ν of S, such that between the elements of S and the elements of $\mathfrak{S}$ there exists a (1-1)-correspondence, we can effectively define a subset S' of S that is not an element of $\mathfrak{S}$—i.e., not an S_ν. The method we are going to describe constitutes what we may call a "general diagonal procedure."

3.2.1.1 First let us consider a very simple example. Let S be a set having 3 elements, a, b, and c. Let $\mathfrak{S}$ be the following set of three subsets:

† See any good book on the Theory of Equations.

$\{\{a, b\}; \{b, c\}; \{c\}\}$. Both S and $\mathfrak{S}$ have 3 elements, so that the stipulation of a (1-1)-correspondence between the elements of S and the elements of $\mathfrak{S}$ is satisfied; suppose that we denote the elements of $\mathfrak{S}$, in the order named above, by S_a, S_b, S_c (thus $S_a = \{a, b\}$ for example). Now the *method* is expressed in the following:

(3a) Rule. *Form a set S' which contains an element x of S if and only if S_x does not contain it.*

In the present case, S' will be, precisely, the set $\emptyset$, for S_a contains a, S_b contains b, and S_c has c as its only element; so S' does not contain a, b, or c.

3.2.1.2 The general case presents no difference. The rule (3a) is stated in sufficiently general terms if we agree that S_x is to denote the element of $\mathfrak{S}$ that is paired with $x \varepsilon S$ in the given (1-1)-correspondence. And we shall prove the basic theorem:

3.2.1.3 Theorem. *If S is any set and $\mathfrak{S}$ is the collection of all subsets of S, then S and $\mathfrak{S}$ do not have the same cardinal number.*

Proof. Suppose there exists a (1-1)-correspondence T between the elements of S and the elements of $\mathfrak{S}$. Let S_x denote the element of $\mathfrak{S}$ that is paired with $x \varepsilon S$ in this (1-1)-correspondence. Form the set S' according to (3a). Then S' is a subset of S. But then $S' \varepsilon \mathfrak{S}$ and accordingly must be an S_x—say S_a. This is absurd, since, by (3a), S' contains a if and only if S_a does not contain a. We must conclude, then, that T does not exist.

3.2.1.4 A corollary of 3.2.1.3 is the well-known elementary formula of arithmetic: $n < 2^n$ for every natural number n. For, if a set S has n elements, then S has 2^n subsets† (counting the null subset); and clearly $2^n \nless n$, since each element x of S constitutes the element $\{x\}$ of $\mathfrak{S}$. As soon as we have defined "<" for cardinal numbers, we shall have a similar inequality for them (Theorem 4.2.3.1).

3.2.2 Cantor's diagonal procedure as a special case

To see the relation between the above and Cantor's diagonal procedure, we shall represent real numbers, the elements of R, in the binary scale instead of the decimal.

3.2.2.1 For theoretical work in analysis and function theory, the binary representation has long been used. In recent years, in the development of high-speed

† In elementary texts, these are called "selections" instead of "subsets"; thus from a set of 3 things, $2^3 = 8$ selections may be made. In view of the discussion of III 6, we avoid the use of the term "selection" for obvious reasons when dealing with infinite sets.

computing machines and even in the study of the human nervous system, the binary scale has assumed great importance.

By the binary scale we mean the representation of real numbers on the base 2, using only the digits 0 and 1. Thus, whereas 101.01 is the number $10^2 + 1 + 10^{-2}$ in the decimal scale, in the binary scale it becomes $2^2 + 1 + 2^{-2} = 4 + 1 + \frac{1}{4} = 5\frac{1}{4}$. Other bases may be used, of course; thus, in the *ternary* scale, whose base is 3, the above number denotes $3^2 + 1 + 3^{-2} = 10\frac{1}{9}$ (there exists, incidentally, a society for the propagation of the duodecimal scale, whose base is 12). When more than one base is being used to represent numbers, the base used to represent a number may be indicated by enclosing the number in parentheses followed by a subscript denoting the base. Thus, $(101.01)_2 = 5\frac{1}{4}$ and $(101.01)_3 = 10\frac{1}{9}$. In short, the actual value of a number is computed in powers of the base, the powers being determined by the positions of the digits just as in the decimal system.† Note that 10. always denotes the base number.

3.2.2.2 In a manner similar to that given in 1.4, we represent each real number by a form (1c) (Section 1.4), where now, however, each k and each a are either 0 or 1. Since the set

$$E^1 = \{x \mid (x \varepsilon R) \,\&\, (0 < x \leqq 1)\}$$

has the cardinal number c (see Problem 14), we may restrict ourselves to the case where all k's are zero, i.e., to *binary fractions*. And this time we allow finite binary fractions; indeed, such numbers will be represented by *both* their finite and infinite binary fractions (thus, 0.1 and $0.011 \cdots 1 \cdots$, although both representing $\frac{1}{2}$, will be present). Since the number of finite binary fractions is countable (see Problem 2), it follows from Theorem 2.5.2 that the set R' of all such finite and infinite fractions still has the cardinal number c.

Instead of the argument as phrased in 1.4, we proceed as follows: Each binary fraction $.a_1a_2a_3 \cdots a_n \cdots$ determines a subset of N, namely, the subset which contains a natural number n if and only if $a_n = 1$. Thus the number $0.11 \cdots 1 \cdots$ in which every a_n is 1 determines the set N itself; the number $0.1011 \cdots 1 \cdots$ in which only a_2 is zero determines the subset consisting of all elements of N except the natural number 2; the subset of N consisting of all odd numbers is determined by the number $0.1010 \cdots$ in which only a_n's with odd subscripts are 1's; etc. Conversely, every subset N' of N determines a binary fraction $0.a_1a_2 \cdots a_n \cdots$, in which $a_n = 1$ if and only if $n \varepsilon N'$. Thus the cardinal number of R' is the same as that of the set, $\mathfrak{N}$, of all subsets of N.

By Theorem 3.2.1.3, $\mathfrak{N}$ and N do not have the same cardinal number. Hence R' and N do not have the same cardinal number.

† The arithmetic of the binary scale is extremely simple, and the elementary school child would certainly welcome the adoption of the scale. For example, the only "multiplication table" he would need to learn would be $0 \times 0 = 0$, $1 \times 0 = 0$, and $1 \times 1 = 1$!

3.2.2.3 The appeal to Theorem 3.2.1.3 has concealed the use of the diagonal procedure in the above argument. If we recall the proof of 3.2.1.3, however, we see that the argument depends, as in 1.4, on supposing that there does exist a (1-1)-correspondence between the elements of R' and the elements of N, and, if r_n' is the element of R' paired with n in this correspondence, on choosing 0 or 1 for the nth digit of a new number r according as the nth digit of r_n' is 1 or 0 respectively; the latter is the exact analogue, as explained in 3.2.2.2, of defining a subset N' of N which contains n or does not contain n according as the nth digit of r_n' is 0 or 1, respectively.

3.2.3 An alternative procedure

The use of the binary scale in 3.2.2 suggests an alternative way of presenting the general diagonal procedure. First, however, we introduce the following definitions:

3.2.3.1 If S and A are sets (not necessarily disjoint), then a *mapping* of S *into* A is a function $f: S \rightarrow A$; however, if the range of f is A itself, then we call f a *mapping* of S *onto* A. The set of all mappings of a set S into a set A is usually denoted by the symbol "A^S."

In the case of the so-called functions of a real variable, both S and A are identical with R (or, more generally, subsets of R); in the case of complex functions, S and A are subsets of the complex number system. For example, using the equations of plane analytic geometry, the function $y = x$ gives a mapping of R *onto* R; the function $y = e^{-x^2}$ gives a mapping of R *into* R and *onto* the set $\{y \mid 0 < y \leqq 1\}$.

3.2.3.2 Now, if S is any set, then a subset S' of S determines, uniquely, a function $f(x)$, $x \in S$, whose values are in the set S_1 consisting of the numbers 0 and 1. This function is defined as follows: If $x \notin S'$, we let $f(x) = 0$; otherwise $f(x) = 1$. (That is, in the "mapping" terminology, a subset S' determines a mapping of S into S_1; this is a mapping of S *onto* S_1 in every case except when $S' = 0$ or $S' = S$.) If $\mathfrak{F}$ is the set of all such functions (i.e., $\mathfrak{F} = S_1{}^S$), and $\mathfrak{S}$ the set of all subsets of S, then between the elements of $\mathfrak{F}$ and the elements of $\mathfrak{S}$ there exists a (1-1)-correspondence, in which to each S' corresponds the function $f(x)$ defined above, and conversely.

3.2.3.3 Alternative proof of Theorem 3.2.1.3. Supposing that T exists as in the earlier proof of Theorem 3.2.1.3, it follows that there exists a (1-1)-correspondence T' between the elements of S and the elements of $\mathfrak{F}$. Denote the element of $\mathfrak{F}$ that is paired with a given element a of S by "$f_a(x)$." But we can now define (diagonal procedure!) a function $f(x)$ such that, for each element a of S, $f(a) \neq f_a(a)$.† But, as a mapping of S into $S_1 = \{0, 1\}$, $f(x)$ is an element of $\mathfrak{F}$; it therefore is some $f_m(x)$, where $m \varepsilon S$. But this is impossible, since $f(m) \neq f_m(m)$ by definition.

3.2.3.4 Remark. The substitution of the set $\mathfrak{F}$ for the set $\mathfrak{S}$ as in 3.2.3.2 has led to the use of the symbol 2^S for the set of all subsets of a set S; here the "2" denotes the set $\{0, 1\}$.

4 Cardinal numbers and their ordering

In 1.5.1 we defined what we mean by saying that two sets *have the same cardinal number*. We did not, however, define *cardinal number*. In introducing the symbols $\aleph_0$ and c we remarked that they would come, with practice, to have "the same significance for us as the number 15 for example; for we never have defined the latter and we never *define* any of the natural numbers in the primary schools," since by the time a child enters school he has usually learned to count and knows (we sometimes say "knows intuitively") the meaning of the numbers used in counting. That is, the natural numbers already have an "intuitive meaning" for him, and to attempt a definition of them would seem senseless to him.

4.1 It is possible, however, to define these numbers as special cases of what we call cardinal numbers. If we analyze the psychology of the "intuitive meaning" of the number 2, we shall probably conclude that "2 apples" brings up to the mind of the hearer an image of a *pair*, here a pair of apples. A similar remark might hold for the phrase "20 apples"; but it would hardly hold for "200 apples." From the psychological viewpoint, it seems probable that 200 is simply one of the numbers one ultimately gets by starting with the numbers whose mental images are distinct—1, 2, 3—and applying consecutively the operation of adding 1, as taught in the elementary schools. (This is certainly the case with a number like 3,762,147; it is conceivable that, owing to some special circumstances of our occupation, our experience with 200 may induce a special intuitive knowledge of 200.) But numbers such as $\aleph_0$ and c are hardly to be attained in any such manner (by adding 1, that is).

If we return to the mental image brought up by "2," we may recognize that it constitutes a sort of abstract *norm*—a *pair* of not any special objects, but what we might call a "pair in the abstract." It was this that we had in mind when we remarked that, through frequent use, the number $\aleph_0$ would come to have a similar significance for us, possibly like the abstract image of the natural numbers in their natural order—which is

† That is, $f(a) = 0$ if $f_a(x)$ has the value 1 at $x = a$; otherwise $f(a) = 1$.

probably our usual concept of N (it being difficult to dissociate the individual numbers from the *order* induced by the counting process). Similar remarks might be made about the number c.

4.1.1 In 1879, G. Frege proposed a definition of "cardinal number" which was later (1901) and independently proposed by B. Russell. This so-called *Frege–Russell definition* is usually stated as follows: The cardinal number of a set S is the set of all sets that have the same cardinal number (in the sense of this relation as given in 1.5.1) as the set S.

Another way of putting this is to observe that the relation "have the same cardinal number as" is an equivalence relation (II 8.1) in the set U of all sets, and cardinal numbers are the classes of the corresponding class decomposition of U (II 8.5). Thus 2, as a cardinal number, would be the class of all pairs; and to say that a given set A "has 2 elements" is simply saying that A is an element of this class. Similarly, 1 would be the class of all sets having a single element (the class of all "singletons").

Unfortunately, unless we restrict the manner in which we operate with sets, or the set notion itself, these notions involve contradiction. For example (see Problem 28), the set of all sets is self-contradictory; and similarly the set of all sets having a single element (which, according to the Frege–Russell definition is the number 1) and the set of all pairs (the number 2) and so on, can be shown to be self-contradictory. Unless we are prepared, then, either to give up the idea of adopting some definition of cardinal number, or to pause to set up a formal set theory which will be restricted sufficiently to avoid known contradictions, we must get along tentatively with a provisional definition. For this we shall use the "genetic method"—i.e., we shall give a definition which simulates the actual historical evolution of the number concept. In 1.5.1 we defined $\aleph_0$ and c by means of the sets N and R, respectively. These sets we might call the *norms* for $\aleph_0$ and c; for these "numbers" they serve the same purpose that the standard yard in Washington, D.C., serves for the yard length of measure. The test for whether a given set has the number $\aleph_0$ is, does it have the same cardinal number as N? We might, then, adopt the device of establishing a norm for each cardinal number. Then a cardinal number, α, would always be associated with a set, A, which would serve as a criterion as to whether any given set B has cardinal number α or not. This would serve to define the symbol α; thus in Section 1.5.1 what we really defined was the symbols $\aleph_0$ and c.

Some authors consider cardinal numbers purely as symbols (Sierpinski, for instance; [S_1; 20], [S_3; 132]). It is doubtful whether mathematicians in general take this point of view, although number symbols preceded the formation of number concepts (cf. Wilder [Wi]). Most mathematicians consider numbers as *concepts*, relating to the "size" of sets. The most basic aspect of the form of a set—disregarding all other aspects such as color, shape, substance and the like if the set be a collection of physical objects; and disregarding order and other relations, operations and the like if the set be a collection of mathematical entities—is the "number" of its elements. Consequently, some have proposed that "number" be defined as the only property of a set that remains after rejecting all such properties as were just outlined. There results a negative definition which leaves one feeling that after rejecting all such properties there is possibly nothing left!

It is easy to set up norms for natural numbers, if we first define $\emptyset$ to be the norm for 0. For the first natural number, 1, the set $\{\emptyset\}$, whose single element is the set $\emptyset$, may serve as a norm. In particular, then, a set A would be said to "have 1 element" or, equivalently, "the number of elements in A is 1" if A has the same cardinal number as $\{\emptyset\}$. Similarly, the norm for 2 would be the set $\{\emptyset, 1\}$ whose elements are $\emptyset$ and 1; the norm for 3 would be $\{\emptyset, 1, 2\}$; and so on. In the ordinary affairs of life, this corresponds closely to our counting practices. True, when we wish to ascertain the number of elements in a set S of three elements, say, we don't precisely use the norm $\{\emptyset, 1, 2\}$, counting "null set, one, two," but we do say "one, two, three," assigning these words one at a time to the elements of S; so, except for change of words, we do virtually the same thing. And if we were interested in only defining natural numbers we would proceed in this fashion, perhaps. However, since we want to define "transfinite" cardinal numbers like $\aleph_0$ and c as well, this would necessitate taking up ordinal numbers first. So we shall content ourselves at present with an even more "naive" approach, the genetic.

4.1.2 If we study extant records of the evolution of the natural number concept, it becomes clear that measurement of the size of a collection was the prime motive (cf. Wilder [Wi]). Tally sticks from paleolithic times have been found; tablets from the ancient Egyptian culture recording large numbers raise suspicions that the sizes of armies, groups of captives, and the like, have been magnified for purposes of glorification of the ruling class—thus taking advantage of the degree of symbolism which had been attained. Not until *symbols* for numbers—e.g., the remarkable Babylonian numerals—were devised, did a *concept* of number develop. It is typical of cultural evolution that the origins of symbols become obscured, and that they take on meanings that they did not originally have. And the symbols for numbers seem to have passed from a descriptive (adjective) character to a nominal (noun) status. The symbols persisted from one culture to another and ultimately grew to stand for "something," viz., the number concept.

Concepts, once formed, as products of our interactions with the physical and social environments, are just as "real" as the objects in the physical world. And this applies to the number concept. Analysis of the concept leads inevitably to the conclusion that it consists of an idealization of the elementary notion of "size." So instead of trying to give the concept an illusory "reality" by identifying it with the "class of all sets" having the same size, we may prefer to let the symbol be a name for the concept itself. The viewpoint of the working mathematician and the scientist occupied with the applications of mathematics is precisely this, so far as the uses of the natural numbers are concerned. For some purposes, we may formalize the notion in an axiomatic framework just as the early conception of physical space was formalized in Euclidean geometry. But a satisfactory formalization is not as easily achieved as in the case of geometry. And now that we wish to extend the notion to infinite collections, it appears advisable to make the initial definitions as "natural" as possible. Formalization can (and has) come later.

Consequently, we shall assume that with each set we associate the concept of its size and call this the *cardinal number* of the set. This concept will be denoted by a symbol; in case more than one symbol is assigned to a cardinal number (for instance, the Arabic "2" and the Roman "II"), we consider these equivalent; this equivalence will be expressed by the "=" sign, and is to be interpreted as logical identification of the concepts denoted. Thus "2 = II" means that the same cardinal number is denoted by "2" and "II". And if between the elements of a set A and those of a set B there exists a (1-1)-correspondence, we shall say that they have the same "size"—i.e., that their cardinal numbers are the same. For finite sets, this is precisely the situation with the natural numbers. Thus the natural numbers become cardinal numbers in this context; i.e., insofar as they measure size alone (their role as ordinal numbers will be discussed in Chapter VI). And the natural number symbols $1, 2, 3, \cdots$, will be used as symbols for cardinal numbers of finite sets; to them we add 0 as the cardinal number of $\emptyset$.† Notice that our convention regarding the meaning of "cardinal number" conforms to the expression "has the same cardinal number as." For we may now "dissect" the latter phrase, interpreting "cardinal number" as having the new meaning, and the result is the same as before. Thus, as shown in 1.2, N and F have the same cardinal number, and the standard symbol for this cardinal number is $\aleph_0$.

For individual sets it is convenient to use a symbolism introduced by Cantor; if A is a set, the symbol $\overline{\overline{A}}$ denotes its cardinal number. And if A and B are sets having the same cardinal number, the expression "$\overline{\overline{A}} =$

† Thus the symbols for "finite cardinals" are $0, 1, 2, \cdots$.

$\overline{\overline{B}}$" symbolizes the fact (in conformity with the convention regarding "=" stated above). In particular $\overline{\overline{N}} = \overline{\overline{F}} = \aleph_0$.

It should be noticed, regarding our convention, that three distinct entities are involved in the notion of a particular cardinal number; (1) certain sets, (2) the common size of these sets (the actual cardinal number), and (3) a symbol (or equivalent symbols). In practice, (2) and (3) are commonly employed as though they were the same; thus we speak of "the number 2" without mentally differentiating between the symbol "2" and the cardinal number 2. If no inconsistency results from such confusion, it is not serious, and we shall frequently conform to this practice. As a matter of fact, in applications this confusion of symbol and thing symbolized is well known to be the rule rather than the exception. (In our daily life, it is one of the functions of symbols to substitute for the things symbolized.) Thus, in balancing accounts, we use symbols in reckoning without regard for their meanings, only returning to the things symbolized (such as the state of our finances) in the final act of interpreting the results of the reckoning.

In theoretical considerations, such as in the proving of theorems, a consideration of cardinal numbers may start with either (1) or (2). In case we commence with (3), we may wish to transfer attention to (1). For example, if α is a cardinal number (precisely, "α" is the symbol associated with a cardinal number), then we may wish to consider a set A having this cardinal number; in this case we may call A a *representative set* for (or set representative of) α; in symbols, $\overline{\overline{A}} = \alpha$. If α is frequently used, we may wish to fix upon a representative set as a norm, such as the set N for $\aleph_0$ and R for c.

4.2 "Size" is relative, and the reason we consider cardinal numbers at all is that we wish to compare the sizes of sets. So what shall we mean by saying that A is of smaller size than B or, in cardinal number symbols, what shall we mean by " $\overline{\overline{A}} < \overline{\overline{B}}$"? We define this relation as follows:

(4a) **Definition.** If α and β are cardinal numbers, then $\alpha < \beta$ if, for arbitrary sets A, B such that $\overline{\overline{A}} = \alpha$, $\overline{\overline{B}} = \beta$, A has the same cardinal number as a subset of B but the converse fails.

4.2.1 That $2 < 5$ by Definition (4a) is clear (take A as a set of two coins and B as a set of five books, for example). That is, (4a) corresponds exactly to the "<" indicated for the natural numbers when we speak of their "natural order" (sometimes called "order of magnitude"; see III 4.1f).

Consider $\aleph_0$ and c. Here we may use N and R as the representative sets A and B of (4a). As N is a subset of R, the identity correspondence of

N with N as a subset of R shows that N has the same cardinal number as a subset of R. However, R cannot have the same cardinal number as a subset of N, since this would imply that R is denumerable (see Problem 12). Hence, by definition (4a), $\aleph_0 < c$.

4.2.2 The cardinal number f

Let E denote the set of all mappings (3.2.3.1) of R into $\{0, 1\}$. Then, if E_1 denotes the set of all subsets of R, $\overline{\overline{E}} = \overline{\overline{E_1}}$ (by 3.2.3.2). Let us denote $\overline{\overline{E}}$ by f. By Theorem 3.2.1.3, $c \neq f$. Also, $\aleph_0 \neq f$ since the contrary would imply $\aleph_0 = c$.† Hence f is a new cardinal number. Is $c < f$?

For each $r \in R$, let f_r be the mapping of R into $\{0, 1\}$ obtained by pairing each $x \in R$ with 0 if $x \neq r$, and by pairing r with 1. For $r, r' \in R$ and $r \neq r'$, f_r and $f_{r'}$ are different mappings, and the collection $\{(r, f_r)\}$ constitutes a (1-1)-correspondence between the elements of R and the elements of a subset $E' = \{y \mid (y \in E) \;\&\; (y \text{ is an } f_r)\}$ of E. Thus R has the same cardinal number as a subset of E. To show that $c < f$, we may repeat the type of argument used in the proof of Theorem 3.2.1.3. But, this time, if T represents a (1-1)-correspondence between E_1 and a subset R_1 of R, and for each $x \in R_1$ we denote the element of E_1 corresponding to x by "S_x," then the set

$$S' = \{x \mid x \in R_1 \;\&\; x \notin S_x\}$$

is a subset of R leading to the same absurdity as in 3.2.1.3.

4.2.3 We remarked in 3.2.3.1 that the set of all mappings of a set S into a set A is usually denoted by the symbol A^S. When A is the set $\{0, 1\}$, we usually use the symbol 2^S (3.2.3.4); and the cardinal number of 2^S is denoted by "$2^{\overline{\overline{S}}}$." Thus the cardinal number of the set of all mappings of N into $\{0, 1\}$ is denoted by "$2^{\aleph_0}$"; this is also the cardinal number of the set of all subsets of N. Since we showed in 3.2.2.2 that this cardinal number is c, we have the classical relation

$$2^{\aleph_0} = c. \tag{4.2.3a}$$

And, since we showed in 4.2.1 that $\aleph_0 < c$, we have

$$\aleph_0 < 2^{\aleph_0}. \tag{4.2.3b}$$

Similarly, since $f = 2^c$, by 4.2.2, and $c < f$, we derive the relation

$$c < 2^c. \tag{4.2.3c}$$

† As shown below, E has a subset E' which has the cardinal number c. Hence, a (1-1)-correspondence between the elements of N and the elements of E would induce a (1-1)-correspondence between the elements of an infinite subset N' of N and the elements of E'; the latter would then have cardinal number $\aleph_0$ (cf. Problem 12).

Relations (4.2.3b) and (4.2.3c) suggest that, perhaps for every cardinal number α, $\alpha < 2^{\alpha}$.

4.2.3.1 Theorem. *For every cardinal number* α,

$$\alpha < 2^{\alpha}.$$

Proof. Let A be a set having cardinal number α. We need only show that (i) A has the same cardinal number as a subset of 2^A, and (ii) no subset of A has the same cardinal number as 2^A. The (1-1)-correspondence in which, for each fixed $a \varepsilon A$, a is paired with the element of 2^A which maps a into 1 and all other elements into 0, shows that (i) holds. That (ii) holds may be shown by the same type of argument as that used in proving 3.2.1.3 (as adapted to the proof of "$c < f$" in 4.2.2 for instance).

4.2.3.2 A special case of Theorem 4.2.3.1 is the well-known formula $n < 2^n$ of arithmetic (see 3.2.1.4).

4.2.3 Lemma. *The relation* $<$ *defined in* (4a) *is transitive; i.e.,* $\alpha < \beta, \beta < \gamma$ *imply* $\alpha < \gamma$. *And, if* $\alpha < \beta$, *then* $\alpha \neq \beta$.

[We leave the proof to the reader.]

4.2.4 Existence of infinitely many transfinite cardinal numbers

Cardinal numbers of infinite sets, such as the numbers $\aleph_0$, c, and f, are called *transfinite* cardinal numbers. The *finite* cardinal numbers are 0 and the natural numbers $1, 2, \cdots, n, \cdots$.

Theorem 4.2.3.1 gives a means of demonstrating the existence of infinitely many transfinite cardinal numbers. Starting with $\aleph_0$, we get $c = 2^{\aleph_0}$ and $f = 2^c$. And, in addition, we may now consider 2^f. By virtue of Lemma 4.2.3, this is not one of the numbers $\aleph_0, c, f$, since $f < 2^f$. And evidently by this process of "exponentiation" we exhibit an infinite collection of different transfinite cardinal numbers (which would not necessarily constitute *all* the transfinite cardinals, however).

It may be argued that we can go even further. Let $\alpha_1, \alpha_2, \cdots, \alpha_n, \cdots$ be cardinal numbers, such that $\alpha_n < \alpha_{n+1}$ for all n, and let $\overline{\overline{A}}_n = \alpha_n$. Then the cardinal number α of the set $A = \bigcup_{n=1}^{\infty} A_n$ satisfies the relation $\alpha_n < \alpha$ for all n (Problem 26).

Inasmuch as the set N of finite cardinals is simply ordered (II 7) relative to its natural order, and we can now extend N by adding $\aleph_0, c, f, 2^f, \cdots$, and obtain a larger simply ordered collection, we can ask: Do the cardinal numbers (finite and transfinite) form a simply ordered collection relative to the order relation defined in (4a)?

4.2.5 By Lemma 4.2.3, the relation $<$ for cardinal numbers satisfies axioms (2) and (3) of *simple order* (II 7) for any collection C of cardinal numbers. We cannot conclude, however, that axiom (1) of simple order holds, for there is no guarantee, *a priori*, that, given two sets A and B, there will exist a (1-1)-correspondence between one of these sets and a subset of the other.† *A priori*, given two sets A and B, there are four possibilities. Let us use the arrow "$\rightarrow$" to denote the existence of a (1-1)-correspondence; i.e., $A \rightarrow B$ or $B \leftarrow A$ will denote that there exists a (1-1)-correspondence between A and a subset of B. *Non-existence* of such a correspondence will be denoted by $\not\rightarrow$. Then the four possibilities are:

(4.2.5a) $$\begin{matrix} A \rightarrow B \\ A \leftarrow B \end{matrix}$$

(4.2.5b) $$\begin{matrix} A \rightarrow B \\ A \not\leftarrow B \end{matrix}$$

(4.2.5c) $$\begin{matrix} A \not\rightarrow B \\ A \leftarrow B \end{matrix}$$

(4.2.5d) $$\begin{matrix} A \not\rightarrow B \\ A \not\leftarrow B \end{matrix}$$

Cases (4.2.5b) and (4.2.5c) correspond, respectively, to $\overline{\overline{A}} < \overline{\overline{B}}$ and $\overline{\overline{B}} < \overline{\overline{A}}$. In the case of two finite sets A and B, (4.2.5a) would certainly imply that $\overline{\overline{A}} = \overline{\overline{B}}$; for the only way (4.2.5a) could hold in this case would be for the correspondences indicated to be a (1-1)-correspondence between the elements of A and the elements of B (see III 5.2.1). Therefore, it seems reasonable to ask if the same conclusion holds when A and B are infinite sets.

This was one of the earliest questions to arise (and to be answered) after the Cantor theory of sets was announced, and below we shall give the theorem constituting the affirmative answer. Consequently, if we can show that case (4.2.5d) cannot occur, no matter what the sets A and B may be, we shall then be able to conclude that the cardinal numbers (finite and transfinite) do form a simply ordered set relative to the "$<$" defined in (4a) and the "$=$" defined in 4.1.2. As we shall see in the next chapter, non-occurrence of (4.2.5d) is equivalent to acceptance of the Choice Axiom (III 6.3). Thus, without assumption of the Choice Axiom, the most we could assert is that the cardinal numbers form a partially ordered collection relative to the relation "$\leq$" formed from combination of the above "$<$" and "$=$" (see Problem II 16). The central importance of the Choice Axiom in the theory of infinite sets becomes clearer as we proceed.

We now state and prove the theorem referred to above. First, let us

† Contrary to the assertion in Young [Y; 80].

note that (4.2.5a) implies that A has the same cardinal number as some subset B_1 of B, and B has the same cardinal number as some subset A_1' of A. From this as hypothesis we shall show that it follows that A and B have the same cardinal number. Second, it should be pointed out that, from the correspondences given to satisfy (4.2.5a), the (1-1)-correspondence establishing the equal cardinality of A and B will be *effectively* given by the proof of the theorem. As we point out later, this fact may be used to give a new enumeration of the rational numbers, for instance.

4.2.6 **Bernstein equivalence theorem.**† *If A and B are sets such that A has the same cardinal number as a subset B_1 of B, and B has the same cardinal number as a subset A_1' of A, then A and B have the same cardinal number.*

Proof of 4.2.6. Let f denote a (1-1)-correspondence between A and B_1, and g a (1-1)-correspondence between B and A_1'. In the figure, which is purely schematic, we have represented A and B as the sets of all points in two vertical rectangular bands which run off at the bottom of the figure. Each symbol denotes the set of all points of the band below the line (solid or broken) on which it rests. Thus A_1' denotes the set of all points below the first cross-hatched rectangle.)

If we think of f as a mapping of A onto B_1, and g as a mapping of B onto A_1', then in the notation of II 4.4.1, the expression

$$y = g(f(x))$$

denotes a (1-1)-correspondence T between A and some subset A_1 of A_1'. (For if $x \varepsilon A$, then $f(x) \varepsilon B_1$; and, consequently, $g(f(x)) \varepsilon A_1'$ since g maps B onto A_1'.) Now T is also a (1-1)-correspondence between the elements of A_1 (as a subset of A) and a subset A_2 (of A_1) (see figure); and, for general $n \varepsilon N$, we have in T a (1-1)-correspondence between the elements of a set A_n (as a subset of A) and a subset A_{n+1} of A_n (definition by induction). If, then, we let $A = A_0$, and for each n let A'_{n+1} be the subset of A_n that corresponds to A_n' in T, we can define a (1-1)-correspondence T' between the elements of A and the elements of A_1' as follows:

(a) For each n, T' agrees with T for elements of $A_{n-1} - A_n'$ (cross-hatched in figure); i.e., between the elements of $A_{n-1} - A_n'$ as a subset of A, and the elements of $A_n - A'_{n+1}$ as a subset of A_1', the (1-1)-correspondence T' is the same as T.

† Sometimes called "Cantor-Bernstein theorem," since it was conjectured by Cantor (it was proved by F. Bernstein in Cantor's seminar); also called "Schröder-Bernstein theorem," since it was independently proved by E. Schröder (on the basis of a logical calculus). This theorem is a special case of a theorem of Banach; see Sierpinski [S_1; 90ff]. (An interesting and simple proof of Banach's theorem may be obtained if we are familiar with the theory of linear graphs; see König [Kö; 85].)

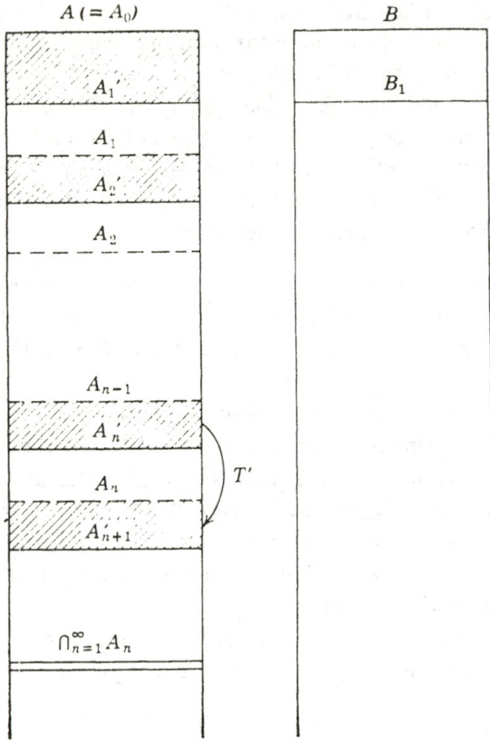

(b) T' is the identity correspondence for elements of $A_n' - A_n$ (all n), and for elements of $\bigcap_{n=1}^{\infty} A_n$. In the figure, T' makes each cross-hatched portion correspond to the next cross-hatched portion below it, and leaves the other portions fixed.

Finally, the combination of T' and g^{-1} gives a (1-1)-correspondence between the elements of A and the elements of B. In symbols, $h(x) = g^{-1}(T'(x))$ for each $x \varepsilon A$.

4.2.6.1 It should be noted, as predicted above, that the (1-1)-correspondence established in the above proof between the elements of A and the elements of B is effectively defined (3.1.8) if the given correspondences f and g are effectively defined.

4.2.6.2 Note, too, that from 4.2.6 it follows that, if A has the same cardinal number as a subset of B, then either $\overline{\overline{A}} = \overline{\overline{B}}$ or $\overline{\overline{A}} < \overline{\overline{B}}$, inasmuch as the first half of the condition for $\overline{\overline{A}} < \overline{\overline{B}}$ (4a) is already fulfilled. We shall express this relation by $\overline{\overline{A}} \leqq \overline{\overline{B}}$.

4.2.7 In concluding this chapter we raise another question—one which, unfortunately, we shall not be able to answer: Consider the natural numbers. Between 1 and 2, in the natural order, there is no cardinal number; 2 is the "immediate successor" of 1. And, in general, for any $n \varepsilon N$, $n + 1$ is the immediate successor of n.

4.2.7.1 Question. Is there any cardinal number α such that $\aleph_0 < \alpha < c$? Or is c the immediate successor of $\aleph_0$?

4.2.7.2 In a way, this is not a good question to ask unless we rule out (4.2.5d). However, it is legitimate, even though we do not know that the cardinal numbers are simply ordered. If we grant the simple ordering, it will be noted that 4.2.7.1 is equivalent to asking: Does every infinite subset of R have either the cardinal number $\aleph_0$ or the cardinal number c? We shall return to this question in V 3.6.3.

SUGGESTED READING

Cantor [C_1]
Fraenkel [F_3; I, II]
Grelling [Gr; 16–25]
Hobson [Ho; §§ 6, 58–62]
Kamke [Ka_1; I–II]
Richardson [R; XV]
Sierpinski [S_1; II–V], [S_3; II–V]
Young [Y; VI, VIII]

PROBLEMS

1. If $S = \emptyset$, why does 3.2.1.3 hold?
2. Show that the set of all finite subsets of N is countable.
3. Show that in the coordinate plane the set $K = \{(x, y) \mid (x \varepsilon F) \,\&\, (y \varepsilon F)\}$ is denumerable. Then use Theorem 3.1.6 to show that the set of all circles with radii elements of F and centers in K is denumerable. Work the analogous problem for three-dimensional space.
4. Use the results of Problem 3 to show that, if physical space is assumed to be euclidean, then the set of all physical objects (assuming some kind of a division of the material universe into things called "objects") is countable.
5. That 2.5.3 is stated as a corollary of Theorem 2.5.2 implies that the set I must be Dedekind infinite. Justify this by showing I is Dedekind infinite; base your proof on III 5.3.2, thus avoiding use of the Choice Axiom.
6. To carry out the proof implied for Corollary 2.5.5, it is necessary to know that the set of transcendental numbers is Dedekind infinite. Prove this.
7. Show that if B is a Dedekind infinite set, and A a set such that $A \supset B$, then A is Dedekind infinite.
8. Show that F may also be defined as the set of all elements of R whose decimal representations ultimately begin to repeat (such as $\frac{1}{5} = .19999 \cdots$, $\frac{1}{7} = .142857142857142857 \cdots$).
9. Show that if in the proof of 2.4 we made each block a single digit, there would not result a (1-1)-correspondence of the type desired.

10. Modify the proof of 2.4 to show that the subset $E^3 = \{(x, y, z) \mid (0 < x \leqq 1) \& (0 < y \leqq 1) \& (0 < z \leqq 1)\}$ of coordinate 3-dimensional space has the cardinal number c. How about space of four dimensions, in which each point has *four* coordinates (x, y, z, w); five dimensions; etc.?

11. In a Hilbert space, each point is represented by an infinite sequence of coordinates $(x_1, x_2, \cdots, x_n, \cdots)$, where x_n is a real number. Show that the set $E^\omega = \{(x_1, x_2, \cdots, x_n, \cdots) \mid 0 < x_n \leqq 1 \text{ for all } n\}$ has the cardinal number c.

12. Show that every infinite subset of N has the cardinal number $\aleph_0$ (cf. Problem 24 of Chapter III).

13. Show that the set of positive real numbers has the cardinal number c. As a corollary, the set of non-negative real numbers has the cardinal number c.

14. Show that the set of all real numbers between 0 and 1 has the cardinal number c. As a corollary, the subset $\bar{R}^1 = \{x \mid 0 \leqq x \leqq 1\}$ of R has the cardinal number c.

15. In the set $\bar{R}^1 = \{x \mid (x \, \varepsilon \, R) \& (0 \leqq x \leqq 1)\}$, let $x \approx y$, where $x, y \, \varepsilon \, \bar{R}^1$, mean that $x - y \, \varepsilon \, F$. Show that $\approx$ is an equivalence relation in $\bar{R}^1$. How many elements are there in each class of the class decomposition of $\bar{R}^1$ corresponding to $\approx$? What would you guess is the cardinal number of the set of all classes in this class decomposition?

16. Contrast (especially as regards effectiveness) the decomposition of $\bar{R}^1$ defined in Problem 15 with the following (due to Sierpinski): We define a function $f(x)$, $x \, \varepsilon \, \bar{R}^1$, as follows: First express x as a "non-finite decimal" in the ternary scale, $x = (0.a_1a_2 \cdots a_n \cdots)_3$, such that for no n are all a_n, $a_{n+1}, a_{n+2}, \cdots$ zeros unless $x = 0$. Then, if $x = 0$, or infinitely many of the digits a_n are 2's, let $f(x) = 0$. Otherwise, let n be the smallest (Problem 24 of Chapter III) natural number such that all digits $a_n, a_{n+1}, a_{n+2}, \cdots$, are *all* 0 or 1; then let $f(x) = (0.a_na_{n+1}a_{n+2} \cdots)_2$. Finally, for every t such that $0 \leqq t \leqq 1$, let $R_t = \{x \mid f(x) = t\}$.

17. Show that if a set S has a proper subset S_1, such that there exists a (1-1)-correspondence between the elements of S and the elements of S_1, then S_1 is Dedekind infinite.

18. Make the extension of the mapping defined in 1.2.1 to define a (1-1)-correspondence between N and F.

19. Denote the ith prime by p_{i-1}; thus $p_0 = 2$, $p_1 = 3$, etc. Show that if we express each rational number (in its lowest terms) in the form $p_0^{a_0} \cdot p_1^{a_1} \cdot \cdots \cdot p_n^{a_n}$ where p_n is the largest prime for which $a_n \neq 0$, then we can effectively define a (1-1)-correspondence between F and the set P of all polynomials in x with integral coefficients.†

20. Convert the mapping defined in Problem 19 to a (1-1)-correspondence between N and the set P.

21. Show that the set of all rational powers of rational numbers is countable. Note that we can conclude that not all irrational numbers are rational powers of rationals (such as $\sqrt{2}$, for instance).

22. Show that the class of all numbers x of the form a^b, where a is a rational number and b a rational power of a rational, is countable.

23. It is known that all numbers of the form a^b, where a is an algebraic number different from 0 or 1, and b is an irrational algebraic number, are transcendental. Show that not all transcendental numbers are of this form. (Give a constructive definition of one such.)

† I am indebted to Professor Philip Obreanu for this problem.

24. Let f map N into F by the formula $f(n) = n$. Let g map F into N by the formula $g(p/q) = 2^p \cdot 3^q$ if $p \geqq 0$ and $g(p/q) = 2^{|p|} \cdot 3^q \cdot 5$ if $p < 0$, where it is assumed that $q > 0$ and p and q are integers prime to one another. If h is the resulting (1-1)-mapping of N onto F worked out in the proof of the Bernstein equivalence theorem (4.2.6), find $h(24)$, $h(360)$ and $h^{-1}(33/49)$.

25. Apply the Bernstein equivalence theorem to prove: *If* $A \subset B$, *then* $\bar{\bar{A}} \leqq \bar{\bar{B}}$. Conversely, show that this theorem has the Bernstein equivalence theorem as a consequence.

26. Prove that, if $\alpha_1, \alpha_2, \cdots, \alpha_n, \cdots$ are cardinal numbers such that $\alpha_n < \alpha_{n+1}$ for all n, then the cardinal number α of the set $\bigcup_{n=1}^{\infty} A_n$, where $\bar{\bar{A}}_n = \alpha_n$, satisfies the relation $\alpha_n < \alpha$ for all n.

27. Show that, if E' is the set of all single-valued functions defined over $\bar{R}^1$ with values restricted to 0 and 1, then the cardinal number of E' is f.

28. Show that, if we assume that there exists a "set of all sets" U and that α is its cardinal number, we can apply Theorem 4.2.3.1 to show that our assumption leads to contradiction.

29. Suppose we assume that there exists a set, U, whose elements are all those sets having exactly one element (i.e., U is the set of all "singletons"). Then if S is a subset of U, the set $\{S\}$ having the single element S must be an element of U. Show that we can apply Theorem 4.2.3.1 to arrive at a contradiction.

30. Is the set of all sets each of which has exactly two elements self-contradictory?

31. Compare the material in 3.1, especially 3.1.8.1, with the following ("Richard paradox"): If a specific English dictionary, D, is used to form sentences, some of them may designate natural numbers; e.g., "Let N be the number of moons of the earth" would designate the number 1. However, consider the sentence, "Let N be the smallest natural number not definable in twenty words or less from the dictionary D."

32. Let r be any real number. In the sequence (F) of 1.2 (any other such ordering of the rational numbers, for instance, that of 1.2.1, would do as well, however), let f_1 be the first rational number in the sequence such that $f_1 < r$; let f_2 be the first rational number in the sequence such that $f_1 < f_2 < r$; and, in general, having defined f_n, let f_{n+1} be the first rational number in (F) such that $f_n < f_{n+1} < r$. Show that the limit of the sequence $f_1, f_2, \cdots, f_n, \cdots$ in the real numbers is r.

33. Is *the set of all cardinal numbers* a self-contradictory notion?

V

Well-Ordered Sets; Ordinal Numbers

In Chapter II we defined, incidental to exemplifying the use of the axiomatic method, two kinds of order: simple order (II 7) and partial order (II, Problem 16). In the present chapter we are mainly concerned with a special kind of simple order, called "well-ordering." To obtain it, we shall augment the three axioms given in II 7 by a fourth axiom. Before doing so, however, we consider some special examples of simply ordered sets, by way of introduction to the new notion.

1 Order types

Although there are *two* ways of simply ordering the elements of a set $\{a, b\}$ which has only two elements—$a < b$ or $b < a$—the result in either case is just an *ordered pair*. Similarly, although a set having three elements can be simply ordered in *six* ways, the result in any case is an *ordered triple*. And, in general, although a set having n elements, $n \varepsilon N$, can be simply ordered in $n!$† ways, the result is always an *ordered n-tuple*.

1.1 Let us make this precise; we haven't even defined the italicized terms. In the terminology of II 4.4.4, all that we have said in the preceding paragraph is that *every two simply ordered sets of n elements are isomorphic with respect to the simple order axioms*. Another way of putting this is to say that, if we add to the simple order axioms (II 7) the axiom, *The set C has exactly n elements*, $n \varepsilon N$, then the resulting system of axioms is categorical (II 4.5.1).

1.1.1 The analogous statements do not hold for infinite sets. In particular, to add to the simple order axioms the axiom, *The set C has exactly* $\aleph_0$ *elements*, does not render the axiom system categorical; for consider the set N. It can be assigned its *natural order*: $n < n + 1$ for all n. Let us call this ordered set ON. Another simple ordering of N is obtained as follows: (i) if $a, b \varepsilon N$, and a is odd, b is even, then $a < b$; (ii) if a and b are both odd, then $a < b$ denotes the natural order $(1 < 3,$

† Where $n!$ denotes *n factorial*; i.e., $1 \cdot 2 \cdot 3 \cdot \ldots \cdot n$.

etc.); (iii) if a and b are both even, then $a < b$ means that $b < a$ in the natural order (i.e., $4 < 2$, $6 < 4$, etc.). Schematically, the order just defined is

$$1, 3, 5, \cdots, 2n + 1, \cdots; \cdots, 2n, \cdots, 6, 4, 2.$$

Let us call this ordered set ON_1. The sets ON and ON_1 are not isomorphic with respect to the simple order axioms (Problem 1). Generally, we abbreviate this statement to: ON and ON_1 do not have the same order type.

(1a) Definition. If A and B are two simply ordered† sets which are isomorphic relative to the simple order axioms, then we say that A and B *have the same order type.*

1.1.2 We are now in a position similar to that in which we were when we had defined the relation "have the same cardinal number," where we had no definition of the term "cardinal number"; for now we have the relation "have the same order type" between simply ordered sets, but no definition of "order type." With a sufficiently restricted set theory we might give the analogue of the Frege-Russell definition of cardinal number (IV 4.1.1), and define the order type of a simply ordered set S as the collection of all simply ordered sets which have the same order type as S. But, again, contradiction would result unless we do incorporate the definition in such a set theory. So, lacking the latter at the present stage of our discussion, we ask: What is it that we compare in two simply ordered sets when we set up this relation "have the same order type" between them?

Evidently, we again measure "size," since the isomorphism relation is a (1-1)-relation and hence we are certainly comparing the cardinal numbers of the sets. But we obviously compare something in addition to "size," since "size" has nothing, *per se*, to do with order. We should recall here Definition II 4.4.4, according to which isomorphism with respect to an axiom system Σ means (1-1)-correspondence with preservation of Σ-statements. Here we are dealing with isomorphism with respect to the simple order axioms, and the Σ-statements preserved are concerned with order. However, it is of importance to notice that "order" can be conceived of in two senses. In one sense the *individuality* of elements is taken into account; thus $a\ b\ c$ and $b\ a\ c$ are different orders ("permutations") of the three letters a, b, c. In another sense—and it is this, along with the "size," which is designated by the term *order type*—the individuality of elements is ignored; more precisely, *if two elements are exchanged, the order type remains unchanged.* Another way of putting this is to

† Obviously we could say "partially ordered" (Problem 16 of Chapter II) here instead of "simply ordered." The latter is more significant for our purposes, however.

state that the order type is that aspect of the arrangement of the elements of a simply ordered set, which remains unchanged when any two elements are exchanged. Thus $a\ b\ c$ and $b\ a\ c$ are both instances of the same order type, the ordered triple. And it is this that is preserved by those mappings which are isomorphisms with respect to the simple order axioms.

As in the case of cardinal numbers, order types may be denoted by suitable symbols called *ordinal numerals.* Symbols for the same order type are considered equivalent, and this equivalence is again denoted by "=," thereby symbolically signifying the logical identity of the concepts denoted by the symbols. Since, as observed in 1.1, a simply ordered set of n elements has a unique order type, it is customary to use the natural number numerals as ordinal numerals for order types of finite sets. Thus a simply ordered set having n elements will be said to be of, or to have, *order type n* or, alternatively, to be an *n-tuple.* (By convention, "ordered pair" is used as synonymous with "order type 2," and "ordered triple" as synonymous with "order type 3".) The order type of $\emptyset$ may be denoted by "0."

Finally, we note that our convention regarding an individual order type involves the three distinct entities: (1) certain simply ordered sets; (2) the common order type of these sets; and (3) a symbol (or equivalent symbols). Remarks concerning these, similar to those made above regarding cardinal numbers, may be made here. In particular, (2) and (3) are commonly employed as though they were the same, so that in arithmetic we speak of "the cardinal number 2" and "the ordinal number 2"; in the latter case, what we actually mean is "the order type denoted by '2'". ["Ordinal number" is a term that denotes a special kind of order type, to which the order type n happens to belong; it will be defined later (3.5)].

For an individual simply ordered set A, we can use the bar symbol, "$\overline{A}$," to denote the order type so that, if two simply ordered sets A and B have the same order type, this fact may be expressed by the symbols "$\overline{A} = \overline{B}$." And if α is a given order type, any set A such that $\overline{A} = \alpha$ may be called a set representative of the order type α. An alternative symbolism, "$A \varepsilon \alpha$," is frequently used to denote that $\overline{A} = \alpha$; it derives from the concept of an order type as a collection of sets all of the same order type (analogue of the Frege-Russell definition).

1.1.3 We have observed above that a simply ordered set of n elements has only one order type, i.e., the order type n. But an infinite set, such as N, may have many order types. Not all of these are sufficiently important to have been assigned special symbols. However, the order type of the set ON of 1.1.1 is denoted by "ω." The symbol "$^*\omega$" was used by Cantor to indicate the order type of the negative integers in their "natural order" ($-2 < -1$, etc.); this order type is the same as that of the ordered set of natural numbers with "$<$" meaning the reverse of the natural order. The order type ON_1 may then be denoted by "$\omega + {}^*\omega$"; the use of "$+$" here is in accord with the following:

1.1.4 Addition ($+$) of order types is defined as follows: If Ω_1 and Ω_2 are order types, A_1 and A_2 are disjoint sets of order types Ω_1 and Ω_2, respectively, then $\Omega_1 + \Omega_2$ is the order type determined by $A = \bigcup_{i=1}^{2} A_i$, ordered so that $<$ is the same as originally for each A_i, $i = 1, 2$, but if $a_1 \varepsilon A_1$ and $a_2 \varepsilon A_2$, then $a_1 < a_2$. Note that ${}^*\omega + \omega$ would not be the same as $\omega + {}^*\omega$, so that this addition is not commutative. See Sierpinski [S_1; VIII], [S_3; XII].

We shall not go into the "arithmetic" of order types, but the addition of order types is an aid to the understanding of the ordinal number symbols to be introduced in the sequel. If n is used to denote the order type n, notice that $n + \omega$ is the same as ω; however, this is not the case with $\omega + n$.

2 The order type ω

The order type ω, because of its central importance, has been subjected to axiomatic treatment. (When we speak of "infinite sequence" in elementary algebra, we mean an ordered set having the order type ω.)

2.1 An axiomatic definition of order type ω

Let N be a non-empty set and $<$ a binary relation between elements of N. Then N is called of order type ω if

(a) N is simply ordered with respect to the binary relation $<$.†

(b) If $a \varepsilon N$, then the set $\{x \mid x < a\}$ is finite.

(c) N has no last‡ element.

Any set of order type ω obviously satisfies these axioms. They can be shown to form a categorical system. (We leave this to the reader as an exercise.)

2.2 An axiomatic definition of ω in terms of cuts

The notion of "cut" is widely used for defining certain basic order types.

† This means, of course, that the single axiom (a) implies all three axioms of simple order as given in II 7.

‡ In a simply (or partially) ordered set, an element x is called the *last* element of a set A if $x \varepsilon A$ and if, for every $y \varepsilon A$, $y \leqq x$. "First" is defined analogously. Uniqueness of first and last elements (when such exist) follows from the simple (or partial) order axioms.

(2a) Definition. If A and B are subsets of an ordered set S, then $A < B$ means that $A \neq \emptyset \neq B$, and that $a \in A$, $b \in B$ imply $a < b$. If $S = A \cup B$ and $A < B$, then we say that A and B form a *cut*† of S; we shall denote such a cut by the symbol $[A, B]$ (always implying $A < B$).

We can then define a set N to be of order type ω if

(a) N is simply ordered with respect to the binary relation $<$.
(b) N has a first element.
(c) N has no last element.
(d) If $[A, B]$ is a cut of N, then A has a last element and B has a first element.‡

2.2.1 Incidentally, it is interesting to note that the sets which are of the same order type as the elements of the set F of rational numbers (IV 1.1) in their natural order (order of magnitude) may be characterized, as to their order relations, by the assumption of (1) denumerability, (2) axioms (a) and (c), (3) denial of (b), and (4) the assertion that (d) fails for *every* cut $[A, B]$.§

2.3 An axiomatic definition of ω in terms of spanning sets

A very useful notion, in dealing with ordered sets, is that of spanning set.

2.3.1 Definition. If S is a simply ordered set, then a subset X of S will be said to *span* S, or to be *cofinal with* S, if $s \in S$ implies the existence of an $x \in X$ such that $s \leqq x$. We also call X a *spanning set* or *cofinal subset* of S.

2.3.1.1 Evidently, if a simply ordered set S has a last element x, then the set $X = \{x\}$ is a spanning set of S. Also, S spans itself. In the case where S is N and $<$ is the natural order, the set X of all odd numbers is a spanning set. In the case where S is the set R of all real numbers with $<$ meaning the order of magnitude, the set N as a subset of R is a spanning set.

2.3.2 In terms of spanning sets, we can define a non-empty set N to be of order type ω if

(a) N is simply ordered with respect to the binary relation $<$.
(b) Every infinite subset of N is a spanning set of N.
(c) N has no last element.

† Frequently called "Dedekind cut" because of its use by Dedekind in the study of irrational numbers. See Dedekind [D_1].

‡ See Sierpinski [S_1; 143ff], [S_3; 208, Exercise 3].

§ See Sierpinski [S_1; 145ff], [S_3; 209, Th. 1]; Sierpinski denotes this order type by η.

2.4 Recognition of type ω among the well-ordered sets

We shall now define the fundamental notion of well-ordered set upon which the concept of ordinal number depends.

(2b) Definition. A simply ordered set is called *well-ordered*† if every non-empty subset of it has a first element.

Well-ordered sets are frequently called "sequences" or "well-ordered sequences."

An axiomatic definition of well-ordered set is obtained by adding, to the simple order axioms of II 7, the axiom (4): "Every non-empty subset of C has a first point."

2.4.1 A *well-ordering* of a set S is the result of an assignment of a binary relation $<$ between the elements of S in such a way that the set S becomes simply ordered with respect to $<$ and satisfies Definition (2b). A *well-ordering type* is an order type which forms a well-ordering.

2.4.2 *Every finite simply ordered set is well-ordered* (with respect to the order relation $<$ defining the simple order).

2.4.3 The set N, ordered so as to be of order type ω (the natural order; ON of 1.1.1) is well-ordered. However, when ordered so as to be of order type $^*\omega$, it is not well-ordered, since a well-ordered set has a first element (as a subset of itself). Thus an infinite set, when simply ordered, may or may not be well-ordered (with respect to the order defining the simple order). The set ON_1 of 1.1.1 is not well-ordered. The set F, ordered according to magnitude, is not well-ordered; the subset consisting of its positive elements has no first element. The null set, $\emptyset$, is well-ordered since it satisfies Definition (2b) vacuously.

2.4.4 Consider the following example: The order type $\omega + 1$, as defined in 1.1.4, is a well-ordering type. One of its representations is obtained by adding to ON a new element, p, such that, for all $x \in ON$, $x < p$ by definition. (It can, of course, be exemplified by redefining $<$ for N so that, for all $x \in N$ for which $x \neq 1$, $x < 1$; and, for $x,y \in N - \{1\}$, $x < y$ is the natural order.)

† It may sometimes be necessary to say "well-ordered with respect to $<$," where $<$ is the binary relation with respect to which the set is simply ordered. This will always be the case when more than one ordering of a set is under consideration. English writers frequently use the term "normally ordered." This has the advantage of allowing the noun "normal ordering," whereas the (universally used in the United States) term "well-ordered" leads to "well-ordering," which sounds suspiciously like a grammatical error.

2.4.5 *If S is a well-ordered set, and $s \varepsilon S$, then the set $\{x \mid s < x\}$, if not empty, has a first element s'.* We call s' the *immediate successor* of s.

2.4.6 We can now define a non-empty set N to be of order type ω if the following axioms hold:

(a) N is well-ordered.
(b) N has no last element.
(c) If N_1 is a subset of N such that the first element of N is in N_1, and such that, if $s \varepsilon N$ is in N_1 then the immediate successor s' of s is in N_1, then $N_1 = N$.

Axiom (c) is, of course, the mathematical induction principle discussed in III 5. As we shall see below, (c) is intimately related to well-ordering, through a weakened† type of induction principle called "transfinite induction." Its necessity in the above axiom system is shown by the order type $\omega + \omega$, which satisfies (a) and (b) but not (c).

3 The general well-ordered set

In Section 2 we studied some of the properties of sets of type ω, which is the most important, from the viewpoint of its wide variety of applications, of the types of well-ordered sets. In this section we turn to the general types of well-ordered sets. That they must constitute a wide class is probably already apparent from the above discussion; for example, each $\omega + n$ is a well-ordering type. So also is $\omega + \omega$. And, in general, if W_1 and W_2 are well-ordering types, then $W_1 + W_2$ (see 1.1.4) is a well-ordering type. Thus we can generate new well-ordering types from old by the process of addition.

The importance of the notion of well-ordered set is chiefly due to the fact that (1) it represents an extension of the order types $0, 1, 2, \cdots, n, \cdots$ and ω; (2) a method—transfinite induction—is applicable to every well-ordered set in a manner similar to the application of mathematical induction to order types n and ω; and (3) by virtue of (1), numbers called "ordinal numbers" can be introduced that represent a natural extension of the natural numbers.

However, if these attributes of well-ordered sets are to be of advantage, particularly in the case of infinite sets which may come to our attention with no order assigned at all (or perhaps with an order which is not a well-ordering, as in the case of the rational numbers in their natural order), it would be advantageous to know when and how sets may be well-ordered.

† In the sense that its hypothetical "if" requires more than the corresponding part of (c).

In the case of some sets we have already shown how they may be well-ordered. In particular, we showed in IV 1.2 and IV 1.3 how to well-order the sets F (rational numbers) and A (algebraic numbers); in both cases we set up an ordering of type ω. However, we did not well-order the set R of all real numbers. We showed in IV 1.4 that R is uncountable, and hence could not be given a well-ordering of type ω. This poses the question: Does there exist a well-ordering of the set R of all real numbers?

If we tried to well-order the set R, we might commence by a method like that which we used in IV 3.1, where we first arranged a type ω sequence of real numbers and then, by using diagonal methods, rearrangement rules, etc., constructed new numbers. And if we ordered them (as we did) as they were constructed, they formed, at each stage of the construction, a well-ordered set. However, aside from the fact that we ran into such difficulties as exhaustion of possible rearrangement rules, we would have no guarantee that all numbers in R would be obtained by the process even if such difficulties could be overcome. In view of such considerations, we would have to look for another method. And since the process used in IV 3.1 leads to a collection not effectively countable, we might suspect that any process of construction would lead to a similar result, before all numbers in R had been accounted for. Of course, we might give up the task as hopeless, concluding that there cannot exist any well-ordering of R.

It is not improbable that most mathematicians who had given thought to the problem had arrived at this conclusion when the German mathematician Ernst Zermelo published his famous Well-Ordering Theorem in 1904 (Zermelo [a]). We state this theorem now, but at present we give only an indication of the proof, since we are chiefly interested in showing how the Choice Axiom enters into it.

3.1.1 Definition. If S is a simply ordered set and $s \in S$, then by S/s we denote the set $\{x \mid x < s\}$, and by s/S the set $S - S/s$; each of these being considered sets ordered relative to the same binary relation $<$ with respect to which S is simply ordered. We call S/s the *section of S determined by s.*

Axiom (b) of 2.1 stipulates that all sections of the set N are finite.

If the set S of the above definition has a first element a, then evidently $S/a = \emptyset$ and $a/S = S$. And if S is well-ordered, then both S/a and a/S are well-ordered sets. The latter statement is a particular case of "*If W is a well-ordered set with respect to some binary relation $<$, then every subset W' of W is well-ordered with respect to the same relation $<$.*"

In the sequel, when we speak of a *well-ordered subset W' of a well-ordered set W*, we shall mean the above set W' unless another ordering is specified.

3.1.2 **Well-Ordering Theorem (Theorem of Zermelo).** *If S is any set whatsoever, then there exists a well-ordering*† *of S.*

Indication of proof. Let $\mathfrak{S} = 2^S - \{\emptyset\}$; then, as the elements S_ν of $\mathfrak{S}$ are non-empty sets, there exists by the General Choice Axiom (III 6.4) a set $\mathfrak{R}$ whose elements are pairs (S_ν, x_ν) in which $x_\nu \varepsilon S_\nu$, and such that each $S_\nu \varepsilon \mathfrak{S}$ occurs in one and only one pair. Denote each x_ν by $x(S_\nu)$; thus $x(S)$ is the representative element of S itself.

Now let us call a non-empty well-ordered set W, whose elements are also elements of S, a "WS-sequence," if, for every element w of W, $x(S - W/w) = w$.‡ In particular, $w_1 = x(S)$ is a WS-sequence and, moreover, the first element of every WS-sequence is w_1 [since, for the first element w of a WS-sequence, $W/w = \emptyset$ and hence $x(S - W/w) = x(S) = w_1$]. Furthermore, of two different WS-sequences, one is always a section of the other, so that, if $x,y \varepsilon S$ that lie in some WS-sequence W, we can define $x < y$ to mean that $x < y$ in W. Then it can be shown that the collection of all elements of S that lie in WS-sequences is well-ordered with respect to this binary relation $<$, and, finally, that *every* element of S lies in a WS-sequence, thus giving a well-ordering of S.

3.1.3 Comment. Note how strongly the Choice Axiom dominates the proof of the Well-Ordering Theorem. And if we have convinced ourselves that there exists no well-ordering of R, then certainly the Well-Ordering Theorem, which asserts the existence of well-orderings for *all* sets, even sets of inconceivably higher cardinality than R, must raise doubts in our mind regarding the validity of its proof. But there seems to be no element of uncertainty regarding the proof except for the use of the Choice Axiom. Consequently, if we would reject the Well-Ordering Theorem, we would also feel compelled to reject the Choice Axiom. (As we shall see later (Section 5), the two—Choice Axiom and Well-Ordering Theorem—are actually equivalent.)

With few exceptions, the feeling today regarding the principle in mathematical circles may be summarized by the statement: It is all right to use the principle, provided prominent mention is made of the fact. Then anyone interested enough to do so may seek ways of avoiding use of the principle. Of many theorems, it is known that they cannot be proved without use of the principle, or are equivalent to it; we shall see

† Notice that we do not say "one can well-order S," which might be taken to imply that there exists an effective procedure for the well-ordering; since the latter does not at present exist in the literature, we use the "non-committal" "there exists . . ."

‡ Cf. 3.1.1. Although W/w is more than just a set—namely, a set together with an order relation—we shall use it as though it were just a set in set operations such as $S - W/w$, rather than to introduce a new symbol to denote the collection of its elements.

cases of this later in the equivalences of Choice Axiom, Well-Ordering Theorem, and Comparability (Section 5). In other cases, it remains an open question whether a theorem proved by using the Axiom can be proved without it. In such cases, the finding of a proof which does not use the Axiom merits publication of the fact.

Generally speaking, it would be impossible to do mathematics in the modern sense without using the Axiom. This is undoubtedly due to the great generality of much modern mathematics, necessitating the use of sets that do not permit statement of specific rules of selection. Much work has been done to find ways of avoiding use of the Axiom. Therefore, it is not surprising that many equivalent "principles" have been found, some of which have turned out to be easier of application in some situations than the Well-Ordering Theorem.

First, however, let us see what use can be made of the Well-Ordering Theorem. If it should turn out that the Theorem leads to many fruitful results, and yet never leads to contradiction, we may choose to overcome any prejudices that we harbor and admit it to the society of respectable mathematics. After all, the real power of the Theorem is in its application to uncountable sets, and in dealing with these, new methods that were not necessary for handling problems involving finite and denumerable sets may be needed.

3.2 The transfinite induction principle

One of these new methods is that of *transfinite induction*, already mentioned above as bearing a relation to well-ordered sets analogous to the relation that mathematical induction bears to the well-ordered set N of natural numbers.

(3a) **Transfinite induction principle.** Let W be a simply ordered set and let us, in analogy to III 5.1, call a subset W_1 of W "inductive," provided that, if it contains (as subset) a section W/w, it contains w. Then W is said to *satisfy the transfinite induction principle* if (1) W has a first element w_0 and (2) if W_1 is an inductive subset of W containing w_0 then $W_1 = W$. The property (2) is called the *transfinite induction principle*. When W is of order ω, the transfinite induction principle becomes identical with the mathematical induction principle.

3.2.1 **Theorem.** *In order that a non-empty simply ordered set W should be well-ordered, it is necessary and sufficient that W satisfy the transfinite induction principle.*

Proof of necessity. That W has a first element, w_1, follows from Definition (2b). Let W_1 be a subset of W satisfying the "if" part of (2) in (3a). Suppose $W - W_1 \neq \emptyset$. Then by Definition (2b), $W - W_1$

has a first element, w. As w is the first element not in W_1, $W/w \subset W_1$. But then $w \varepsilon W_1$, since W_1 is inductive, which contradicts the definition of w. We conclude, then, that $W = W_1$.

Proof of sufficiency. Let W be a non-empty simply ordered set satisfying the transfinite induction principle. To show W well-ordered, we have to show that, if W' is a non-empty subset of W, then W' has a first element. Suppose not. Then let us define a subset W_1 of W as follows: $W_1 = \{x \mid (x \varepsilon W) \ \& \ (x < W')\}$. Then W_1 contains w_1, the first element of W, else w_1 is the first element of W'. And, if $w \varepsilon W$ is such that $W/w \subset W_1$, then $w \varepsilon W_1$, since otherwise w would be the first element of W'. But then W_1 is inductive and $W_1 = W$. But this implies $W' = \emptyset$, contradicting the non-empty character of W'. We conclude, then, that W' has a first element.

3.2.2 Proof by transfinite induction

Just as the mathematical induction principle is used to demonstrate that *all* elements of a type ω sequence have a certain property (see III 5.2.1, for instance), so is the transfinite induction principle used to show that all elements of a well-ordered set possess a given property.

If it is desired to prove that all elements of a given *denumerable* set S have a certain property P, then the proof may take the following form: As S is denumerable, its elements may be put in the form of a type ω sequence:

$$x_1, x_2, \cdots, x_n, \cdots.$$

We then prove that x_1 has property P, and that if, for any n, an x_n has property P, so has x_{n+1}. That all elements of S have property P then follows from the mathematical induction principle; we called this *proof by mathematical induction* in III 5.1.

Now suppose that W is any set whatsoever, and we wish to prove that all its elements have a certain property P. Even though W may be uncountable, it may be that for certain reasons (which we do not go into at present) W can be assigned a well-ordering (2.4.1). And then, with W well-ordered, it may be that we can prove that: (1) the first element, w_1, of W has property P; (2) if for any element w of W, all elements of W/w in the well-ordering have property P, then w has property P. We may then conclude that *all* elements of W have property P. For, by Theorem 3.2.1, the well-ordering of W satisfies the transfinite induction principle, and hence, if we form a set W_1 by placing an element of W in W_1 if it has property P, it will follow from the transfinite induction principle that $W_1 = W$. This type of proof is called *proof by transfinite induction*, and it evidently has proof by mathematical induction for type ω sequences as a special case. (See Problem 17.)

3.2.3 Definition by transfinite induction

If for every $n \varepsilon N$ we wish to define a mathematical entity E_n, we frequently (1) define E_1, and (2) define E_{n+1} in terms of E_n (see III 5.1); often E_{n+1} is defined in terms of several or all of the terms $E_1, E_2, \cdots, E_n$ [as for example in sequences $\{u_n\}$ where $u_{n+1} = (u_n + u_{n-1})/2$]. We used this type of definition in III 5.3, for instance; it is called *definition by induction.*

More generally, we can use the transfinite induction principle to justify *definition by transfinite induction.* Let W be a well-ordered collection, and suppose that we define some mathematical entity $E(w_1)$, w_1 being the first element of W; if we then give, for each $w \varepsilon W$, a definition $E(w)$ in terms of the section W/w or its elements, we may conclude from the transfinite induction principle that we have defined an $E(w)$ corresponding to *every* $w \varepsilon W$. This procedure is called definition by transfinite induction, and again definition by induction for sequences of type ω is a special case.

3.3 Example of an existence proof based on the Well-Ordering Theorem

Let us consider the set, R, of all real numbers. We have already shown (IV 1.2) that the set, F, of all rational numbers is a denumerable subset of R. We shall now show that every real number can be expressed in a unique way as a linear polynomial with coefficients in F. For this purpose we ask: Does there exist a subset B of R having the following property: Every real number $r \neq 0$ is uniquely expressible in the form

$$r = f_1 b_1 + f_2 b_2 + \cdots + f_n b_n, \tag{3.3a}$$

where $f_1, f_2, \cdots, f_n$ are all non-zero numbers in B (it is expected that different numbers "r" may require different values of "n," of course)?

If such a set B exists, we can show that it does not contain 0; for if it did, then a term $f_{n+1}b_{n+1}$, where $f_{n+1} = 1$ and $b_{n+1} = 0$, could always be added to the expression (3.3a) so that (3.3a) would not be unique. Also, B will not contain more than one rational number; for suppose f_1 and f_2 were two different rational numbers in B. As neither can be 0, we can express f_1 in the form mf_2 (where $m = f_1/f_2$). Then if r is a third rational, $r = pf_1$ and $r = pmf_2$ (where $p = r/f_1$), and r is not uniquely expressible in terms of the elements of B. So if we were to try to construct B, we might decide to put the number 1 in B and no other rational number; every rational number f would then be expressible in the form $f = f \cdot 1$. In particular, if we asked for a set B which would do for F what we have asked that it do for R, then not only would such a B exist, but would contain exactly one element.

A set B, as defined above for the real numbers, is called a "Hamel basis," being named after G. Hamel who first proved its existence (see

Hamel [a]), or simply a "basis." The existence of a Hamel basis leads to important consequences in the study of real functions and their applications.

3.3.1 Theorem (Hamel). *There exists a basis for the set, R, of all real numbers.*

Proof. By Theorem 3.1.2 there exists a well-ordering, W, of R. We then define a subset B of R as follows: Using the method of transfinite induction, we first define for each element w of W a set $B(w)$ as follows: (1) if the first element of W is the number zero, we delete it from W; then, for the first element, w_1, of (the new) W, we define $B(w_1) = \{w_1\}$; (2) having defined $B(w)$ for all elements of a section W/w_ν, let $B'(w_\nu) = \bigcup_{w < w_\nu} B(w)$; then we define $B(w_\nu) = 0$ or $\{w_\nu\}$ according as there does or does not exist a relation of the type

$$\sum_{i=1}^{n} f_i b_i = 0$$

between elements b_i of the set $B'(w_\nu) \cup \{w_\nu\}$ in which not all $f_i = 0$. This defines $B(w)$ for all elements of W. Finally, let $B = \bigcup_{w \varepsilon W} B(w)$. We assert that B is a basis for R. The proof of this will be left to the reader (Problem 16).

3.3.2 Theorem 3.3.1 was introduced by Hamel in order to show the existence of discontinuous solutions $f(x)$ of the functional equation

$$f(x + y) = f(x) + f(y) \qquad (3.3.2a)$$

where $f(x)$ is a single-valued real function of a real variable (i.e., a mapping of R into R in the sense of IV 3.2.3.1). Cauchy had noted that the only *continuous* functions satisfying (3.3.2a) are the functions $f(x) = cx$, where c is a constant. However, without the continuity restriction, Hamel showed that the equation has infinitely many other solutions. For example, we may define $f(x)$ quite arbitrarily for the elements of the basis B, and let $f(0) = 0$. Then, for an $r \varepsilon R$ not in B, there is a relation of type (3.3a) which is unique because of the properties of a basis. By placing $f(r) = \sum_{i=1}^{n} f_i f(b_i)$, the function $f(x)$ is defined for all $x \varepsilon R$ and is easily shown to satisfy the required condition (3.3.2a).†

3.4 Application to comparability of well-ordered sets

We shall next show, by transfinite induction, that, given two well-ordered sets, either there exists a (1-1)-correspondence between their elements that preserves order, or there exists an order-preserving (1-1)-correspondence between the elements of one and the elements of a *section* of the other; and, moreover, that this correspondence is unique.

† This result has interesting applications in topology, showing the existence of certain "pathological" configurations; see Jones [a], for instance.

3.4.1 Theorem. *Let W and W' be well-ordered sets. Then either W and W' are of the same order type, or one is of the same order type as a section of the other.*

If W is empty, the theorem holds trivially. Otherwise, let w_1 be the first element of W; then either all elements of W' are already in order-preserving (1-1)-correspondence with elements of W/w_1 (meaning $W' = \emptyset$) or not; if they are not, pair w_1 with the first element w_1' of W'. In general, if $w \varepsilon W$ and each element of W/w is already paired with some element of W', then either all elements of W' are already in order-preserving (1-1)-correspondence with elements of W/w or not; if they are not, we pair w with the first element of W' not already paired with elements of W. By the transfinite induction principle, either all elements of W' are paired in this manner with elements of W, or conversely. We leave the details of the proof to the reader.

3.4.2 Theorem. *If W/w_1 and W/w_2 are different sections of a well-ordered set W, then they are of different order types.*

Proof. Suppose $w_1 < w_2$. If w_1 is the first element of W, the theorem follows trivially. Otherwise, suppose that it holds for each element, w_1, of a section W/w', where w' is not the first element of W. Then it holds for w'. For suppose not, and hence that there exists a (1-1)-correspondence T, preserving order, between the elements of W/w' and the elements of some section W/w_2 such that $w' < w_2$. Let the element of W/w' that is paired with w' (as an element of W/w_2) in T be denoted by v'. But then W/v' and W/w' are of the same order type (according to T), in contradiction of the supposition that the theorem holds for all elements of W/w'. The theorem now follows by the transfinite induction principle.

3.4.3 Corollary. *No well-ordered set W is of the same order type as one of its sections.*

Proof. Augment W by a new element e, and define $w < e$ for all $w \varepsilon W$; apply 3.4.2.

3.4.4 Theorem. *If W and W' are non-empty well-ordered sets, then the order-preserving* (1-1)*-correspondence implied by Theorem* 3.4.1 *is unique.*

3.5 Ordinal numbers

(3.2b) Definition. The order types of *well-ordered sets* are called *ordinal numbers*. The ordinal number of a well-ordered set W is the order type of which W is a representation, and may be denoted, following Cantor, by $\overline{W}$ (cf. the last paragraph of 1.1.2).

3.5.1 Symbols for ordinal numbers

Since a *finite* set, when simply ordered, is of a unique order type, and this is a well-ordering type, we may use the symbols for the cardinal numbers of the finite sets as ordinal numbers also. Thus, to say that the ordinal number of a set is n, $n \in N$, is to imply that it is of order type n (1.1.2) (hence the term *n-tuple*). The ordinal number of the null set is denoted by "0."

However, in view of the fact that an infinite set may be so ordered as to belong to more than one well-ordering type (cf. 1.1.1 and 2.4.4 for example), the cardinal number symbols for *infinite* sets cannot be used to denote ordinal numbers. It would be meaningless, for instance, to speak of a set "having the ordinal number $\aleph_0$." It was in recognition of this fact that a special symbol, ω, was introduced above for the order type of the natural numbers in their natural order. Order types $\omega + n$, $n \in N$, have also been mentioned above (Section 3), as well as $\omega + \omega$. The latter may be denoted by "$\omega \cdot 2$," however, since, if 2 is understood as the ordinal number 2, $\omega \cdot 2$ becomes $\omega + \omega$ in accordance with the following definition:

3.5.1.1 Definition. If ω_1 and ω_2 are ordinal numbers, then the *product* $\omega_1 \cdot \omega_2$ is the order type of the (well-ordered) set obtained by replacing each element of a well-ordered set O_2 of ordinal number ω_2 by a well-ordered set O_1 of ordinal number ω_1; the order in the new set agreeing with the original order between elements of the replacement sets O_1, and with that of O_2 between elements of different replacement sets.†

Hence, $2 \cdot \omega$ would be ω, since by the above definition it is the order type obtained by replacing each element n of N by an ordered pair (n_1, n_2). Likewise, $n \cdot \omega$ is the same as ω, but $\omega \cdot n$ is the order type obtained from n successive sequences of type ω. The symbols $\omega \cdot \omega$, $\omega \cdot \omega \cdot \omega, \cdots$, are usually replaced respectively by $\omega^2, \omega^3, \cdots$.‡

3.5.1.2 A simple example of order type ω^2 is obtained by ordering the natural numbers in the following manner: (1) Arrange the numbers that are *not* positive integral powers of primes in their natural order; thus, $1, 6, 10, 12, \cdots$. Call this sequence S_1. (2) Arrange the powers of 2 in their natural order; thus, $2, 4, 8, \cdots$. Call this sequence S_2. $(n + 1)$ In general, arrange the powers of the nth prime, p_n, in their natural order,§

† Thus, if $x < y$ in O_2, then, for every element x_1 of the O_1 replacing x and every element y_1 of the O_1 replacing y, $x_1 < y_1$.

‡ We shall not go into the complete arithmetic of ordinal numbers, however; as in the case of cardinal numbers, the reader is referred to Sierpinski [S_1; VIII, X], [S_3; XIV].

§ We assume the fundamental number-theoretic theorem that the number of primes is infinite; see Courant and Robbins [C-R; 22–23].

and call the resulting sequence S_{n+1}. Retaining the natural order in each S_n, if $n, n' \varepsilon N$ such that $n < n'$, we let $S_n < S_{n'}$; thus $12 < 4, 4 < 3$, etc.

3.5.2 Ordering of the ordinal numbers

Suppose ω_1 and ω_2 are ordinal numbers, and let $O_1 \varepsilon \omega_1$, $O_2 \varepsilon \omega_2$. By $\omega_1 = \omega_2$ we mean "=" as defined in 1.1.2. If $\omega_1 \neq \omega_2$, then O_1 and O_2 are not of the same order type, so that (Theorem 3.4.1) O_1, say, is of the same order type as a section of O_2. Then we write $\omega_1 < \omega_2$. That this definition of < for any set of ordinal numbers satisfies the simple order axioms (II 7) follows from the results stated in 3.4.

Furthermore, if Φ is any collection of ordinal numbers, then there exists a smallest element in Φ. For let $\overline{W} \varepsilon \Phi$, and suppose $\overline{W}$ not the smallest element of Φ. Then, with $W \varepsilon \overline{W}$, every element $\overline{W}'$ of Φ such that $\overline{W}' < \overline{W}$ is represented by a section W/w' of W. But, as W is well-ordered, the elements w' appearing in such sections W/w' have a first element w_1' in W and the corresponding W/w_1' is of an order type $\overline{W}_1'$ which is the smallest in Φ. And we can state

3.5.2.1 *Every collection of ordinal numbers is well-ordered by the above definition of* <.

We may glimpse the "march of the ordinal numbers" by considering them in the order defined above:

$$0, 1, 2, \cdots, n, \cdots; \omega, \omega + 1, \cdots, \omega + n, \cdots; \omega\cdot 2, \omega\cdot 2 + 1, \cdots, \omega\cdot 2 + n, \cdots; \cdots; \omega\cdot n, \cdots; \omega^2, \omega^2 + 1, \cdots; \omega^2 + \omega, \omega^2 + \omega + 1, \cdots; \cdots; \omega^n, \omega^n + 1, \cdots; \cdots \tag{3.5.2a}$$

If the reader will refer to IV 3, he will now see the origin of the subscripts that were used to differentiate the various sequences of irrational numbers constructed there. Notice that *each ordinal number is the order type of the well-ordered set of ordinal numbers that precede it.*

3.5.2.2 *Every ordinal number* α *is the order type of the well-ordered set of all ordinal numbers* β *such that* $\beta < \alpha$.

Proof. Let $W \varepsilon \alpha$. Consider any ordinal number $\beta < \alpha$, and let $B \varepsilon \beta$. Then, by definition, B is of the same order type as a section W/b of W. Conversely, an element b of W determines a section W/b having an ordinal number β such that $\beta \neq \alpha$ (by Corollary 3.4.3) and hence such that $\beta < \alpha$.

The pairs (β, b) constitute a (1-1)-correspondence between the elements of B and the elements of W. This correspondence is order-preserving. For example, if $\beta' < \beta < \alpha$, then, since W/b as defined above is of order type β, for any $B' \varepsilon \beta'$ there is a section of W/b determined by an element b' that is of order type β'; this section is also a section of W and $b' < b$. That conversely a relation $b' < b$ implies a corresponding relation $\beta' < \beta$ will follow from the fact that W/b' is a section of W/b.

It should also be noticed that the procedure followed in IV 3 virtually forced the generation of a well-ordered collection of numbers. As a matter of fact, this was the type of procedure which led Cantor to the study of well-orderings,† and it is a common form of application of the well-ordering idea and ordinal number symbols.

Notice, too, that all the symbols actually set forth in (3.5.2a) designate order types of *countable* representative sets (cf. Theorem IV 3.1.6). Indeed, we derive the feeling, as we study the "march of the ordinal numbers," that they never seem to reach the "uncountable stage." Yet we have shown that R is uncountable (IV 1.4) and that there exists a well-ordering of R (3.1.2—the Well-Ordering Theorem). Hence the "march" must continue to ordinal numbers representing uncountable sets! Furthermore, by 3.5.2.1 there must be a *first* such ordinal, which we may designate by ω_1. Does ω_1 correspond to a well-ordering of R? We return to this question below.

3.5.3 The Burali-Forti antinomy

In III 1.1 we mentioned the announcement by Burali-Forti in 1897 of the existence of contradiction in the "unrestricted" theory of sets.‡ As usually formulated, the Burali-Forti antinomy consists of the argument that, if we allow a "set Γ of all ordinal numbers," then Γ must be well-ordered by 3.5.2.1 and hence itself have an order type; but this is impossible, since, if ω' is any order type, then there is an order type $\omega' + 1$ (obtained by adding a new element to any set of order type ω', as in 2.4.4 for instance). Thus if γ is the ordinal number of Γ, then, since $\gamma + 1 \varepsilon \Gamma$, we should have $\gamma + 1 < \gamma$!

3.6 The alephs; the continuum problem

Because of the Well-Ordering Theorem (3.1.2), and because of the existence of uncountable infinite sets (R, $2^R, \cdots$) of various cardinal

† The procedure referred to here was the generation of successive *derived sets* or *derivatives* of point sets; the derivative of a point set is the set of all its limit points (see VII 3.5.2 and VII 4). An ω-type sequence of such derivatives generally does not exhaust the set of different derivatives, so that it is necessary to go on to "transfinite derivatives." The successive derivatives, in any case, form a well-ordered collection. See Hobson [Ho; §§ 52–54, 57, 68].

‡ In [a], Bernstein stated that Cantor was already aware of this contradiction in 1895 and communicated it by letter to Hilbert in 1896.

numbers, the ordinal numbers are divided into distinct categories according to the "cardinality" of the sets occurring as representative of the various order types. The cardinal numbers corresponding to these categories are called "alephs," aleph being the Hebrew "capital A"—$\aleph$. We have already used the symbol $\aleph_0$ to designate the cardinal number of denumerable sets; and, since all the transfinite (meaning $\geqq \omega$) ordinal numbers specifically set forth in (3.5.2a) correspond to denumerable sets, the latter numbers must all be in the $\aleph_0$ category, or *class* (which is the term usually employed in this connection).

By convention, the so-called *finite* ordinals, 0, 1, 2, $\cdots$, n, $\cdots$, are put into one class called the *first class* of ordinals; this is the only class containing ordinals corresponding to sets of different cardinality. The *second class* of ordinals consists of all those ordinals that correspond to sets containing $\aleph_0$ elements, namely, ω, $\omega + 1, \cdots; \omega \cdot 2, \cdots; \cdots$; that is, if ω' is an ordinal in the second class, and $W \varepsilon \omega'$, then $\overline{\overline{W}} = \aleph_0$.

Now (cf. 3.5.2.1) let ω_1 denote the first (in the order defined in 3.5.2) ordinal such that, if $W \varepsilon \omega_1$, then W is uncountable. Let $\aleph_1$ denote the cardinal number of such a W. Then all ordinal numbers ω' such that, for $W \varepsilon \omega'$, $\overline{\overline{W}} = \aleph_1$ constitute the *third* class of ordinals. We call $\aleph_1$ the cardinal number corresponding to the third class of ordinals (and $\aleph_0$ the cardinal number corresponding to the second class of ordinals). Continuing in this manner, we obtain a series of alephs, $\aleph_0$, $\aleph_1$, $\aleph_2, \cdots$ corresponding to distinct classes of ordinal numbers.

We shall not go into all the ramifications of this theory and its relation to the ordinal numbers,† but shall be content to point out some of its consequences that are of greatest significance for our purposes.

Not all mathematicians, as for instance the intuitionists (cf. Chapter X), admit the existence of ordinal numbers of arbitrarily large class; not even, indeed, the set of all ordinals of the second class. See Lusin [L; 27], and Borel's preface to Sierpinski [S_1], for example.

3.6.1 In the first place, because of the Well-Ordering Theorem *every infinite cardinal number is an aleph*. For, if S is any set, then by the Well-Ordering Theorem there exists a well-ordered set W whose elements are the elements of S. Hence the cardinal number of W, which is an aleph, is the same as the cardinal number of S.

3.6.2 In the second place, the possibility (4.2.5d) of Chapter IV cannot occur, since, of two alephs, one must be $\leqq$ the other (this follows directly, of course, from the Well-Ordering Theorem and Theorem 3.4.1, with the A and B of IV (4.2.5d) forming the W and W' of the latter

† See, however, Sierpinski [S_1; XI], [S_3; XV].

theorem). We express this fact by saying that every two sets are *comparable*; and that the principle involved, namely that IV (4.2.5d) is impossible, is *comparability*. It should be noticed that, if we use $\Rightarrow$ for implication—"$A \Rightarrow B$" meaning "A implies B"—then we have

(3.6.2a) Choice Axiom $\Rightarrow$ Well-Ordering Theorem $\Rightarrow$ Comparability.

The power of the choice axiom becomes more and more evident!

3.6.3 The continuum problem; continuum hypothesis

Thirdly (recalling the question raised in IV 4.2.7.1), where, in relation to the other cardinal numbers, does the cardinal number c stand, from the standpoint of $<$ (IV 4a)? In particular, is $c = \aleph_1$? This is known as the "continuum problem," and may also be formulated as the query: Has every infinite subset of R either the cardinal number $\aleph_0$ or the cardinal number c? Two types of proof that the answer is affirmative were proposed,† but were not generally accepted.‡

Cantor, himself, conjectured that $c = \aleph_1$, and it has been found that a considerable number of theorems can be based on the hypothesis that the conjecture holds. It has become customary to call the conjecture "the continuum hypothesis." Some theorems turned out to be *equivalent* to the continuum hypothesis;§ others which originally were proved on the basis of the hypothesis have subsequently been proved without it. In no case has a theorem based upon it ever been shown to be false. In spite of this, it was considered by some mathematicians that the hypothesis is probably false.‖ Others deemed it impossible to prove it or disprove it on the basis of ordinary set theory; that it constituted an independent axiom.

Since $c = 2^{\aleph_0}$ (IV (4.2.3a)), the continuum hypothesis may be stated in the form $\aleph_1 = 2^{\aleph_0}$. One may, then, state a generalized continuum hypothesis in the form of the assumption that $\aleph_{\alpha+1} = 2^{\aleph_\alpha}$ for all cardinal numbers $\aleph_\alpha$. Without the Well-Ordering Theorem, the hypothesis would of course take the form: If e is a cardinal number $\geqq \aleph_0$, then there exists no cardinal number m such that $e < m < 2^e$.

† D. Hilbert [a] and F. Bernstein [b].

‡ In the reprinting of Hilbert [a] as Anhang VIII of [H_2], it is significant that the proposed plan of proof is omitted. For a critique of Bernstein [b], see Rosser [e].

§ For a discussion of relations to other mathematical principles and theorems, the reader is referred to Sierpinski [S_2]. For a hypothesis alternative to the continuum hypothesis, namely $2^{\aleph_0} = 2^{\aleph_1}$, see Lusin [a] and Sierpinski [a].

‖ See Gödel [a] for comment and references. On p. 524, loc. cit., Gödel remarks ". . . one may on good reason suspect that the role of the continuum problem in set theory will be this, that it will finally lead to the discovery of new axioms which will make it possible to disprove Cantor's conjecture."

In 1940 Gödel showed† that, if the generalized continuum hypothesis be added to the well-known axiom system given by Zermelo for set theory (to be discussed in VIII 8), along with the choice axiom, then any contradiction that might be implied by these axioms could be transformed into a contradiction implied by the axioms for set theory alone. This took the form of a relative consistency proof; a model for the extended system was found within the Zermelo system, so that if the latter is consistent, so must the former also be consistent. In 1947, it was shown by Sierpinski [b] that the generalized continuum hypothesis implies the choice axiom.‡

Finally, in 1963, Paul Cohen announced (see Cohen [a], [b]) that denials of both Choice Axiom and generalized continuum hypothesis are consistent with the axioms of a set theory such as those of Zermelo cited above. In conjunction with Gödel's 1940 result, then, it appears that these principles have an independence in an axiomatic set theory which makes their use, or rejection, just as much a matter of arbitrary selection as that between a euclidean or non-euclidean geometry. But it seems highly probable that the fruitfulness resulting from the choice axiom, already frequently pointed out above, will dictate its acceptance wherever needed in applications of the theory of sets. The situation regarding the continuum hypothesis seems not so clear; a decision will perhaps have to wait upon further investigation of its consequences beyond what is already known.

4 The second class of ordinals

If the order type ω is the most important of the order types of well-ordered infinite sets, then it is probably true that the order type ω_1 is the second most important.§ For instance, it is a fruitful source of examples, such as models for axiom systems (we see an instance of this below in VI 1.2.6.2). And it may be expected that, in any method of proof which makes use of the continuum hypothesis, the collection of all ordinals of the first and second classes may be involved. For, inasmuch as the order type of this set is ω_1, the set R may (using the continuum hypothesis) be well-ordered so as to be of type ω_1.||

It is useful, therefore, to know some of its chief properties. One of these is expressed in the following:

† K. Gödel, [G].

‡ Sierpinski remarks that this result was previously announced, without proof, by A. Lindenbaum and A. Tarski, in 1926.

§ That is, important from the standpoint of frequency of application, number and importance of theorems dependent thereon, etc.

|| Just as, for example, when we wish a well-ordering of a denumerable set (such as F), we usually select a type ω sequence for it, rather than an $\omega + 1$, $\omega \cdot 2$, or ω^2 sequence.

4.1 Theorem. *No well-ordered set of type ω_1 is spanned* (2.3.1) *by a countable subset.*

Proof. Suppose that W is a well-ordered set of type ω_1 and that W_1 is a countable subset of W that spans the latter. As W_1 is countable, we may order its elements in a type ω sequence

$$x_1, x_2, \cdots, x_n, \cdots, \tag{4.1a}$$

although this is of course *not* necessarily the order in which these elements occur in W.

Now, for each n, the set $W(x_n) = \{x_n\} \cup W/x_n$ (3.1.1) is countable since the order type is less than ω_1. Also, the W's satisfy the relation $W = \bigcup_{n=1}^{\infty} W(x_n)$. But this implies, by virtue of Theorem IV 3.1.6, that W is denumerable. As W is of type ω_1, however, it is uncountable.

4.2 An axiomatic characterization of type ω_1

The result of the preceding section can be used to give an axiomatic characterization of sets of type ω_1:

Let W_1 be any set and $<$ a binary relation between elements of W_1 such that:

(a) W_1 is well ordered with respect to the relation $<$.
(b) If $a \in W_1$, then the section W_1/a is countable.
(c) No countable subset of W_1 spans W_1.

(Compare this with the characterization of type ω in 2.1) We leave to the reader (Problem 23) the proof that this is a categorical axiom system characterizing sets of type ω_1.

4.3 We can define limits for ordinal numbers as follows:

4.3.1 Definition. A (type ω) sequence of ordinal numbers,

$$\alpha_1, \alpha_2, \cdots, \alpha_n, \cdots \tag{4.3.1a}$$

is called an *increasing sequence* if $\alpha_n < \alpha_{n+1}$ for all n.

4.3.2 Definition. An ordinal number α is called the *limit* of an increasing sequence (4.3.1a) if, for every ordinal number $\beta < \alpha$, all but a finite number of the α_n satisfy the relation $\beta < \alpha_n < \alpha$.

And we can then state:

4.3.3 Theorem. *Every increasing sequence of ordinals of the first and second class has a limit of the second class.*

Proof. Consider any increasing sequence (4.3.1a). By Theorem 3.5.2.2, the ordinals of the first and second classes form a well-ordered set of type ω_1, and, by Theorem 4.1, there exist ordinals of the first or second class which are preceded by *all* elements of (4.3.1a). Of these there is a first in the order defined in 3.5.2; denote it by α. Evidently $\omega \leqq \alpha$ [cf. axiom (b) of 2.1], so that α is of the second class.

If β is an ordinal number such that $\beta < \alpha$, then there exists an α_n such that $\beta \leqq \alpha_n$; for otherwise α would not fulfill the condition stated in its definition. Then, for all $m > n$, $\beta < \alpha_m < \alpha$.

4.3.4 Definition. If an ordinal number has an immediate predecessor, we call it an ordinal number *of the first kind*; by convention we also term the number 0 of the first kind. All numbers not of the first kind we call *of the second kind.*

4.3.5 Theorem. *Every ordinal number α of the second class which is of the second kind is the limit of an increasing sequence of type* (4.3.1a).

Proof. The number α, being of the second class, designates the order type of a well-ordered set of cardinal number $\aleph_0$; in particular, the ordinals β such that $\beta < \alpha$ form a denumerable set, and may be rearranged in a sequence

(4.3.5a) $$\beta_1, \beta_2, \cdots, \beta_n, \cdots$$

of type ω (the order in this sequence being *not* necessarily that of the numbers in the sense of 3.5.2).

Form a sequence $\{\alpha_n\}$ as follows: Let α_1 be β_1. As α has no immediate predecessor, there is an ordinal α' such that $\alpha_1 < \alpha' < \alpha$; and, as α' is in the sequence (4.3.5a), we can define a number α_2 as the β_i, say $\beta_{n(1)}$, of *smallest* subscript $n(1)$, which satisfies the relation $\alpha_1 < \beta_{n(1)} < \alpha$.

In general, having defined α_k, we let α_{k+1} be the β_i, say $\beta_{n(k)}$, of smallest subscript $n(k)$, such that $\alpha_k < \beta_{n(k)} < \alpha$. This defines, by induction, a sequence $\{\alpha_n\}$ of type (4.3.1a).

We assert that α is the limit of this sequence. For suppose not. Then there is an ordinal number $\beta < \alpha$ such that, for all n, $\alpha_n < \beta$. Now β is an element of (4.3.5a), say β_m. Let $\alpha_{n'}$ be the last β_n that precedes β_m in the sequence (4.3.5a)—then $\alpha_{n'} = \beta_{n(n'-1)}$, and the number $\alpha_{n'+1} = \beta_{n(n')}$ comes *after* β_m in the sequence (4.3.5a). However, since $a_{n'} < \beta_m < \alpha$, the choice of $\beta_{n(n')}$ for $\alpha_{n'+1}$ is contrary to the rule that it have the smallest subscript satisfying the required relation to $\alpha_{n'}$ and α. Hence the supposition that α is not the limit of the sequence $\{\alpha_n\}$ leads to contradiction.

5 Equivalence of Choice Axiom, Well-Ordering Theorem, and Comparability; other equivalent principles

In 3.1.2 we gave an outline of a proof that the Choice Axiom implies the Well-Ordering Theorem. And in 3.6.2 we saw that the latter implies comparability; i.e., of two arbitrary sets A and B, one at least has the same cardinal number as a subset of the other. In this section we show the actual equivalence of all three of these properties of sets.

5.1 Let us first notice the more obvious implications. Thus in 3.6.2 we observed that comparability is a direct result of the Well-Ordering Theorem, since, if we take the two given sets A and B in well-ordered form and apply Theorem 3.4.1 (which states that one of these sets is then of the same order type as the other or as a section of the other), we obtain immediately a (1-1)-correspondence between the elements of one and the elements of a subset of the other.

5.1.1 Similarly, the implication

$$\text{Well-Ordering Theorem} \Rightarrow \text{Choice Axiom}$$

is virtually immediate.† For (using the symbols of III 6.3), if $\mathfrak{S}$ is a collection of disjoint non-empty sets S_ν, let $S = \bigcup_\nu S_\nu$. By the Well-Ordering Theorem there exists a well-ordered set W whose elements are the elements of S. Since W is well-ordered, every non-empty subset of W has a first element, so that now we can *define* the representative element of S_ν to be the *first* element of S_ν in the well-ordered set W.

5.1.2 We already have, then, the implications

(5.1.2a) Choice Axiom $\Leftarrow$ Well-Ordering Theorem $\Rightarrow$ Comparability.

And in order to complete the proof of the equivalence of all three properties we have to show that the implication arrows in (5.1.2a) can be reversed.

5.1.3 Comparability $\Rightarrow$ Well-Ordering Theorem.‡ We let the reader prove the following lemmas, which follow quickly from the definition of well-ordering.

5.1.3.1 Lemma. *If W is a simply ordered set such that for every $w \in W$ the set W/w is a well-ordered subset of W, then W is well-ordered.*

5.1.3.2 Lemma. *If W and W' are well-ordered sets such that there exists an order-preserving* (1-1)-*correspondence between the elements of W and the sections of W' (where $W'/w_1 < W'/w_2$ if $w_1 < w_2$), then W and W' are of the same order type.*

† This was pointed out by É. Borel [a] upon the appearance of Zermelo's first paper [a]. And as a consequence Borel expressed the opinion that Zermelo's Well-Ordering Theorem was therefore only to be considered as part of the proof of the *equivalence* of the two properties, not as a proof *establishing* the possibility of well-ordering every set; the latter would demand, in Borel's opinion, an effective procedure for well-ordering.

‡ The proof given in this section is due to Hartogs [a].

5.1.3.3 Let S be any non-empty set, and let us consider all the possible well-orderings of subsets W of S. These form a collection $\mathfrak{W}$, non-empty, since, if $s \varepsilon S$, then $\{s\}$ is a well-ordered subset of S. In $\mathfrak{W}$, order-preserving (1-1)-correspondence is an equivalence relation $\approx$ (II 8). Let K be the collection of classes which are elements of the class decomposition (II 8.5) of $\mathfrak{W}$ corresponding to this $\approx$.

The collection K becomes a well-ordered set (which we shall continue to call K) if, for $\mathfrak{k}_1,\mathfrak{k}_2 \varepsilon$ K we let $\mathfrak{k}_1 < \mathfrak{k}_2$ if and only if, for $W_1 \varepsilon \mathfrak{k}_1$, $W_2 \varepsilon \mathfrak{k}_2$, there is a section W_2/x_2 such that $W_1 \approx W_2/x_2$; that is, if there is a $W_2/x_2 \varepsilon \mathfrak{k}_1$. For consider any $\mathfrak{k}_2 \varepsilon$ K. If $\mathfrak{k}_1 < \mathfrak{k}_2$, and $W_2 \varepsilon \mathfrak{k}_2$, the W_2/x_2 defined above is unique by virtue of Theorem 3.4.2. Conversely, a W_2/x_2 determines a unique class $\mathfrak{k}_1$ such that $W_2/x_2 \varepsilon \mathfrak{k}_1$. Hence the collection $\{\mathfrak{k}_1 \mid \mathfrak{k}_1 < \mathfrak{k}_2\}$ is of the same order type as W_2, hence well-ordered. Then since $\mathfrak{k}_2$ was arbitrary, the collection K is well-ordered by Lemma 5.1.3.1.

Each $W_2 \varepsilon \mathfrak{k}_2 \varepsilon$ K is of the same order type as the section $\mathrm{K}/\mathfrak{k}_2$. For, as we have just shown, for any $\mathfrak{k}_1 < \mathfrak{k}_2$ there is a corresponding section $W_2/x_2 \varepsilon \mathfrak{k}_1$, etc., and that $\mathrm{K}/\mathfrak{k}_2$ and W_2 are of the same order type follows from Lemma 5.1.3.2. It follows that K *cannot be of the same cardinal number as S or as any subset of S.* For, if K did have the same cardinal number as a subset S_1 of S, then S_1 would have a well-ordering which we denote by W_1, and hence there would exist a $\mathfrak{k}_1 \varepsilon$ K such that $W_1 \varepsilon \mathfrak{k}_1$. But W_1, as just shown, is of the same order type as the section $\mathrm{K}/\mathfrak{k}_1$. It would then follow, from the transitivity of the relation of having the same order type, that K and $\mathrm{K}/\mathfrak{k}_1$ have the same order type, violating Corollary 3.4.3. In symbols, then

$$\overline{\overline{\mathrm{K}}} \neq \overline{\overline{S}} \text{ and } \overline{\overline{\mathrm{K}}} \not< \overline{\overline{S}}.$$

5.1.3.4 However, if we now assume the comparability property of sets, it follows that

$$\overline{\overline{\mathrm{K}}} > \overline{\overline{S}}.$$

This implies that, between the elements of S and the elements of a subset of K, there exists a (1-1)-correspondence; and this induces a well-ordering of S.

5.1.3.5 It is worthy of note that in 5.1.3.3, without any use of either the Choice Axiom or the comparability property, it was shown that:

Theorem. *If S is any set, then there exists a well-ordered set* K *such that neither of the relations* $\overline{\overline{\mathrm{K}}} = \overline{\overline{S}}$, $\overline{\overline{\mathrm{K}}} < \overline{\overline{S}}$ *holds.*

5.1.4 Choice Axiom $\Rightarrow$ Well-Ordering Theorem.† Continuing with the symbols used in 5.1.3, let us denote the order type of K by β, and

† Cf. Sierpinski [S_1; 230ff], [S_3; XVI].

suppose that, for every ordinal number $\alpha < \beta$, $\bar{\bar{S}} \neq \bar{\alpha}$, where $\bar{\alpha}$ is the cardinal number of a set of order type α.

Now, as we saw in III 6.4, the choice axiom implies that for every subset S' of S there may be assumed to exist a definite representative element which we denote by $x(S')$. Let $x_0 = x(S)$. Then, for any ordinal $\alpha < \beta$, if we have defined an x_γ for all $\gamma < \alpha$, we may define x_α to be $x(S - S_\alpha)$, where $S_\alpha = \bigcup_{\gamma < \alpha} x_\gamma$ (we may assume $S - S_\alpha \neq \emptyset$, since by our supposition $\bar{\bar{S}}_\alpha \neq \bar{\bar{S}}$ and $S_\alpha \subset S$). Then by the transfinite induction principle [which is applicable (3.2.1) to all ordinals of the well-ordered set of ordinals representing sections of K] we have defined for every $\alpha < \beta$ an element x_α of S, and, for $\alpha' < \alpha < \beta$, $x_{\alpha'} \neq x_\alpha$. But this implies that $\bar{\beta} = \bar{\bar{K}} \leqq \bar{\bar{S}}$, which we showed above not to be the case. Hence the supposition that, for all $\alpha < \beta$, $a \neq \bar{\bar{S}}$ leads to contradiction.

We conclude, then, that, for some $\alpha < \beta$, $\alpha = \bar{\bar{S}}$, and since α is an ordinal number this implies that there is a well-ordered set A whose elements are in (1-1)-correspondence T with the elements of S. The correspondence T induces a well-ordering of S.

5.2 Zorn's Lemma

During the years since Zermelo's proof brought out the equivalence of the Choice Axiom and the Well-Ordering Theorem, various alternative principles have been proposed. Some of these were motivated by a desire to avoid the direct use of transfinite numbers and transfinite induction, while others were conceived in connection with proofs which seemed more easily and naturally carried out by means of the alternatives.

One of these, which evolved in various forms and came finally to be called "Zorn's Lemma" may be stated as follows:

5.2.1 Zorn's Lemma. *If every simply ordered subset of a partially ordered set S has an upper bound, then S has at least one maximal element.*

(The notion of partially ordered set S was defined in Problem 16 of Chapter II. An element u of S is called an upper bound of a simply ordered set T if, for all $x \in T$, $x \leqq u$. An element m of a partially ordered set S is called a maximal element of S if for no element y of S distinct from y is $m \leqq y$.)

To illustrate the use of Zorn's Lemma, we give the proof of Hamel's Theorem (3.3.1) as based upon it, so that contrast with the proof given previously may be brought out. We shall call a set A of real numbers *independent* if there exists no relation of the form

$$f_1 a_1 + f_2 a_2 + \cdots + f_n a_n = 0 \qquad (5.2.1a)$$

where the a's are elements of A and the f's are rational numbers not all zero.

5.2.2 Proof of Hamel's Theorem on the basis of Zorn's Lemma

Let $\mathfrak{A}$ be the collection of all sets of independent elements of R; $\mathfrak{A} \neq \emptyset$, since $\{1\}$ and $\{1, \sqrt{2}\}$ are elements of $\mathfrak{A}$. Order (partially) the elements of $\mathfrak{A}$ by inclusion; thus $\{1\} < \{1, \sqrt{2}\}$, since $\{1\} \subset \{1, \sqrt{2}\}$. Then every simply ordered subset $\mathfrak{B}$ of $\mathfrak{A}$ will have an upper bound in $\mathfrak{A}$; one such will consist of the union of all elements of $\mathfrak{B}$ (inasmuch as every finite collection of the elements of such a union will be in some element of $\mathfrak{B}$). Applying Zorn's Lemma, there is a maximal element A of $\mathfrak{A}$, and A will be a basis for R. For if r is a real number not in A, $A \cup \{r\}$ cannot form an independent set of real numbers (since A is maximal), and therefore there will exist a relation

$$fr + f_1a_1 + \cdots + f_na_n = 0$$

in which the f's are rational, $f \neq 0$, and the a's are in A. This yields a relation of the form (3.3a).

SUGGESTED READING

Beth [Be; § 117]
Birkhoff, G. [Bi; III]
Fraenkel [F_3; III]
Gödel [G], [a]
Grelling [Gr; 35–40]
Halmos [Hal; 59–80]
Hobson [Ho; §§ 64–66]
Huntington [Hu; VII]
Kamke [Ka_1; IV]
Rosser [Ross; XIV]
Sierpinski [S_1; IX], [S_2; I]; [S_3; XI–XV]

PROBLEMS

1. Show that the sets ON, ON_1 of 1.1.1 are not isomorphic with respect to the simple order axioms.

2. Show that the set ON, and the set OF which consists of the elements of F ordered in their natural order of magnitude, are not isomorphic with respect to the simple order axioms.

3. Show that the order types of R, $R^+ = \{x \mid x \varepsilon R, 0 < x\}$, and the set $R^1 = \{x \mid x \varepsilon R \ \& \ 0 < x < 1\}$, each arranged in the natural order of the real numbers, are the same. [Compare Problem 14 of Chapter IV.]

4. Show that the set of all points in the coordinate plane can be so ordered as to be of the same order type as the set R (the latter ordered in its natural order of magnitude).

5. Show that if $\mathfrak{S}$ is a collection of disjoint, non-empty sets S_ν of rational numbers, then the choice axiom is not needed to assert the exist of a representative set (III 6.3) for $\mathfrak{S}$.

6. Show that, if we assume the Choice Axiom, then a simply ordered set S is well-ordered if and only if it has no subset A of order type $^*\omega$ (the order relations between elements of A being the same as in S).

7. Let A be a well-ordered set and B a simply ordered set that is not well-ordered. If $\overline{\overline{A}} = \overline{\overline{B}}$, and $f : A \rightarrow B$ is a (1-1)-correspondence between A and B, show that there exist $x, y \in A$ such that $x < y$ in A but $f(y) < f(x)$ in B.

8. Let f be a mapping (IV 3.2.3.1) of a set S onto a set T. Prove that $\overline{\overline{S}} \geqq \overline{\overline{T}}$, and indicate how the choice axiom comes into the proof.

9. Let $\mathfrak{A} = \{A_\nu\}$, where the A_ν are disjoint, non-empty sets. Prove that $\overline{\overline{\mathfrak{A}}}$ is less than or equal to the cardinal number of $\bigcup_\nu A_\nu$. Is the choice axiom used in the proof?

10. Show that a partially ordered set in which every non-empty subset has a first element is well-ordered.

11. On the basis of the axioms of 2.1, show that (1) N is well-ordered, (2) if A is a non-empty subset of N containing the first element of N and such that if $a \in A$ then the immediate successor of a in N is in A, then $A = N$ ("mathematical induction principle").

12. Show that the finite well-ordered sets are characterized by the properties of having a last element and of satisfying the mathematical induction axiom (c) of 2.4.6.

13. Given a set N and binary relation $<$ between elements of N, show that N is of order type ω with respect to this $<$ if and only if

(a) N is simply ordered with respect to $<$.

(b) N has a first element.

(c) Between† any two elements of N there exist at most a finite number of elements of N.

(d) N is Dedekind infinite.

14. Given N, $<$, as in Problem 13, show that N is of order type ω if and only if

(a) N is well-ordered.

(b) Every element of N except the first has an immediate predecessor.

(c) N is infinite.

15. Prove directly from the axioms for a well-ordered set (2.4) the validity of the transfinite induction method of showing that all elements of a well-ordered set have a certain property.

16. Show that the set B defined in the proof of 3.3.1 is a basis for R.

17. Show that when a well-ordered set is of type ω, then the mathematical induction principle (III 5) follows from the transfinite induction principle (3a).

18. If S is a given set, show that by assuming Zorn's Lemma we may prove that there exists a well-ordering of S. [*Hint:* Let $\mathfrak{A}$ be the collection of all well-ordered sets of elements of S; order these by inclusion as in 5.2.2.]

19. Show that a set W and binary relation $<$ between elements of W form a set of order type ω_1 if and only if

(a) W is well-ordered with respect to $<$.

(b) W is uncountable.

(c) Between any two elements of W there exist only a countable set of elements of W.

20. Show that a set of order type $\omega_1 \cdot \omega_2$ as defined in Definition 3.5.1.1 is well-ordered.

21. A well-ordering W of a set S is a *minimal* well-ordering of S if no well-ordering of S is of the same order-type as a section W/w of W. For example,

† In any simply ordered set, if $x < y$ and $y < z$, then y is said to be *between* x and z.

a type ω ordering of N (its natural order) is a minimal well-ordering of N. Show that every set has a minimal well-ordering.

22. Show that if S is any set and α is the order-type of its minimal well-ordering, then α is the smallest ordinal number in its class.

23. Prove that the axiom system of 4.2 is a categorical system characterizing the sets of order type ω_1.

VI

The Linear Continuum and the Real Number System

Our discussion of the axiomatic method in Chapters I and II led to a consideration of sets and the theory of the infinite. In the latter connection we generalized the number concept as it is represented by the natural numbers, finding two complementary types—the cardinal and the ordinal numbers.

In the present chapter we use the notions thus developed to study the concepts which form the foundation of modern mathematical analysis, namely, the linear continuum and the real number system. We defer to Part II our discussion of the philosophical and technical criticisms that have of late been directed at these concepts.

In IV 1.4–1.5, we discussed the cardinality of the set R of real numbers, using their decimal representation. Of course, a real number may be represented by other symbols; thus, "$\sqrt{2}$," "π," "e," are commonly used symbols since they identify the (frequently used) numbers concerned, while their decimal representations can only be observed to a finite number of digits and do not, therefore, serve to completely identify them. Generally, however, we do not have a special symbol for a real number, and in discussing the general real number we must ultimately resort to some kind of infinite symbol like the decimal representation.

Another mode of representing real numbers is by geometric segments. This was the only mode of representation used by the Greeks, who called real numbers "magnitudes." Operations with these "magnitudes," all done geometrically, are described in Euclid's *Elements*. It was not until the publication of a detailed discussion of the decimal representation of "fractions" by Stevin in 1585 that this mode of representation began to find favor (it was known to earlier scientists, however). The Babylonian astronomers used finite sexagesimals similar to what we call "finite decimals" like 0.25, and these were known to the Greeks. But although they were used by Greek astronomers, their possibilities were never realized by the Greek mathematicians. On the other hand, the conception of "magnitude" formed a satisfactory substitute for the abstractions of Greek mathematics, and we continue to use it today in the manner in which we frequently develop elementary Calculus from Analytic Geometry. It is not until we proceed to more advanced subjects such as Function Theory that we attempt to acquaint the student with "arithmetic" modes of symbolizing and conceptualizing real numbers.

Before describing the manner in which the nineteenth century analysts "arithmetized" the real numbers on the basis of the natural numbers, we shall give a somewhat analogous but simpler presentation which should not only serve to bring out the nature of the real number system, but furnish a means of axiomatizing its structure. We shall confine our discussion to the set R^+ of positive real numbers for the present; extension to the negative numbers and zero offers no difficulty.

We suggested in IV 4.1.2 that we can use the intuitive notion "size" of a set as a means of conceptualizing the natural numbers. Now, once we have the natural numbers, we can conceptualize the (positive) fractions p/q in which p and q are prime to one another, as *ordered pairs* of natural numbers, as an alternative to the Greek concept of length. We can go further: $\frac{3}{5}$, for example, can be associated with a set of 5 elements and one of its subsets having 3 elements. In the case of "improper" fractions like $\frac{5}{3}$, we may consider different sets of 3 elements, associating therewith all the elements of one and only two from another. This is, indeed, the concept behind the use of fractions as applied to collections instead of linear magnitudes. To complete this line of thought, we then make the convention that we will associate the same concept with a fraction p/q in which p and q are not prime to one another, as one does with the equivalent fraction in its "lowest terms"; thus $\frac{6}{10}$ and $\frac{3}{5}$ are different symbols for the same concept. Moreover, since finite decimals reduce to fractions in the usual manner—0.37 represents the same rational number as $\frac{37}{100}$—one has a mode of conceptualizing these also. It should be pointed out, however, that the additional step of reduction to the set concept is not necessary to conceptualize fractions, since the concept of an ordered pair of natural numbers is already sufficient; the further step in terms of sets may or may not be made, as one prefers.

The (positive) real number, then, can be conceived as an infinite sequence (type ω) of finite decimals; thus $\pi - 3$ may be conceived as the sequence

$$(1) \qquad .1, .14, .141, .1415, .14159, \cdots;$$

that is, the sequence of "decimal approximations." And of course any such sequence may, conversely, be conceived as a real number. Finally, if we agree that the symbol $.14159\cdots$ is only an abbreviation of (1), we arrive at the symbols used in IV 1.4—with the additional feature that we now have a concept which these symbols denote. As in IV 1.4, we make the convention that each element of R is to have a unique representation, and we achieve this by replacing each finite decimal by a decimal ending in an infinite sequence of 9's as before.

Let us recapitulate, both for clarity and additional comment. First, we had the set N of natural numbers. On the basis of N we formed the set, F^+, of (positive) rational numbers which we considered as ordered pairs of natural numbers. Now, to avoid logical difficulties, we should keep N and F^+ distinct, recognizing, however, that each element n of N has a representative of form $n/1$ in F^+, a representative of the same concept as for n. Then on the basis of F^+ we form the set R^+ of positive real numbers, each of which is an infinite sequence of elements of F^+ of the form

(2) $$k_1k_2\cdots k_m\cdot a_1a_2\cdots a_n$$

where $k_1k_2\cdots k_m$ is the integral part of the element and $a_1a_2\cdots a_n$ its decimal part. We call this infinite sequence an "infinite decimal" and abbreviate it to the usual form:

(3) $$k_1k_2\cdots k_m\cdot a_1a_2\cdots a_n\cdots$$

And again we recognize that F^+ and R^+ must be kept distinct, although each element p/q of F^+ has a representative in R^+ which is an infinite decimal. In particular, we notice that we now have three concepts of a natural number n: (1) as a cardinal number, (2) as an ordered pair of natural numbers (i.e., cardinal numbers), and (3) as an infinite decimal. In practice, we may, and usually do, think of it in its simplest form (1), but in theoretical work with R^+ we should adhere to (3).

Remark. The reader should be reminded that the above discussion is intended only to provide a conceptual model for what follows, namely, the search for an axiom system which will characterize the order type which we now proceed to set up in R^+. In addition, it will make the process to be described in Section 3 (the building up of the real numbers on the basis of the Peano axioms) seem more natural.

1 Analysis of the structure of the real numbers as an ordered system

In several instances we have referred, in the past, to the "natural order" of the collection R, with the remark that we meant the order according to magnitude. Here we shall be more explicit about this, since the type of order with which we are dealing has some very surprising aspects which can be appreciated only by a more careful examination of the order relation itself. Since we do not at present consider operations in R, being interested only in its ordering, we restrict our discussion to R^+, the set of all positive real numbers (cf. Chapter V, Problem 3).

1.1 Definition of < for R^+

Given two positive real numbers r and r', just how can we give an explicit definition of what we mean by the relation $r < r'$? Suppose that we use the form (3) to represent r, with a similar form to represent r', but with primes on the k's, a's, and m in the latter case. Since the integral parts of these numbers are natural numbers, we can first specify that, if

$$k_1k_2\cdots k_m < k_1'k_2'\cdots k_{m'}' \tag{1a}$$

where the < in (1a) is that of the natural order in N, then $r < r'$. If the converse of (1a) holds—$k_1'k_2'\cdots k_{m'}' < k_1k_2\cdots k_m$—then $r' < r$. But, in case these numbers are equal, we then consider the decimal parts of r. The first inequality (in the order of the subscripts), $a_n < a_n'$ (or $a_n' < a_n$), will now determine the order. Thus, if $a_1 < a_1'$, then $r < r'$. Or, if $a_1 = a_1'$, $a_2 = a_2'$, and $a_3 < a_3'$, then $r < r'$. If, however, $a_n = a_n'$ for every n, then the relation $r = r'$ holds.

1.2 Separability property of R^+

There is little difficulty in showing that the order relation defined in 1.1 satisfies the simple order axioms (II 7), and we leave this to the reader. What we should like to do next is to consider the relation of the set F^+ of positive rational numbers to the set R^+ in terms of this order. We shall use the symbols for elements of F^+ which represent them as elements of R^+, i.e., infinite decimals.

1.2.1 Theorem. *If r and r' are (positive) real numbers, and $r < r'$, then there exists $f \in F^+$ such that $r < f < r'$.*

Proof. Let r and r' be represented as in 1.1, and suppose that $r < r'$. If $k_1k_2\cdots k_m < k_1'k_2'\cdots k_{m'}'$, let a_k' be the first non-zero digit in the sequence $a_1', a_2', \cdots, a_n', \cdots$. Let f be the number $k_1'k_2'\cdots k_{m'}'.a_1'a_2'\cdots(a_k' - 1)99\cdots$, which has integral part $k_1'k_2'\cdots k_{m'}'$ and digits $a_1'a_2'\cdots a_{k-1}'$, the same as in r', but with a_k' replaced by $a_k' - 1$ and all a_n' for $n > k$ replaced by the digit 9. Then $f \in F$ and $r < f < r'$ by the rule given in 1.1.

If the integral parts of r and r' are the same, let i be the first value of n such that $a_n < a_n'$. Then let k be the first subscript greater than i such that a_k' is not zero. Let $f = k_1'k_2'\cdots k_{m'}'.a_1'\cdots a_i'\cdots(a_k' - 1)999\cdots$, a number which agrees with r' in its integral part and in the first $k - 1$

digits of its decimal part, but replaces a_k' by $a_k' - 1$ and all successive digits by 9. Then $f \varepsilon F^+$ and $r < f < r'$.

If in a simply ordered set we say that an element y *separates* elements x and z if y is between x and z, then we can state:

1.2.2 Corollary. *Every two elements of R^+ are separated by an element of R^+.*

1.2.2.1 The property of R^+ stated in 1.2.2 is frequently called the *density* property of R^+. A simply ordered set satisfying 1.2.2 is called *dense.*

1.2.3 If a real number can also be represented in the decimal system in the form $k_1k_2\cdots k_m.a_1a_2\cdots a_n\cdots$, where *all* a_n from some subscript n on are all zero, then we call it a *finite decimal.* (Thus, $2.499\cdots 9\cdots$ is a finite decimal since it can be represented in the form 2.5. On the other hand, $\frac{1}{3}$ is not a finite decimal, nor is any irrational real number.)

Now the numbers f defined in the proof of 1.2.1 are finite decimals; hence we can state:

1.2.4 Theorem. *If r and r' are real numbers, $r < r'$, then there exists a finite decimal d such that $r < d < r'$.*

1.2.5 We showed in IV 1.2 that F has the cardinal number $\aleph_0$; in IV 1.4 that the cardinal number of R is a number c different from $\aleph_0$; and in IV 4.2.1 that $\aleph_0 < c$. Thus, although the number of elements in R is greater than the number of elements in F, it turns out that between every two elements of R there is an element of F. For finite sets there is nothing surprising about such a property, but in the present case we might erroneously reason that, since each *two* elements of R are separated by *one* element of F, the number of elements in R is at most *twice* as great as that in F, i.e., $\aleph_0 + \aleph_0$, which, we showed in IV 2.5.1, is $\aleph_0$.† The "paradox" is explained if we notice that the *same* element of F separates many different pairs of elements of R.

1.2.6 The set F^+ is called a *countable separating set* for R^+, and the *existence* of such a set is designated the *separability* property of R^+. In general, if O is a simply ordered set, then O is called *separable* if it has a countable subset S such that between each two elements of O there exists an element of S; and S is called a *countable separating set* for O. Although the separability property implies density, not all simply ordered sets having the density property have the separability property; not even if, in addition, they have cardinal number c. For consider the following examples:

† We did not define addition or multiplication in Chapter IV for cardinal numbers; but, as the reader would probably surmise, $2\cdot\aleph_0 = \aleph_0 + \aleph_0$, and to obtain the latter we would select disjoint sets $A, B \varepsilon \aleph_0$ and find the cardinal number of $A \cup B$.

1.2.6.1 Example (Huntington [Hu; 45]). In the coordinate plane, let S denote the set $\{(x, y) \mid (0 \leqq x \leqq 1) \ \& \ (0 \leqq y \leqq 1)\}$. Define $<$ for S as follows: If $p = (x, y)$, $q = (x', y')$, let $p < q$ mean that either (1) $x < x'$, or (2) $x = x'$ and $y < y'$. That $p = q$ means $x = x'$ and $y = y'$. Then S is simply ordered with respect to this $<$. Also, S has the density property (1.2.2.1). We proved in IV 2.4 that the cardinal number of S is c. But S does not have the separability property.

For consider any real number r such that $0 \leqq r \leqq 1$. Let $I_r = \{(r, y) \mid 0 \leqq y \leqq 1\}$. Also let $p_r = (r, 0)$, $q_r = (r, 1)$. Then $p_r < q_r$, and a point x_r between p_r and q_r would have to be an element of I_r. Since the cardinal number of the collection of sets I_r is c, any set $M \subset S$ having the property that $M \cap I_r \neq \emptyset$ *for all* r would have to have cardinal number equal to c. Hence, any set $M \subset S$ having the property that, for every $p, q \,\varepsilon\, S$ there exists $x \,\varepsilon\, M$ such that x is between p and q must have the cardinal number c!

1.2.6.2 Example. Let Ω denote the collection of all ordinal numbers of the first and second classes (V 3.6). For each successive pair $\alpha, \alpha + 1 \,\varepsilon\, \Omega$, introduce a set R_α of the same order type as the set of real numbers $\{r \mid 0 < r < 1\}$, defining $a < x_\alpha < \alpha + 1$ for all $x_\alpha \,\varepsilon\, R_\alpha$. And, if $\alpha < \beta$, where $\alpha,\beta \,\varepsilon\, \Omega$, let $x_\alpha < x_\beta$, for all $x_\alpha \,\varepsilon\, R_\alpha$, $x_\beta \,\varepsilon\, R_\beta$. And finally, if $\alpha < \beta < \gamma$, where $\alpha,\beta,\gamma \,\varepsilon\, \Omega$, let $\alpha < x_\beta < \gamma$ for all $x_\beta \,\varepsilon\, R_\beta$. The set $\Gamma = \Omega \cup \bigcup_{\alpha\varepsilon\Omega} R_\alpha$ is then a simply ordered set having the density property.

The set Γ does not have the separability property. For, if Γ' were a countable separating set for Γ, then for every $\alpha \,\varepsilon\, \Omega$ there would exist a $\gamma \,\varepsilon\, \Gamma'$ such that $\alpha < \gamma$. Then Γ' would be a countable spanning set (V 2.3.1) of Γ. However, this would imply the existence of a countable spanning set of Ω, which is impossible by Theorem V 4.1 (cf. Problem 5).

It is easy to prove that $\overline{\overline{\Gamma}} = c$ by using the Bernstein equivalence theorem. In the symbols of IV 4.2.5, it is trivial that there exists a mapping $\bar{R}^1 \to \Gamma$, where $\bar{R}^1 = \{x \mid x \,\varepsilon\, R, 0 \leqq x \leqq 1\}$. To show that there is a mapping $\Gamma \to \bar{R}^1$, we may first appeal to the relation $\aleph_1 \leqq c$ to assert the existence of a mapping $f\colon \Omega \to \bar{R}^1$. Then with S as in 1.2.6.1, and $\bar{R}^1$ as its subset $\{(x, y) \mid 0 \leqq x \leqq 1, y = 0\}$, we may extend f to a mapping $\Gamma \to S$ simply by mapping each R_α into $I_{f(\alpha)}$. Hence $\overline{\overline{\Gamma}} \leqq \overline{\overline{S}} = c$.

Note that the ordered sets S (1.2.6.1) and Γ are not of the same order type (Problem 6).

1.3 The possible types of cuts of R^+

Our investigation of the order type of R^+ would not be complete without a study of the cuts of R^+. The notion of a cut of a simply ordered system was introduced in V 2.2, Definition (2a), and used there to characterize sets of order type ω. In a set of order type ω, every cut $[A, B]$ is of a single type, viz., the type in which A has a last element and B has a first element.

Cuts of R^+ are of *two* types: that in which A has a last element and B no first element, and that in which A has no last element and B has a first element. The former is exemplified in the case where $A = \{x \mid (x \leqq 1)\}$, $B = \{x \mid 1 < x\}$; the latter by $A = \{x \mid x < 1\}$, $B = \{x \mid 1 \leqq x\}$. The type of cut found in sets of order type ω is *never* found in R^+, in view of Corollary 1.2.2.

1.3.1 Theorem. *If* $[A, B]$ *is a cut of the collection* R^+, *ordered as in* 1.1, *then either* A *has a last element or*† B *has a first element.*

Proof. It will be sufficient to prove the theorem for the set $R^1 = \{r \mid (r \in R) \& (0 < r < 1)\}$ instead of R^+, since R^+ and R^1 are of the same order type (Chapter V, Problem 3). Hence we assume that $[A, B]$ is a cut of R^1.

Suppose that A has no last element and that B has no first element. By Theorem 1.2.4, the set, F', of all positive finite decimals, with integral part zero, is a countable separating set (1.2.6) for R^1. The set F' is denumerable, and its elements can therefore be represented by symbols

$$(1.3.1a) \qquad d_1, d_2, \cdots, d_n, \cdots, \qquad n \in N,$$

where the order in (1.3.1a) is, of course, not that in R^1. Since the d's do occur in two orders, let us agree that "$<$" denotes the order relation in R^1, and that "$<'$" denotes the order in (1.3.1a)—thus $d_n <' d_{n+1}$ for all n, although, for some values of n, $d_{n+1} < d_n$ may hold.

Let $d_{n(1)}$ denote the first‡ (in terms of $<'$) element of (1.3.1a) that is an element of A. Having defined $d_{n(k)}$ for a $k \in N$, let $d_{n(k+1)}$ be the first (in terms of $<'$) element of (1.3.1a) such that $d_{n(k+1)} \in A$ and $d_{n(k)} < d_{n(k+1)}$. That such an element exists for all k follows from the supposition that A has no last element. We have defined, then, by induction a sequence

$$(1.3.1b) \qquad d_{n(1)}, d_{n(2)}, \cdots, d_{n(k)}, \cdots$$

of distinct elements of A such that (1) $d_{n(k)} < d_{n(k+1)}$ for all $k \in N$ and (2) if $a \in A$, then there exists k such that $a < d_{n(k)}$.

That (2) holds may be shown as follows: Suppose that there exists an $a \in A$ such that for no k is $a < d_{n(k)}$. Let $b \in A$ such that $a < b$.

†As stated before [II 7f], we always use "or" in the sense of "and/or," not as a complete disjunction. The complete disjunction actually holds here, however, as stated in the preliminary remarks above. Incidentally, "or" is another of our logical "universals" which we constantly find ourselves explaining. Cf. I 3.4.

‡ V 2.1, second footnote.

Then since the finite decimals form a countable separating set, there exists a d_n of the sequence (1.3.1a) such that $a < d_n < b$; let n' be the smallest value of n satisfying these inequalities. Now $d_{n(1)} <' d_{n'}$ since otherwise $d_{n'}$ would have been the first element of (1.3.1a) in A. Let $n(i)$ be the largest subscript in (1.3.1b) such that $d_{n(i)} <' d_{n'}$. Then $d_{n'} <' d_{n(i+1)}$. But this is impossible, since $d_{n(i)} < d_{n'}$ and $d_{n(i+1)}$ was supposed, by the induction definition, to be the *first* element of (1.3.1a) to lie in A and satisfy the relation $d_{n(i)} < d_{n(i+1)}$! We must conclude, then, that (2) holds.

To simplify symbols, let us denote $d_{n(k)}$ by d_k'. Sequence (1.3.1b) then becomes

(1.3.1b) $$d_1', d_2', \cdots, d_k', \cdots$$

And since each d_k' is a finite decimal, we may exhibit the elements of (1.3.1b) in a vertical array:

$$\begin{aligned} d_1' &= .a_1{}^1 a_2{}^1 a_3{}^1 \cdots \\ &\cdot\ \cdot\ \cdot\ \cdot\ \cdot\ \cdot\ \cdot \\ d_k' &= .a_1{}^k a_2{}^k a_3{}^k \cdots \\ d_{k+1}' &= .a_1^{k+1} a_2^{k+1} a_3^{k+1} \cdots \\ &\cdot\ \cdot\ \cdot\ \cdot\ \cdot\ \cdot\ \cdot\ \cdot \end{aligned} \tag{1.3.1c}$$

there being only finitely many a's in each row.

Since $d_k' < d_{k+1}'$ for every k, $a_1{}^k \leqq a_1^{k+1}$. And since the vertical array (1.3.1c) is infinitely long, from some point on (as k increases), it must be the case that $a_1{}^k = a_1^{k+1}$; designate this common value by a_1. Then for the numbers in (1.3.1c) that begin with this a_1, there will be a point reached as k increases where always $a_2{}^k = a_2^{k+1}$; designate this common value by a_2. Continuing in this way, we may define (by induction) a sequence $a_1, a_2, a_3, \cdots$ which we may use as the digits of an infinite decimal

(1.3.1d) $$d = .a_1 a_2 \cdots a_i \cdots$$

The digit a_i is the ultimate common ith digit of the d_k'. Evidently $d \,\varepsilon\, R^1$.

We assert that $d \notin A$. To see this, let $a \,\varepsilon\, A$ and recall that we showed above that there is a k such that $a < d_k'$; also notice that as d is defined, $d_k' < d$ for every k. We therefore conclude that $d \,\varepsilon\, B$. But B has no first element, so there must be $b \,\varepsilon\, B$ such that $b < d$. Let the decimal denoting b be $.b_1 b_2 \cdots b_n \cdots$. Now $b < d$ implies that there is a smallest subscript h such that $b_h < a_h$. But this also implies that $b_h < a_h{}^k$ for k large enough and therefore that $b < d_k'$. But this is impossible, because $b \,\varepsilon\, B$, $d_k' \,\varepsilon\, A$, and $A < B$.

Thus the supposition that A has no last element and B has no first element leads to contradiction.

1.3.2 Remarks. We pointed out in IV 2.1 that, on the basis of an axiom system for ordinary euclidean geometry, such as that of Hilbert, it can be shown that the "Cantor axiom" holds. In the terminology that we have since introduced, this "axiom" states that, if L is the set of all points on any euclidean straight line, and $<$ is introduced in the usual manner for points of L, then *L and the set R of all real numbers are of the same order type.*

It follows that the theorems proved above concerning order in R^+ (which hold, of course, in R also) can also be stated for L. So far as Theorem 1.2.1 is concerned, this results independently from the fact that on L every finite interval has a midpoint and it is not difficult to set up a process of "bisecting" intervals on L so as to define, by induction, a set of points on L analogous to the countable separating set F (see Problem 7). However, to obtain the analogue of Theorem 1.3.1 requires a continuity axiom of some kind, which may well take the form of 1.3.1 itself. Indeed, as we shall see directly, these properties (separability, and the "cut property" stated in 1.3.1) virtually characterize the structure of the euclidean line.

1.3.2.1 An interesting by-product of the proof of Theorem 1.3.1 is a constructive procedure by which, given any real number d, and the sequence (1.3.1a), we may set up a sequence (1.3.1b) of finite decimals "approximating" d in a sense easily defined. However, if $0.a_1a_2\cdots a_n\cdots$ is any given infinite decimal, the finite decimals $0.a_1$, $0.a_1a_2, \cdots, 0.a_1a_2\cdots a_n, \cdots$ are the most obvious candidates for such a sequence. Of course, in the proof of 1.3.1 the number d had to be constructed from the sequence, rather than the reverse!

1.4 Axiomatic definition of the order type of R

We can use the properties of R which we have found above to give a categorical system of axioms which characterize the order type of R.†

1.4.1 Let C be a set of undefined elements and $<$ an undefined binary relation between elements of C such that

(1) *C is simply ordered with respect to $<$.*

(2) *If $[A, B]$ is a cut of C, then either A has a last element or B has a first element.*

(3) *There exists a non-empty, countable subset S of C such that, if $x,y \in C$ such that $x < y$, then there exists $z \in S$ such that $x < z < y$.*

(4) *C has no first element and no last element.*

† Compare Huntington [Hu; V], Sierpinski [S_1; 151], [S_3; 217, Th. 1].

These axioms can be briefly designated as, respectively,

(1) *Simple order axiom.*
(2) *Dedekind cut axiom.*
(3) *Separability axiom.*
(4) *Unboundedness axiom.*

1.4.1.1 **Definition.** Any set C with relation $<$ satisfying the axioms (1) to (4) of 1.4.1 will be called a *linear continuum* with respect to this $<$.

We shall designate the above axiom system by Λ.

1.4.2 Consistency of the axiom system Λ

Is it possible to show the axiom system Λ consistent? Here is a case where we cannot, for a model, exhibit any finite collection of physical objects, such as we did for the system Γ of I 2.2. As a matter of fact, for our consistency proof we resort to the concept which underlies the axioms (see I 5, II 6)! The collection R^+ and $<$ defined as in 1.1 form a model of the system Λ. (Cf. Problem 8.)

1.4.3 Independence of the axioms of Λ

The following interpretations, together with the system $(R, <)$, show the independence of the axioms of Λ (cf. II 3, Definition 3.1.1).

1.4.3.1 Interpretation of $[\Lambda - (1)] + \sim(1)$. Let M_1 be an ordered set defined as follows: The elements of M_1 are the points in the coordinate plane of the sets $S_1 = \{(x, y) \mid (x \leqq 0) \ \& \ (y = 0)\}$, $S_2 = \{(x, y) \mid (x > 0) \ \& \ (x = y)\}$, $S_3 = \{(x, y) \mid (x > 0) \ \& \ (x = -y)\}$. Definition of $<$: For $p, q \,\varepsilon\, S_i$, $i = 1, 2, 3$, where $p = (x_1, y_1), q = (x_2, y_2), p < q$ means $x_1 < x_2$. For all $p \,\varepsilon\, S_1$, $q \,\varepsilon\, S_i$, $i = 2, 3$, $p < q$. For $p \,\varepsilon\, S_2$, $q \,\varepsilon\, S_3$, there is no $<$ relation. For M_1, axiom (1) fails, axioms (2) to (4) are satisfied.

1.4.3.2 Interpretation of $[\Lambda - (2)] + \sim(2)$. Let M_2 consist of the elements of F with $<$ defined as in 1.1.

1.4.3.3 Interpretation of $[\Lambda - (3)] + \sim(3)$. Let M_3 be the set S and relation $<$ of 1.2.6.1, with the points (0, 0) and (1, 1) deleted. Or let M_3 be the set Γ and relation $<$ of 1.2.6.2, with the first element of Γ deleted. (Cf. Problems 9 and 10.)

These interpretations also show, incidentally, that the assumption of density (1.2.2.1) alone in axiom (3) would not have been sufficient to characterize the order type of R. (The integers in their usual order suffice as a model of $[\Lambda - (3)] + \sim(3)$, but they do not have the density property.) For both S and Γ have the density property, yet neither is of the same order type as R, nor is either of the same order type as the other (cf. Problem 6).† The "lines" S and Γ are "too long" to put in

† The example S was also used by Young [Y; 84ff] for the same purpose.

euclidean spaces (to "straighten out" S, imagine that the position of S in the plane is the result of a folding process, accordion fashion, and perform the unfolding process on it).

1.4.3.4 Interpretation of $[\Lambda - (4)] + \sim(4)$. Let M_4 be any one of the following ordered subsets of R: $\{x \mid 0 \leqq x \leqq 1\}$, $\{x \mid 0 \leqq x < 1\}$, $\{x \mid 0 < x \leqq 1\}$.

1.4.3.5 Remark. It is of some interest to observe that the system $\Lambda - (4)$ has only four possible types of interpretations. That is, with isomorphism with respect to $\Lambda - (4)$ as an equivalence relation $\approx$ in the set K of all interpretations of $\Lambda - (4)$, there are exactly four classes in the class decomposition of K corresponding to $\approx$ (II 8.4, 8.5).

1.4.4 Categoricalness of the system Λ

To show that Λ is a categorical system, we first show that, if M is any model of Λ, then M and R are isomorphic with respect to Λ. Then, if M_1 and M_2 are any two models of Λ, the transitivity of isomorphism renders M_1 and M_2 isomorphic with respect to Λ, and consequently Λ is a categorical system.

Let M, then, be any model of Λ. Denote elements of M by the letter x with or without suitable indices, and the order relation in M by $<'$. Denote elements of R by the letter r with or without suitable indices, and the order relation in R by $<$ (as defined in 1.1).

Let us denote the subset of M whose existence is asserted in axiom (3) by S', and the elements of S' by $x_1, x_2, \cdots, x_n, \cdots$. And let us denote the elements of F by $r_1, r_2, \cdots, r_n, \cdots$. It will aid our exposition if we let the latter sequence be the sequence (F) of IV 1.2; then $r_1 = 0$, $r_2 = -1$, $r_3 = 1$, etc.

The set S' is non-empty, hence contains at least the element x_1. We shall set up an order-preserving (1-1)-correspondence T between the elements of R and the elements of M in which we have the pair (r_1, x_1). Let $T_1 = \{(r_1, x_1)\}$. Now r_2 $(= -1)$ satisfies the relation $r_2 < r_1$ (since $r_1 = 0$). We shall pair with r_2, then, the element $x_{n(2)}$ of S', whose subscript $n(2)$ is the smallest natural number n such that $x_n <' x_1$. That such an element $x_{n(2)}$ exists follows from the reasoning: By axiom (4), M has no first element, hence there exists x such that $x <' x_1$. By axiom (3) there is an element x_n of S' such that $x <' x_n <' x_1$; and, since N in its natural order is a well-ordered (type ω) set, there exists a smallest such natural number $n(2)$ as desired above. Let $T_2 = \{(r_1, x_1), (r_2, x_{n(2)})\}$.

Having defined a collection T_k of pairs $(r_1, x_1), (r_2, x_{n(2)}), \cdots, (r_k, x_{n(k)})$, such that the resulting (1-1)-correspondence T_k is order-preserving, we consider r_{k+1}. We determine the element $x_{n(k+1)}$ of S' whose subscript

$n(k + 1)$ is the smallest natural number n such that, if we augment T_k by the pair $(r_{k+1}, x_{n(k+1)})$, then the new (1-1)-correspondence T_{k+1} is order-preserving. That such an $x_{n(k+1)}$ exists follows from reasoning similar to that used in showing the existence of $x_{n(2)}$. In this manner, we arrive at a denumerable collection T_ω of pairs such that the correspondence T_ω is an order-preserving (1-1)-correspondence between the elements of F and the elements of a subset S'' of S' consisting of x_1, $x_{n(2)}, \cdots, x_{n(k)}, \cdots$.

We assert $S'' = S'$. For suppose not. Then there exists x_h with smallest subscript h such that $x_h \varepsilon S' - S''$. Evidently $x_1 \neq x_h$, so that there exists a k, say k', such that $n(k') < h$, and $x_{n(k')}$ is the last (in the process of selecting elements of S') one of the elements $x_1, x_2, \cdots, x_{h-1}$ to be selected for the sequence S''. (That is, $x_1, x_{n(2)}, \cdots, x_{n(k')}$ is the "*shortest*" section of the sequence S'' to contain *all* the elements $x_1, x_2, \cdots, x_{h-1}$.) And, as soon as $x_{n(k')}$ is selected, x_h becomes the *first* candidate, in all selections thereafter, for any of the elements $x_{n(k'+1)}, x_{n(k'+2)}, \cdots$. And ultimately it will certainly be chosen; for suppose, for instance, that $x_{n(2)} <' x_h <' x_1$, and that none of the elements $x_{n(3)}, \cdots, x_{n(k')}$ is between $x_{n(2)}$ and x_1 in M. Then, as soon as we come, in F, to the first r_n such that $r_2 < r_n < r_1$, x_h will be selected for S''. Other possibilities as to the relative order between x_h and the elements $x_1, x_{n(2)}, \cdots, x_{n(k')}$ yield the same conclusion.

Then T_ω constitutes an order-preserving (1-1)-correspondence between the elements of F and the elements of S'. To extend this to an order-preserving (1-1)-correspondence T between the elements of R and the elements of M, consider any $r^1 \varepsilon R - F$. Define a cut $[A_1, B_1]$ of F as follows: Let $A_1 = \{r_n \mid r_n < r^1\}$, $B_1 = \{r_n \mid r^1 < r_n\}$. Neither A_1 nor B_1 is empty. Then let A_1' be the set of all elements of S' paired with elements of A_1 in T_ω, and $B_1' = S' - A_1'$. And let $A' = \{x \mid x <' B_1'\}$, $B' = M - A'$. Then $[A', B']$ is a cut of M in which A' has a last element x^1, and we place the pair (r^1, x^1) in T.

That this defines T as an order-preserving (1-1)-correspondence between R and M we shall leave to the reader to show.

1.4.4.1 The proof of Theorem 1.4.4 was carried through in such detail partly to show the manner in which the irrational numbers, that is, the elements of $R - F$, may be determined by cuts of F (as in the case of r^1 and $[A_1, B_1]$ above). The reader may notice for himself that these correspond precisely to those cuts $[A_1, B_1]$ of F in which A_1 has no last element and B_1 has no first element. It should be noted also that, by virtue of Theorem 1.2.4, cuts of the set, D, of all *finite* decimals could have been used above; although a cut $[A_1, B_1]$ for which A_1 has no last and B_1 no first element may then determine a rational number.

1.4.5 A notable unsolved problem

Regarding the axiomatization of the order type of R, there is a well-known unsolved problem. Because of its seeming simplicity, and the failure of all attempts at solving it, it has received widespread attention. In order to state the problem, we need the notion of *interval*:

1.4.5.1 Definition. If $(C, <)$ is a simply ordered system (II 7.1) and $a,b \in C$, then the subset $\{x \mid a \leqq x \leqq b\}$ of C is called a *closed interval*, or simply *interval*, of C. If we wish to designate the determining points a and b, we say "the interval ab." The subset $\{x \mid a < x < b\}$ of C is called an *open interval*; to designate the determining points, we say "the open interval ab."

1.4.5.2 Theorem. *Every set of disjoint open intervals of a linear continuum $(C, <)$ is countable.*

Proof. If $\{I_\nu\}$ is a collection of disjoint open intervals, then by axiom (3) of 1.4.1 there exists, for each ν, an $x_\nu \in S$ (where S is the countable set S specified in this axiom) such that $x_\nu \in I_\nu$. As no two open intervals I_ν have a common element, and $\overline{\overline{S}} = \aleph_0$, it follows that the collection $\{I_\nu\}$ is countable.

1.4.5.3 Problem of Souslin. Because Theorem 1.4.5.2 is virtually a direct corollary of axiom (3), it is natural to ask whether it is not equivalent to axiom (3) in the presence of the other axioms. Specifically, if we temporarily designate by axiom (3′) the statement "Every collection of disjoint open intervals is countable," then *are the system of* 1.4.1, *and the system of* 1.4.1 *with C assumed non-empty, dense, and* (3) *replaced by* (3′), *equivalent*? (Cf. Problems 26, 27 of Chapter II.)

This question was propounded in 1920† by the Russian mathematician Souslin, and, although many attempts have been made to answer it and numerous papers have been published relating to it,‡ it remains unanswered to this day.

2 Operations in R

Throughout Section 1 we have not once considered the possibility of adding or multiplying elements of R. Our attention was focused entirely on the order type of R. However, as a *real number system*, R is something more than an *order type*. Its elements are added, subtracted, etc.

† See *Fund. Math.*, vol. 1 (1920), p. 223, Problem 3; also see Sierpinski [S_1; 152–153].

‡ See, for instance, Miller [a], Dushnik and Miller [a], Sierpinski [c], [S_3; 219f].

2.1 Definitions of operations

It will be assumed that the reader is familiar with the arithmetic commonly taught in the elementary schools, at least to the extent of knowing how to add, subtract, multiply, and divide *finite* decimals,† and of familiarity with the Laws of Signs ("the product of two negative numbers is positive," etc.; cf. VII 2.2.5, VII 2.7.2). We must, of course, revert now to the complete set R of real numbers.

2.1.1 To add two positive real numbers $r_1 = k_1k_2\cdots k_m.a_1a_2\cdots a_n\cdots$ and $r_2 = t_1t_2\cdots t_s.b_1b_2\cdots b_n\cdots$, we let $r_1{}^0 = k_1k_2\cdots k_m$ and (for every $n \varepsilon N$) $r_1{}^n = k_1k_2\cdots k_m.a_1a_2\cdots a_n$; and $r_2{}^0 = t_1t_2\cdots t_s$ and $r_2{}^n = t_1t_2\cdots t_s.b_1b_2\cdots b_n$; and finally let $r_1{}^n + r_2{}^n = s_n$. Then, since $r_1{}^n \leqq r_1^{n+1}$, $r_2{}^n \leqq r_2^{n+1}$, we have $s_n \leqq s_{n+1}$.

Now $r_1{}^n$ never exceeds $r_1{}^0 + 1$ and $r_2{}^n$ never exceeds $r_2{}^0 + 1$; hence s_n never exceeds $r_1{}^0 + r_2{}^0 + 2$. Ultimately, then, the integral part of s_n stays fixed; i.e., there exists an integer N such that, for $n > N$, $n' > N$, s_n and $s_{n'}$ have the same integral parts. There will also exist an integer $i_1 > N$ such that, for $n > i_1$, $n' > i_1$, s_n and $s_{n'}$ have the same digit in the first place after the decimal point. Pursuing the same type of reasoning as was used in the proof of Theorem 1.3.1 to show the existence of the number d, we can show here the existence of an $s \varepsilon R$ such that for any $k \varepsilon N$ there exists an i_k such that, for $n > i_k$, s_n and s have the same integral parts and the same digits in their first k decimal places.

We now *define* $r_1 + r_2$ to be s and write $r_1 + r_2 = s = r_2 + r_1$.

2.1.2 Obviously all the above operation amounts to is the usual "approximating" process used by everyone who computes. If we wish to find the digits representing the sum $\sqrt{2} + \pi$, we use $1.4 + 3.1 = 4.5$; or $1.41 + 3.14 = 4.55$; or $1.414 + 3.141 = 4.555$; or $\cdots$, depending upon how good an approximation we wish to the actual decimal representation of $\sqrt{2} + \pi$ (which is actually an infinite decimal, of course).‡ The above description of the process, however, shows that there actually *exists* an $s \varepsilon R$ which is being "approximated," so that we can give a definition of what we mean by the "sum" $r_1 + r_2$.

When r_1 and r_2 are both negative, then if $|r_1| + |r_2| = s$, $r_1 + r_2$ is of course $-s$. If one is negative and the other positive, we find the difference d of their numerical values and attach the sign of the numerically greater of the numbers r_1, r_2, to d. For this purpose, however, we have to define, for positive r_1, r_2, and $r_1 > r_2$, what we mean by $r_1 - r_2$.

2.1.3 For the latter operation we do the obvious thing: For each $n \varepsilon N$, let $d_n = r_1{}^n - r_2{}^n$, where $r_1{}^n$ and $r_2{}^n$ are as defined above. We

† For a general discussion of arithmetic, Klein [Kl; Part I] is strongly recommended.

‡ However, in practice one usually "smooths off" these figures, as in the case of the third sum, where 3.142 would probably be used instead of 3.141.

leave to the reader (Problem 11) to show that there exists a $d \varepsilon R$ such that for any $k \varepsilon N$ there exists an integer i_k such that, for $n > i_k$, d_n and d have the same integral parts and the same digits in their first k decimal places. We define $r_1 - r_2$ to be d and write $r_1 - r_2 = d$.

If $r_1 < r_2$, and r_1 and r_2 are both positive, then $r_1 - r_2$ is, of course, $-(r_2 - r_1)$. The usual elementary rules governing signs for subtraction of integers will carry over to the cases where not both r_1 and r_2 are positive. (Thus $\sqrt{2} - \pi = -(\pi - \sqrt{2})$; $\sqrt{2} - (-\pi) = \sqrt{2} + \pi$; etc.)

2.1.4 Multiplication and division are handled along similar lines, by performing the operations with the "approximations" $r_1{}^n$, $r_2{}^n$, observing the usual Laws of Signs in F. *Existence* of the corresponding elements of R is proved by the same type of argument as used in the case of addition. For our purposes we are more interested, however, in finding what general characteristics of these operations may serve as axioms for an axiomatization of the real number system.

2.2 Arithmetical "laws"

Certain "laws" commonly used in computation with integers and finite decimals, and usually taken for granted, should be noted here.

2.2.1 Commutative laws

In defining addition for elements of R, we remarked that $r_1 + r_2 = r_2 + r_1$. Similarly, we would have for multiplication that $r_1 \times r_2 = r_2 \times r_1$. These are respectively called the *commutative laws* of addition and multiplication. The corresponding laws do not hold for subtraction and division, of course. Later we shall see examples of "number systems" in which the commutative law of multiplication does not hold; these laws are not trivial, then, and they have to be specifically stated or implied in any axiomatic characterization of the real number system.

2.2.2 Associative laws

Everyone is familiar with the common device of adding a column of integers "down" and then "up" to check the sum. Involved here is not only the commutative law but also (assuming at least three integers are being added) the *associative law* of addition:

(2.2.2a) $$(r_1 + r_2) + r_3 = r_1 + (r_2 + r_3).$$

For any three finite decimals r_1, r_2, and r_3, it makes no difference in what order we carry out the addition. We can show from this fact that, if $r_1, r_2, r_3 \varepsilon R$, then in order to find a "sum," $r_1 + r_2 + r_3$, we may first find $r_1 + r_2 = r'$ and then $r' + r_3 = r$; and then we can prove that, if instead we first find $r_2 + r_3 = r''$, it will follow that $r = r_1 + r''$. The associative law (2.2.2a) holds for any elements of R, then.

2.2.2.1 Notice, incidentally, that the definition of addition in 2.1.1 above does *not* tell us how to find a sum of *three* real numbers. Now that we have defined such a sum, and have the associative law (2.2.2a), we can in an obvious manner define by induction the sum of any finite number of real numbers and show that the *order* of addition is immaterial. Thus $r_1 + \cdots + r_n + r_{n+1} = (r_1 + r_2 + \cdots + r_n) + r_{n+1}$, etc.

For multiplication analogous statements may be made; in particular, the associative law of multiplication of real numbers is $(r_1 \times r_2) \times r_3 = r_1 \times (r_2 \times r_3)$.

2.2.3 The distributive law

In elementary algebra the reader probably first encountered the rule:

$$a \times (b + c) = (a \times b) + (a \times c), \tag{2.2.3a}$$

and subsequently spent hours of labor in exercises illustrating "removal of parentheses." Relation (2.2.3a) is called the *distributive law*, and from its validity for integers and finite decimals may be proved its validity for any elements a, b, c (positive or not) of R (using $+$ and $\times$ as they have been defined above).

The distributive law, stated here chiefly for its theoretical importance, is not just an abstract principle useful for tormenting unwilling students of elementary algebra. Unfortunately the latter are all too frequently left with no understanding of its importance in ordinary calculations. No experienced computer, for example, would calculate the cost of 3 articles costing 98 cents apiece by multiplying 3 by .98; he would automatically use the distributive law:

$$3 \times (1.00 - .02) = 3.00 - .06 = 2.94$$

2.2.4 The monotonic law

Probably least manifest of all the rules of common arithmetic is the *monotonic law*:

$$(x < y) \Rightarrow [(x + a) < (y + a)]. \tag{2.2.4a}$$

Interpreted, this means that, if x, y, and a are any three integers or finite decimals, and $x < y$, then $(x + a) < (y + a)$. Its validity for all real numbers easily follows from its validity for finite decimals.

2.2.5 The Laws of Signs we have already mentioned above (2.1, 2.1.4) and extended *by definition* to elements of R.

2.2.6 This completes our discussion of the order type of R and operations with elements of R. Further light will be thrown on these matters in the sequel, not only by the frequently used alternative approach to the real number system which is described below, but also by the axiomatic characterization which will be given (VII 2.7.1). From the latter we shall see how fundamental properties of R may be proved as theorems.

It will be noted that we have not considered *infinite sums* ("series") or *infinite products* of elements of R. To do so would take us into the proper domain of *mathematical analysis*, and from this we have to abstain, although it is a natural continuation of what we have started above.

3 The real number system as based on the Peano axioms

Because it is one of the most commonly used methods of introducing the real number system, we sketch (leaving out many details) in this section the definition of real numbers, and operations with them, on the basis of the "Peano axioms." An interesting feature of this approach is that it simulates, in many respects, the development above. However, instead of commencing with the rational numbers (assumed already known), it is based on the famous system of axioms due to Dedekind and Peano.†

3.1 The positive integers and operations with them

The Peano axioms deal with a set $\mathscr{N}$ of undefined elements called *numbers* and an undefined binary relation, s, between numbers: $x\,s\,y$ is read "x is a successor of y." The axioms are:

(1) $\mathscr{N}$ contains a number, 1, such that $1\,s\,x$ holds for no $x \in \mathscr{N}$.

(2) If $x \in \mathscr{N}$, there exists a unique $y \in \mathscr{N}$ such that $y\,s\,x$; y is called the *successor* of x.

(3) If $y\,s\,x$ and $y'\,s\,x'$, and y is the same number as y', then x and x' are the same number.

(4) (Mathematical induction principle.) If G is a subset of $\mathscr{N}$ such that: (i) $1 \in G$ and (ii) if $x \in G$ and $y\,s\,x$ then $y \in G$; then $G = \mathscr{N}$.

Since the successor of a given number is unique, we can define a symbol $S(x)$ to denote the successor of x; $S(x)$ is a single-valued function over $\mathscr{N}$ with values in $\mathscr{N}$, or a mapping of $\mathscr{N}$ into $\mathscr{N}$ [not *onto* $\mathscr{N}$ because of axiom (1)] in the sense of IV 3.2.3.1.

3.1.1 Definition of addition. To define addition, we first define, for all x, $x + 1$ to be $S(x)$:

(3.1.1a) $$x + 1 = S(x), \qquad x \in \mathscr{N}.$$

Then, for all y, we define:

(3.1.1b) $$x + S(y) = S(x + y).$$

† See Peano [P_2]; also arithmetical portions of Peano [P_3], as for instance [P_3; II, § 2, 1]. Also compare Dedekind [D_2; § 6].

Definitions (3.1.1a) and (3.1.1b) induce a definition, by the mathematical induction principle, of $x + y$ for all $x, y \in \mathcal{N}$.†

3.1.2 Definition of multiplication. With addition defined, multiplication may now be defined in terms of addition: We first define $1 \times y = y$ for all $y \in \mathcal{N}$, and then for all $x \in \mathcal{N}$ we define $S(x) \times y$ to be $(x \times y) + y$. This defines $x \times y$ for all $x, y \in \mathcal{N}$ by the mathematical induction principle.

3.1.3 If we set up a (1-1)-correspondence between the elements of the set $\mathcal{N}$ of Peano's axioms and the natural numbers, in which the "1" of axiom (1) is paired with the natural number 1; and, having paired a number x of the set $\mathcal{N}$ of Peano's axioms with a natural number n, we then pair $S(x)$ with the natural number $n + 1$; it will then be seen that we have an isomorphism between any collection $\mathcal{N}$ satisfying the Peano axioms and the set N of natural numbers not only with respect to the axioms, but with respect to $+$ and $\times$, where for the natural numbers these operations have their usual meanings. This was obviously the motivation of the definitions in 3.1.1 and 3.1.2. For every number x we now define $x < S(x)$; stipulating transitivity for this relation, we extend it to all pairs of numbers in such a way as to satisfy the simple order axioms. The set $\mathcal{N}$ of Peano's axioms then becomes of order type ω, and we could use the usual symbols $1, 2, 3, \cdots$, to denote the elements of $\mathcal{N}$. This is the order which is induced by the correspondence indicated above between the numbers and the natural numbers, and corresponds to what we call the "natural order" of the natural numbers.

3.2 Positive rationals and operations with them

To define a set $\mathcal{F}^+$ of positive rationals on the basis of the set $\mathcal{N}$ of numbers postulated by the Peano axioms, we may proceed as follows: Let M be the collection of all ordered pairs (a, b), $a, b \in \mathcal{N}$. Since we have defined multiplication for elements of $\mathcal{N}$ (3.1.2), we may set up an equivalence relation $\approx$ between the elements of M as follows: $(a, b) \approx (c, d)$ if and only if $a \times d = b \times c$. Then $\mathcal{F}^+$ is defined to be the set of all equivalence classes of M with respect to $\approx$; i.e., the elements of the class decomposition of M corresponding to $\approx$ (see II 8.5). We call each element of $\mathcal{F}^+$ a *positive rational number.*

† As pointed out in Landau [La$_1$], this definition must be justified by (1) showing that for a fixed x there is at most one possibility of defining $+$ for all y so as to satisfy (3.1.1a) and (3.1.1b); and (2) showing that for every x such a possibility does in fact exist. The reader may do this as an exercise, or consult the proof of Theorem 4 in Landau's book, *loc. cit.* For a revealing and penetrating discussion of addition, multiplication, etc., as well as justification of the mode of definition by mathematical induction, see Henkin [c].

3.2.1 It is instructive to compare $\mathscr{F}^+$ with the model F^+ described in the Introductory Remarks. In the latter case, a positive rational number was a pair p/q of natural numbers prime to one another; alternative symbols were of the form np/nq. In $\mathscr{F}^+$, however, the positive rationals are equivalence classes, each of the form $\{(x, y) \approx (a, b) \mid a, b \text{ fixed}, x \times b = a \times y\}$.

3.2.2 We must define addition and multiplication in $\mathscr{F}^+$ as well as a simple order. Let $\{(a, b)\}$ and $\{(c, d)\}$ be elements of $\mathscr{F}^+$. Then we define $\{(a, b)\} + \{(c, d)\}$ to be $\{[(a \times d) + (b \times c), (b \times d)]\}$; of course, it should be checked that this is independent of the choice of the representative elements† (a, b), (c, d) in the two given equivalence classes. Multiplication is defined by letting $\{(a, b)\} \times \{(c, d)\} = \{(a \times c, b \times d)\}$—again checking to see that the definition is independent of the choice of the representative elements (a, b), (c, d). To define the simple order, let $\{(a, b)\} < \{(c, d)\}$ mean that $a \times d < b \times c$ (where in the latter relation "$\times$" and "$<$" are as defined in $\mathscr{N}$).

3.2.3 The reader will note that our guide in making these definitions is the model F^- of the Introductory Remarks; in arithmetic we learned that $a/b + c/d = (ad + bc)/bd$ and $a/b \times c/d = ac/bd$; and $a/b < c/d$ if and only if $ad < bc$.

3.2.4 In the set $\mathscr{F}^+$ with operations and order as just defined, it will be noted that if for each natural number n, one maps the number n into the element $\{(n, 1)\}$ of $\mathscr{F}^+$, then the image of N so obtained in $\mathscr{F}^+$ is isomorphic to N in the sense that the mapping preserves $+$, $\times$, and $<$. Moreover, we can go further, in that if p/q is a positive rational of the model F^+ of the Introductory Remarks, we may map p/q into $\{(p, q)\}$ (where the "p" and "q" of the latter are the images defined by the above mapping of N), and obtain in this fashion an isomorphism between F^+ and $\mathscr{F}^+$ preserving $+$, $\times$, and $<$.

3.3 The positive real numbers and operations

We next describe how we can go from $\mathscr{F}^+$ to a set $\mathscr{R}^+$ of "positive real numbers." Several methods for doing this were developed by the nineteenth century pioneers Dedekind, Weierstrass, Cantor, and Meray. The simplest for our purposes is a slight modification of the "Dedekind Cut" (see V 2.2). Define a *half-cut* of $\mathscr{F}^+$ to be a non-empty proper subset $\mathscr{F}_1$ of $\mathscr{F}^+$ such that (1) $\mathscr{F}_1$ has no last element and (2) every section (V 3.2.1) of $\mathscr{F}^+$ determined by an element of $\mathscr{F}_1$ is a subset of $\mathscr{F}_1$. (It follows from this definition that if $\mathscr{F}_2 = \mathscr{F}^+ - \mathscr{F}_1$, then $[\mathscr{F}_1, \mathscr{F}_2]$ is a

† See the remark following the Corollary to Theorem 2 in II 8.5.

cut of $\mathscr{F}^+$ as defined in V 2.2). Each such half-cut of $\mathscr{F}^+$ is called a *positive real number*; denote the collection of all such by "$\mathscr{R}^+$."

If $\mathscr{F}_1$ and $\mathscr{F}_1'$ are two elements of $\mathscr{R}^+$, then $\mathscr{F}_1 + \mathscr{F}_1'$ is defined to be the set $\{x + x' \mid x \in \mathscr{F}_1, x' \in \mathscr{F}_1'\}$ and $\mathscr{F}_1 \times \mathscr{F}_1'$ the set $\{x \times x' \mid x \in \mathscr{F}_1, x \in \mathscr{F}_1'\}$; it can be shown that each of these sets is a half-cut of $\mathscr{F}^+$ and hence a positive real number, element of $\mathscr{R}^+$. It may be shown that of two half-cuts, one is always a subset of the other. We define $\mathscr{F}_1 < \mathscr{F}_1'$ to mean that $\mathscr{F}_1' - \mathscr{F}_1 \neq \emptyset$. The set $\mathscr{R}^+$ becomes simply ordered by this definition.

3.3.1 If $k_1k_2\cdots k_m.a_1a_2\cdots a_n\cdots$ is a fixed element r of the set R^+ of the Introductory Remarks, and we make a half-cut F_r in the set F^+ by assigning to it every positive rational number which is less than some $k_1k_2\cdots k_m.a_1a_2\cdots a_n$, then we can extend the isomorphism between F^+ and $\mathscr{F}^+$ described in 3.2.4 to an isomorphism between R^+ and $\mathscr{R}^+$. This is accomplished by letting r correspond to the half-cut in $\mathscr{F}^+$ whose elements were made to correspond to the elements of F_r in 3.2.4. Actually, the model so obtained furnishes the motive for the definitions of $+$, $\times$, and $<$ just given above.

3.4 The real number system

The final step to the real number system, as based on the Peano axioms, may be made by a "pairing and equivalence" method analogous to that used in passing from $\mathscr{N}$ to $\mathscr{F}^+$. We have defined addition and multiplication of elements of $\mathscr{R}^+$. Let us define an equivalence relation $\approx$ between ordered pairs (r, s) of elements of $\mathscr{R}^+$ as follows: $(r_1, s_1) \approx (r_2, s_2)$ if $r_1 + s_2 = r_2 + s_1$. The elements of the corresponding class decomposition of the ordered pairs of elements of $\mathscr{R}^+$ will be called *real numbers*, and their totality denoted by "$\mathscr{R}$." A real number $\{(r, s)\}$ for which a representative element (r, s) satisfies (1) $r = s$, is called *zero*; (2) $r < s$, is called *negative*; and (3) $s < r$, is called *positive*. The collection of these new "positive real numbers" of (3) form a system isomorphic with $\mathscr{R}^+$, and the entire collection $\mathscr{R}$ forms a system isomorphic with the model R of Sections 1 and 2.

We leave to the reader the definitions of $+$, $\times$, and $<$ for $\mathscr{R}$ which justify these assertions.

3.5 Remarks

This is as far as we shall go with the foundation of the real number system on the basis of the Peano postulates. The subject and the literature concerning it are large, and we have by no means exhausted the variety of methods that can be used. The methods that we have chosen to discuss are a fair sample, however, and probably the ones most commonly employed in teaching and in books on function theory. For supplementary material, as well as details that space limitations have forced us to omit, the reader is referred to such function theory classics as Hobson [Ho] and Pierpont [Pi]. The more elementary accounts in Klein [Kl] and Young [Y; X] are also recommended.

3.5.1 For one who reads German, an excellent summary of methods and bibliography will be found in F. Bachmann's article in *Encyklopädie der Mathematischen Wissenschaften* [Encyk]. In particular, the several uses of the "pairing and equivalence" method above are set in their proper perspective as special cases of a group-theoretic extension theorem (loc. cit., pp. 12ff). Citations to the original source of the method of cuts (Dedekind), half-cuts (Pasch, Russell, *et al.*), etc., will also be found therein.

For an extended and detailed derivation of the arithmetic and ordering of the real number system (as well as of the arithmetic of the complex number system) from the Peano axioms, see the classical work of Landau [La_2], now fortunately available in English [La_1].

4 The complex number system

We close this chapter with a few remarks about the complex number system. As is well known, complex numbers were first introduced to satisfy the demand for solutions to such equations as $x^2 + 1 = 0$, which cannot be satisfied by real numbers; with them, all algebraic equations attain solutions. The "informal" manner of introducing such new numbers is to set up a new "unit," i, and to call any "number" of the form $a + bi$, where $a, b \in R$, a "complex number." It is stipulated that the new numbers satisfy the same laws (commutative, distributive, etc.) as the real numbers (excepting the monotonic law, which has no meaning in the new, non-ordered system), and that $i^2 = -1$.

4.1 The complex number system as an "extension" of the real number system

We can, however, build on the system $\mathscr{R}$ of Section 3 in the following manner: Define a complex number as an ordered pair (a, b), a,b $\in \mathscr{R}$. Define addition by the rule

(4.1a) $$(a, b) + (c, d) = (a + c, b + d)$$

and multiplication by the rule

(4.1b) $$(a, b) \times (c, d) = [(a \times c) - (b \times d), (a \times d) + (b \times c)].$$

The commutative law, etc., follow from the same law for $\mathscr{R}$. And evidently the complex numbers contain a subcollection, namely $\{(a, b) \mid b = 0\}$, isomorphic with $\mathscr{R}$.

4.2 It will be noticed that if, in the above formulation, we denote (0, 1) by i (where 1 is the element of $\mathscr{R}$ such that r × 1 = r for all r ε $\mathscr{R}$), then, by the multiplication rule (4.1b), $i^2 = i \times i = (-1, 0)$, which corresponds in the above indicated isomorphism with the real number −1. Furthermore, with i so defined, the transition to the usual representation of complex numbers is made by the relations (a, b) = (a, 0) + (0, b) = (a, 0) + [(b, 0) × (0, 1)], where + and × are as defined in (4.1a) and (4.1b); the form (a, 0) + [(b, 0) × (0, 1)] corresponds to a + bi.†

SUGGESTED READING

Eves and Newsom [E-N; VII]
Hobson [Ho; I]
Huntington [Hu]
Pierpont [Pi; I–II]
Stoll [St; II, III]
Young [Y; X–XII]

PROBLEMS

1. Show that the relation < defined for R in 1.1 satisfies the simple order axioms.

2. In the last two sentences of the second paragraph of 1.2.6.1, we made assertions that certain sets M would have cardinal number c. It would have been sufficient to say "$\geqq c$." Why could we assert the former?

3. Show that, if we extended Definition V 3.5.1.1 in suitable manner to a definition of the product of two *simple order* types, then the example of 1.2.6.1 is such a product.

4. Can the example of 1.2.6.2 be represented as a product of two order types according to the definition suggested in Problem 3?

5. Prove the assertions in the last two sentences of the second paragraph of 1.2.6.2.

6. Show that the ordered set S of 1.2.6.1 with (1, 1) deleted and the ordered set Γ of 1.2.6.2 are not of the same order type. Can you show this by proving that for any (1-1)-mapping $f: S \to \Gamma$, there will exist $x, y \in S$ such that $x < y$ and $f(y) < f(x)$?

7. If L is a euclidean straight line ("infinitely long"), then, given any interval ab of L, there is a midpoint p of ab. Use this property, and the fact that L is the union of a denumerable set of such intervals ab, to prove that L has a countable separating set.

8. Would anything be gained, logically or otherwise, by using L as an interpretation of Λ rather than R, in order to prove the consistency of the system Λ?

9. Show that the set S and relation < of 1.2.6.1 satisfy the Dedekind cut axiom.

10. Show that the set Γ and relation < of 1.2.6.2 satisfy the Dedekind cut axiom.

11. Establish the existence of the number d of 2.1.3.

12. Carry out the details justifying the statements made in 3.2.4.

† Complete references, etc., may be found in [Encyk., I 1, Heft 2; 23ff]. See also Young [Y; XII] and Klein [Kl; 55ff], in which will also be found elementary expositions of quaternions.

13. Show that the sets defining $\mathscr{F}_1 + \mathscr{F}_1'$ and $\mathscr{F}_1 \times \mathscr{F}_1'$ in 3.3 are actually half-cuts of $\mathscr{R}^+$; and that the definition of "$<$" in 3.3 satisfies the simple order axioms.

14. Show that by using the (1-1)-correspondence between N and F in IV 1.2, one may arrange the natural numbers in the order type η of the rational numbers (in the natural order of the latter). Hence, show that by using half-cuts in this ordering of N, we can immediately go to the order of $\mathscr{R}$.

15. Prove the categoricalness of the Peano axioms.

16. Set up a (1-1)-correspondence between F (the set of all rational numbers) and A (the set of all algebraic numbers) which preserves order. [*Hint:* Cf. the method used in 1.4.4 to set up a correspondence between F and S'.]

17. Do the set of all irrational real numbers and the set of transcendental real numbers have the same order type (when they are considered as having their natural order as real numbers)?

VII

Groups and Their Significance for the Foundations

Up to this point we have done very little of an algebraic nature. Of course, as a branch of mathematics, algebra is affected by such basic matters as logical laws, number, the infinite, etc., which have been discussed in the earlier chapters. And we have found minor uses for algebra, such as in the definition and enumeration of algebraic numbers (IV 1.3).

In the present chapter we shall take a new road, so to speak, and show how notions that have been developed in modern algebra have influenced and clarified the foundations of mathematics. In particular, the theory of the real number system can not only be placed on a foundation that many consider more satisfactory than is afforded by the methods of Chapter VI, but we shall also see many generalizations that the new approach makes possible and which lead to considerable enrichment of mathematics itself as well as of the particular field of algebra. Furthermore, we shall see geometry in a new perspective, not heretofore possible from the narrower, classical point of view.

1 Groups

The entering wedge, historically, was the notion of a group, which had for a long time been in use in various special forms in algebra and geometry. The ultimate formulation of the properties of an abstract group was an achievement of the axiomatic method, and furnishes an excellent example of the "economic" advantage of the method discussed in II 4.10 and II 6.2. In different branches of mathematics, *operations* such as those of addition and multiplication in elementary arithmetic, and such as those exemplified by the combining of transformations in geometry, had come to be studied from an abstract point of view, and it became apparent that there was an underlying common idea. This led inevitably to the axiomatic definition of group, and thence to a large body of theorems constituting *group theory*, available for application wherever groups could be recognized as playing a role in any field of mathematics. Later demands of modern mathematics led even to generalizations of groups, but we shall mention these more fully later.

1.1 Operations

In Chapter VI we called addition, multiplication, etc., "operations of arithmetic," but thus far we have not discussed the notion of operation for its own sake. As most commonly used, an operation in a set S is nothing but a single-valued function $f(x, y)$, $x,y \in S$. It can be most easily brought within the frame of IV 3.2.3.1 by use of the notion of product set:

1.1.1 Definition. If $S_1, S_2, \cdots, S_n$ are sets, then the *cartesian product* or *product set* (sometimes called simply *product*) $S_1 \times S_2 \times \cdots \times S_n$ is the collection of all ordered n-tuples $(x_1, x_2, \cdots, x_n)$ in which $x_i \in S_i$, $i = 1, 2, \cdots, n$. In case $S_1 = S_2 = \cdots = S_n = S$, then the product set is the collection of all ordered n-tuples $(x_1, x_2, \cdots, x_n)$ in which every $x_i \in S$; in this case the product set may be denoted by the symbol S^n.

1.1.2 Examples. The coordinate plane is the product set R^2, and, in general, coordinate n-space is the product set R^n, with, however, certain additional conditions which give it special "spatial" attributes [notably, the distance formula; thus, in the plane, the distance between (x_1, y_1) and (x_2, y_2) is $\sqrt{(x_1 - x_2)^2 + (y_1 - y_2)^2}$].

1.1.2.1 From the point of view of product set, a binary relation between elements of a set S is simply a subset of S^2; a mapping (IV 3.2.3.1) of a set S into a set S' (or single-valued function of S with values in S') is a special kind of subset of $S \times S'$. It should not be forgotten, however, that the notion of order relation (see II 7 and VIII 8.2.7.1) is basic to that of product set.

1.1.3 Definition. If S is any set, then an *operation* (more specifically, a *binary operation*) in S is a single-valued function $\circ$ defined over S^2. For any element (x, y) of S^2, the value of this function may be denoted by $x \circ y$. And that $x \circ y$ is z may be denoted by the "equation" $x \circ y = z$.

In the case of groups the value of the function is again in S, so that the function becomes a mapping (as defined in IV 3.2.3.1) of S^2 into S.

If, for all $x,y \in S$, $x \circ y = y \circ x$, then the operation is called *commutative*.

1.1.4 Examples. The addition function S defined for $\mathcal{N}$ in VI 3.1.1 is a commutative operation in $\mathcal{N}$ mapping $\mathcal{N}^2$ onto the set $\mathcal{N} - \{1\}$, since always $x + y$ is an $S(z)$ for some z, while 1 is not an $S(z)$.

1.1.4.1 If N is the set of natural numbers and $x \div y = x/y$—i.e., the real fraction $x/y \in F$—then the so-defined division is an operation mapping N^2 into F. But this operation is not commutative.

1.2 Definition of an abstract group†

By a *group* is meant any collection G and operation $\circ$ in G such that:

1. (Closure.) For all $x,y \in G$, $x \circ y$ is an element of G. (In other words, the given operation is a mapping of G^2 into G.)

2. (Associative law.) For all $x,y,z \in G$,

$$(x \circ y) \circ z = x \circ (y \circ z).$$

3. (Existence of an identity.) There exists an element i of G such that, for all $x \in G$, $x \circ i = x$. The element i is called an *identity*, or *identical element*, of the group.

4. (Existence of inverse elements.) If $x \in G$, then there exists an element x^{-1} of G such that $x \circ x^{-1} = i$. The element x^{-1} is called an *inverse*, or *inverse element* of x.

If G has n elements, $n \in N$, then we call G a group of *order* n with respect to the operation, or simply a *finite group* if the specific number of elements is not of importance. If G has infinitely many elements, then we call G of *infinite order*, or simply an *infinite group*. If, for all $x,y \in G$, $x \circ y = y \circ x$, then we call G *commutative*, or *abelian*.

1.2.1 Examples. The ordinary integers of arithmetic with $+$ as operation form a group in which 0 is the identity, the inverse of any integer being the negative of that integer.

1.2.1.1 The rational numbers, without zero, and $\times$ as an operation form a group in which 1 is the identity, the inverse of p/q being q/p.

1.2.1.2 The integers 1, -1 with $\times$ as operation form a group of order 2. As a subset of the ordinary complex number system, 1, i, -1, $-i$ with $\times$ as operation form a group of order 4. And in general, if $r_1, r_2, \cdots, r_n$ are the nth roots of 1, then with $\times$ as operation they form a group of order n. Since the integer 1 with $\times$ as operation is a group of order 1, it follows that a group of order n exists for every natural number n.

† A great deal of work has been done on axiomatic definitions of *group*. The axioms given here form one of the most commonly used systems (except that frequently the commutativity of the identity with all elements and that of an element with its inverse element are assumed—which we prefer to prove by way of exemplification of the derivation of group-theoretic theorems). As they are given here, the axioms are capable of considerable weakening, incidentally. See, for example, Huntington [c].

1.2.1.3 Let $\{R_r\}$ be the set of all rotations of a circle in the plane about its center, where the index r has all values such that $0 \leqq r < 1$ and R_r denotes the rotation through the angle $r \cdot 360°$. Introduce an operation $\circ$ such that $R_r \circ R_{r'} = R_{r+r'}$, where $r + r'$ is to be addition "mod 1" (meaning that, if $r + r' \geqq 1$, then we are to substitute $r + r' - 1$ for the value of $r + r'$; thus, if $r = 0.7$ and $r' = 0.8$, then $R_{r+r'}$ is $R_{0.5}$). Geometrically, $R_r \circ R_{r'}$ means a rotation through $r \cdot 360°$ followed by a rotation through $r' \cdot 360°$ (or vice versa), the end result being the same as if one rotation R_k were made, with k the number "$r + r'$ mod 1." The collection $\{R_r\}$ with the operation so defined forms a group, called the *group of rotations of a circle*.

1.2.1.4 Let G denote the set of all transformations

$$\begin{aligned} x' &= ax + by + e \\ y' &= cx + dy + f \end{aligned} \qquad (a,b,c,d,e,f \in R) \tag{1.2.1.4a}$$

of the coordinate plane, each of which transforms a point (x, y) into a point (x', y') whose coordinates are related to those of (x, y) by (1.2.1.4a). We restrict (1.2.1.4a), however, by requiring $ad \neq bc$. If T_1 and T_2 are two such transformations, then by $T_1 \circ T_2$ we may denote the transformation obtained by first effecting T_2 and then effecting T_1; simple algebraic considerations show that the transformations so combined are equivalent to a single transformation whose coefficients a, b, c, d are determinate. The set G and this operation form a group called the *affine group* of the plane.

1.2.1.5 Thus far the groups exemplified, excepting 1.2.1.4, have all been abelian. A simple example of a non-abelian group is the set of all non-singular square matrices with elements in F, say, of order n ($\geqq 2$), with the usual multiplication of matrices as the operation. The non-abelian character is shown by the case $n = 2$ and the matrices

$$M = \begin{pmatrix} a & b \\ c & d \end{pmatrix} \qquad N = \begin{pmatrix} 1 & 1 \\ 0 & 1 \end{pmatrix},$$

where $M \times N = \begin{pmatrix} a & a+b \\ c & c+d \end{pmatrix}, \qquad N \times M = \begin{pmatrix} a+c & b+d \\ c & d \end{pmatrix}.$

1.2.1.6 Another non-abelian example, important in its applications, is furnished by the so-called *substitution groups*, alternatively called *permutation groups*. Consider, for example, the ordered pair (a, b), and imagine an algebraic expression (e.g., a polynomial) in which a and b occur in various powers. The substitution of b for a, and a for b, in this expression, would presumably change its form unless the form is "symmetric" in a and b; the substitution would itself give a test for such symmetry. Now this substitution may be represented by the matrix

(1.2.1.6a)
$$\begin{pmatrix} a & b \\ b & a \end{pmatrix},$$

in which (a, b) occurs in the first row and (b, a) in the second. There being only *two* letters in the pair (a, b), there are only two possible types of substitutions, barring repetitions, namely (1.2.1.6a) and the so-called identity substitution,

(1.2.1.6b)
$$\begin{pmatrix} a & b \\ a & b \end{pmatrix}.$$

If we use an ordered triple (a, b, c), then *six* substitutions are possible:

(1.2.1.6c)
$$S_1 = \begin{pmatrix} a & b & c \\ a & b & c \end{pmatrix}, \quad S_2 = \begin{pmatrix} a & b & c \\ b & a & c \end{pmatrix}, \quad S_3 = \begin{pmatrix} a & b & c \\ c & b & a \end{pmatrix},$$
$$S_4 = \begin{pmatrix} a & b & c \\ a & c & b \end{pmatrix}, \quad S_5 = \begin{pmatrix} a & b & c \\ c & a & b \end{pmatrix}, \quad S_6 = \begin{pmatrix} a & b & c \\ b & c & a \end{pmatrix}.$$

And, of course, with n letters, $n!$ substitutions are possible, since in the second row of the matrix $n!$ permutations of the n letters are possible. An operation can be set up by the rule that, if S_1 and S_2 are substitutions, then $S_1 \circ S_2$ is the result of the substitution S_1 followed by the substitution S_2. For example,

(1.2.1.6d)
$$\begin{pmatrix} a & b & c \\ b & a & c \end{pmatrix} \circ \begin{pmatrix} a & b & c \\ a & c & b \end{pmatrix} = \begin{pmatrix} a & b & c \\ c & a & b \end{pmatrix}.$$

With such a definition of the operation, the collection of $n!$ substitutions using n letters becomes a group, the so-called *symmetric group of degree n*. That this group is generally non-abelian is exemplified by reversing the order of the substitutions in the left-hand member of equation (1.2.1.6d).

1.2.2 If G is a group, then any subset of G which, using the same operation, is again a group, is called a *subgroup* of G. Thus the pair S_1, S_2 of (1.2.1.6c) form a subgroup of the symmetric group of degree 3; so also do the pairs S_1, S_3 and S_1, S_4. The three substitutions S_1, S_5, S_6 of (1.2.1.6c) form a subgroup, of order 3, of the symmetric group of degree 3.

1.2.3 In a group G we may define, for any $g \varepsilon G$ and $H \subset G$, sets $g \circ H = \{x \mid x = g \circ h, h \varepsilon H\}$ and $H \circ g = \{x \mid x = h \circ g, h \varepsilon H\}$. If H is a subgroup of G such that $g \circ H = H \circ g$ for all $g \varepsilon G$, then H is called a *normal subgroup* or *normal divisor* of G. If G is abelian, every subgroup is, of course, normal. Note, however, that $g \circ H = H \circ g$ does not necessarily imply that $g \circ h = h \circ g$ for every $h \varepsilon H$.

For example, we found in 1.2.1.6 that the substitutions S_1, S_2 form a subgroup of the symmetric group of degree 3; but this subgroup is not a normal subgroup. On the other hand, the substitutions S_1, S_5, S_6 do form a normal subgroup H; note, however, that although $S_4 \circ H = H \circ S_4$, it is not the case that for the individual element S_5 of H, $S_4 \circ S_5 = S_5 \circ S_4$.

The importance of the notion of normal subgroup lies in the fact that if H is a normal subgroup of a group G, then the subsets $H \circ g$, $g \varepsilon G$, form the elements of a new group, denoted by G/H and called the *factor group of G mod H.* ("Mod" is an abbreviation for "modulo.") The group operation in G/H is defined by the relation $(H \circ g) \circ (H \circ g') = H \circ (g \circ g')$. The elements of G/H are called *cosets*, or *cosets of G mod H.*† For example, when G is the symmetric group of degree 3 and H the normal subgroup referred to in the preceding paragraph, then G/H has only two cosets and is another example of the many isomorphic groups of order 2 to be discussed in 1.4.1.

1.2.3.1 Integers mod m. An important and widely used factor group occurs when G is the set of all integers with the operation $+$, and H is the set of all integers of the form mn, where m is a fixed natural number $\geqq 2$. The group G/H has exactly m elements. The integers $0, 1, 2, \cdots, m - 1$ are elements of its respective cosets, and it is customary to denote these cosets by these same numerals. In particular, when $m = 2$, H is the set of all even integers, and G/H consists of two elements "0" and "1" (consequently "$1 + 1 = 0$"); another instance of the abstract group of order 2 (see 1.4.1 below).

1.3 Some fundamental properties of groups

For our purposes, we need only a few elementary properties of groups. These we shall state as theorems to be proved on the basis of the axioms of 1.2. Some of the proofs will be left as exercises, in order that the reader may "get a feeling" for the methods used in abstract group theory. The latter make an excellent example, incidentally, of the methods used in modern algebra, as well as of the rigorous proof of theorems from axioms.

1.3.1 Lemma. *If $a,b,c \varepsilon G$ and $a = b$, then $a \circ c = b \circ c$ and $c \circ a = c \circ b$; and $a \circ c$, $c \circ a$, etc., are all elements of G.*

Proof. By definition, $a = b$ means that a and b denote the same element e of G. By the definition (1.1.3) of operation and axiom 1,

† When H is not normal, the sets $H \circ g$ are called *right cosets of G mod H*, and the sets $g \circ H$ are called *left cosets of G mod H.*

$e \circ c$ is a unique element of G and hence $a \circ c = b \circ c$. The other relation follows similarly.

1.3.2 Lemma (Right-hand cancellation law). *If* $a,b,c \,\varepsilon\, G$ *and* $a \circ c = b \circ c$, *then* $a = b$.

Proof. From $a \circ c = b \circ c$ follows, by Axiom 4 and Lemma 1.3.1, that $(a \circ c) \circ c^{-1} = (b \circ c) \circ c^{-1}$. Axiom 2 allows us to write $a \circ (c \circ c^{-1}) = b \circ (c \circ c^{-1})$, or, using Axiom 4, $a \circ i = b \circ i$. This, by Axiom 3, becomes $a = b$.

1.3.3 Theorem (Commutativity of i). *For all* $a \,\varepsilon\, G$, $a \circ i = i \circ a$.

Proof. With a^{-1} as defined in Axiom 4, the following relations hold by virtue of the axioms cited:

$$\begin{aligned} (i \circ a) \circ a^{-1} &= i \circ (a \circ a^{-1}) && \text{Axiom 2} \\ &= i \circ i && \text{Axiom 4} \\ &= i && \text{Axiom 3} \\ &= a \circ a^{-1}. && \text{Axiom 4} \end{aligned}$$

By Lemma 1.3.2, we then have $i \circ a = a$. But also $a \circ i = a$ by Axiom 3, and, since logical identity is a transitive relation, $a \circ i = i \circ a$.

1.3.3.1 Corollary. *The group G has only one identity.*

Proof. For let i and j be identities. Then $i \circ j = i$ by Axiom 3. Also, $i \circ j = j \circ i$ by Theorem 1.3.3, and $j \circ i = j$ by Axiom 3, thus giving $i \circ j = j$. That $i = j$ now follows.

1.3.4 Theorem (Commutativity of inverses). *For each* $a,a^{-1} \,\varepsilon\, G$, $a \circ a^{-1} = a^{-1} \circ a$.

Proof. The theorem follows from the relations

$$\begin{aligned} (a^{-1} \circ a) \circ a^{-1} &= a^{-1} \circ (a \circ a^{-1}) && \text{Axiom 2} \\ &= a^{-1} \circ i && \text{Axiom 4} \\ &= i \circ a^{-1} && \text{Theorem 1.3.3} \end{aligned}$$

and Lemma 1.3.2.

1.3.4.1 Corollary (Left-hand cancellation law). *If* $a,b,c \,\varepsilon\, G$ *and* $c \circ a = c \circ b$, *then* $a = b$.

1.3.4.2 Corollary (Uniqueness of inverse). *For each* $a \,\varepsilon\, G$, *the inverse* a^{-1} *is unique.*

1.3.4.3 Corollary. *For each* $a \,\varepsilon\, G$, $(a^{-1})^{-1} = a$.

1.3.5 Theorem. *For arbitrary* $a,b \,\varepsilon\, G$, *the relations* $a \circ x = b$, $y \circ a = b$ *are satisfied by unique elements* x, y, *respectively, of* G.

1.4 Isomorphic groups

If G and H are groups which are isomorphic (II 4.4.4) with respect to the group axioms (1.2), then G and H are called *isomorphic* groups. Two groups G and H are isomorphic, then, if there exists a (1-1)-correspondence between their elements which preserves the group operation, i.e., denoting elements of G by symbols g_i and corresponding elements of H by symbols h_i, if $g_1 \circ g_2 = g_3$, then $h_1 \circ h_2 = h_3$, and conversely.

1.4.1 Operation table for a finite group

In the case of finite groups, isomorphism between groups may be shown by examination of their *operation tables*. Suppose that G is a finite group of order n with elements $x_1, x_2, \cdots, x_n$, where x_1 is the identity. We may then make a table of n rows and n columns in which the element in the ith column and jth row is $x_i \circ x_j$. The first row and the first column will contain the elements $x_1, x_2, \cdots, x_n$ in this order (cf. Theorem 1.3.3). Also, in each row, and in each column, no element of G is repeated (as can be shown by use of the cancellation laws), and hence each row and each column contains all elements of the group.

For example, the operation table for a group of order 2 will look like the following, if we drop the "x's" and use only the subscripts:

	1	2
1	1	2
2	2	1

The element in the lower right-hand corner is perforce 1, since each row must contain all elements of the group. It follows that all groups of order 2 are isomorphic. Thus, as abstract groups, the groups of order 2 of 1.2.1.2 and 1.2.2 are the same. A like statement holds for groups of order 3—only one such abstract group is possible—but this is not the case for groups of order 4 (see Problems 12 and 13). The lowest order of non-abelian groups is 6, viz., the order of the group isomorphic with the symmetric group of degree 3.

1.4.2 Relation to substitution groups

The importance of substitution groups is due partially to the following theorem:

1.4.2.1 Theorem. *Every finite group is isomorphic with some substitution group.*

Proof. Let M denote the operation table for a finite group G of order n, drawn up as indicated above. For $j = 1, 2, \cdots, n$, let S_j denote the substitution whose matrix has in its first row the numbers of the columns of M—i.e., $1, \cdots, i, \cdots, n$, and in its second row the elements of the jth row of M taken in order. Schematically, then,

$$S_j = \begin{pmatrix} 1 & \cdots & i & \cdots & n \\ j & \cdots & i \circ j & \cdots & n \circ j \end{pmatrix},$$

where $i \circ j$ indicates the subscript of the symbol for $x_i \circ x_j$.

Now, if S_j and S_k are two such substitutions,

$$S_j \circ S_k = \begin{pmatrix} 1 & \cdots & i & \cdots & n \\ j & \cdots & i \circ j & \cdots & n \circ j \end{pmatrix} \circ \begin{pmatrix} 1 & \cdots & i \circ j & \cdots & n \\ k & \cdots & (i \circ j) \circ k & \cdots & n \circ k \end{pmatrix}$$

$$= \begin{pmatrix} 1 & \cdots & i & \cdots & n \\ j \circ k & \cdots & (i \circ j) \circ k & \cdots & (n \circ j) \circ k \end{pmatrix}.$$

By the associative law, $(i \circ j) \circ k = i \circ (j \circ k)$, and hence $S_j \circ S_k = S_{j \circ k}$. It follows that, if N is the operation table for the substitutions $S_1, \cdots, S_j, \cdots, S_n$, then M and N are identical. These substitutions therefore form a subgroup of the symmetric group of degree n that is isomorphic with G.

1.4.2.2 It follows from Theorem 1.4.2.1 that, as abstract groups, one will find all the properties of finite groups exemplified in the substitution groups.†

1.5 Semigroups

The demands of modern mathematics frequently necessitate the study of sets with operations that do not have all the properties of a group. For example, if a set S and operation in S satisfy Axioms 1 and 2 of 1.2, then they form a so-called *semigroup*. If such a semigroup satisfies the right- and left-hand cancellation laws and has only a finite number of elements, it is also a group (see Problem 8), but for infinite groups this is no longer the case (as for instance in the case where S is the set of natural numbers and the operation is addition).

Other types of systems satisfying only part of the axioms of a group are found of use (as for example in number theory), but will not be discussed here.

† For an elementary exposition of substitution groups, see the series of articles by G. A. Miller [a]. For finite groups in general, see Matthewson [Ma] and Miller, Blichfeldt, and Dickson [M-B-D]. The student who reads German will find the little book of Baumgartner [Ba] easy to understand and very instructive.

2 Applications in algebra and to number systems

In number systems, such as the real and complex number systems, we deal with collections in which *two* operations are given. These operations are not independent, since they are related by the distributive law which we saw exemplified in the real number system (see VI 2.2.3). It is not strange, therefore, that in modern algebra we find great importance attached to the study of sets in which two operations, which we may denote by $+$ and $\times$, are given, satisfying distributive laws.

2.1 If we consider one of the simplest number systems, viz., that of the integers, we notice that with respect to addition, $+$, we have a group, but that with respect to multiplication, $\times$, the collection does not form a group because inverses do not exist. Let us study, then, some of the properties of an abstract system of this sort. Specifically, let X be a collection in which are given two operations denoted by $+$ and $\times$, respectively, and satisfying the following axioms:

1. X is a group with respect to $+$, its identity being denoted by 0 and the inverse of an element x by $-x$.†

2. With respect to the operation $\times$, X satisfies the closure axiom (Axiom 1) and the associative law (Axiom 2) for a group; i.e., X is a semigroup with respect to $\times$.

3. For all $x, y, z \in X$,

$$(3a) \qquad x \times (y + z) = (x \times y) + (x \times z)$$

$$(3b) \qquad (x + y) \times z = (x \times z) + (y \times z).$$

A set X and operations $+$, $\times$ in X satisfying these axioms will be denoted by $(X, +, \times)$. It may also be called a *system* $(X, +, \times)$.

2.2 On the basis of these axioms we can prove certain basic theorems of arithmetic.

2.2.1 Theorem. *If* $a \in X$, *then* $a \times 0 = 0 \times a = 0$.

Proof. Since 0 is the identity for the operation $+$,

$$0 + 0 = 0.$$

And, since X satisfies the closure axiom for a group, Lemma 1.3.1 holds and we can write

$$(2.2a) \qquad a \times (0 + 0) = a \times 0,$$

† We must be careful not to think here of "$-$" as an operation; it is only part of the symbol "$-x$" for an inverse with respect to $+$. We are reserving, in accordance with custom, the symbol x^{-1} for the inverse with respect to $\times$, which will be introduced later.

where both $a \times (0 + 0)$ and $a \times 0$ are elements of X. By (3a) of Axiom 3, (2.2a) becomes

(2.2b) $$(a \times 0) + (a \times 0) = a \times 0,$$

which can be written

$$(a \times 0) + (a \times 0) = (a \times 0) + 0.$$

And, since the left-hand cancellation law (1.3.4.1) holds in every group, we get

$$a \times 0 = 0.$$

The proof that $0 \times a = 0$ may be given analogously, except that (2.2a) becomes $(0 + 0) \times a = 0 \times a$, and the other half (3b) of the distributive law (Axiom 3) is used to obtain, analogously to (2.2b), the relation $(0 \times a) + (0 \times a) = 0 \times a$.

2.2.2 Theorem. *For all $x,y \varepsilon X$, $(-x) \times y = -(x \times y)$.*

Proof. By Theorem 2.2.1, $0 \times y = 0$. In this relation, 0 can be replaced by $x + (-x)$, giving

(2.2c) $$[x + (-x)] \times y = 0.$$

Applying part (3b) of Axiom 3 to relation (2.2c) gives

(2.2d) $$(x \times y) + [(-x) \times y] = 0.$$

Applying Corollary 1.3.4.2, relation (2.2d) shows that $(-x) \times y$ must be the inverse, with respect to $+$, of $x \times y$; i.e., $-(x \times y) = (-x) \times y$.

By a similar argument one can prove:

2.2.3 Theorem. *For all $x,y \varepsilon X$, $x \times (-y) = -(x \times y)$.*

2.2.4 Theorem. *For all $x,y \varepsilon X$, $(-x) \times (-y) = x \times y$.*

Proof. By Theorem 2.2.2, $(-x) \times y = -(x \times y)$ for all $x,y \varepsilon X$. Hence, replacing y by $-y$, we get $(-x) \times (-y) = -[x \times (-y)]$. By Theorem 2.2.3, this gives $(-x) \times (-y) = -[-(x \times y)]$. By Corollary 1.3.4.3, $-[-(x \times y)] = x \times y$.

2.2.5 Remark. Theorems 2.2.2 to 2.2.4 will be recognized as constituting the "Laws of Signs" for any system $(X, +, \times)$ satisfying the above axioms. As these "laws" are understood in elementary arithmetic, however, they are stated in 2.7.2 below.

2.3 Rings

One of the most important basic notions of modern algebra is that of a ring:

2.3.1 Definition. A system $(X, +, \times)$ satisfying Axioms 1 to 3 of 2.1 is called a *ring* if $+$ is commutative (1.1.3).

An important case is that of a ring in which there exists a *right-hand identity* with respect to the operation $\times$; i.e., an element "1" such that $x \times 1 = x$ for all $x \in X$.

2.3.2 Theorem. *If the system* $(X, +, \times)$ *has a right-hand identity with respect to* $\times$, *then* $+$ *is commutative.*

Proof. Denote the identity with respect to $\times$ by 1. Let $x, y \in X$. Then by Axiom 3 of 2.1 we have:

$$(x + y) \times (1 + 1) = [(x + y) \times 1] + [(x + y) \times 1] = x + y + x + y$$

$$(x + y) \times (1 + 1) = [x \times (1 + 1)] + [y \times (1 + 1)] = x + x + y + y.$$

By virtue of these relations and the associative law for $+$, we get

$$x + (y + x) + y = x + (x + y) + y,$$

which by 1.3.2 and 1.3.4.1 gives $x + y = y + x$.

2.3.3 Corollary. *If the system* $(X, +, \times)$ *of* 2.1 *has a right-hand identity with respect to* $\times$, *then it is a ring and is called a ring with a unit.*

2.3.4 Remark. We may prove similarly that both 2.3.2 and 2.3.3 hold if "right-hand" is replaced by "left-hand," a "left-hand" identity being an element 1 such that $1 \times x = x$ for all $x \in X$.

2.3.5 Examples. Examples of rings occur almost everywhere in mathematics.

2.3.5.1 The most common example is afforded by the integers and their elementary arithmetic; here we have a ring with a unit. The like holds also for the arithmetic of the set F, the set R, and the set of complex numbers. The even numbers $0, \pm 2, \cdots, \pm 2n, \cdots$ with $+$ and $\times$ in the arithmetic sense form a ring without a unit.

2.3.5.2 Various sets of functions, with $+$ and $\times$ suitably defined, also form rings. For example, the set of all real single-valued functions forms a ring with a unit. And, since sums and products of continuous functions are continuous, the set of all continuous single-valued real functions forms a ring—a "subring" of the set of all real single-valued functions. Another such subring is the set of all polynomials in a single real variable x with real coefficients.

2.3.5.3 For fixed natural number n, the set of all square matrices of order n with elements integers and $+$, $\times$ in the ordinary sense forms a ring with unit in which $\times$ is not commutative (compare 1.2.1.5).

A ring in which $\times$ is commutative is called a *commutative ring*.

2.3.5.4 From the group of integers mod 2 (1.2.3.1) we may obtain the commutative *ring of integers* mod 2 by defining "$1 \times 1 = 1$." This is the smallest possible type of non-degenerate ring with a unit, in that it contains only the identities for $+$ and $\times$. In a similar way we obtain the *ring of integers* mod m, $m > 2$; a product $i \times j$ $(i, j = 0, 1, 2, \cdots, m - 1)$ is defined by reducing the ordinary product $i \times j$ (in the ring of integers) by multiples of m to bring within the range $0 \leqq i \times j < m$. For example, the ring of integers mod 4 has elements 0, 1, 2, 3; and $2 \times 2 = 0$, $2 + 2 = 0$, $2 \times 3 = 2$, $2 + 3 = 1$, etc.

2.3.6 Ideals. With G and H as in 1.2.3.1, H is not only a subgroup of the *group G relative to the operation* $+$, but is an *ideal* in the *ring* $(G, +, \times)$. In general, any non-empty subset A of a commutative ring $(X, +, \times)$ such that (1) A is a subgroup of the group X relative to $+$, and (2) $a \in A$, $x \in X$ imply $(a \times x) \in A$, is called an *ideal* of the ring X. Ideals occupy a position in the theory of rings analogous to that of normal subgroups in the theory of groups; for just as a normal subgroup H of a group G yields the factor group G/H, so does an ideal A in a ring X yield cosets forming a ring called a *quotient ring*. To get the elements of the latter, we simply form the factor group X/A of the group X relative to $+$; addition of cosets is as before (for the factor group), and $(A + x) \times (A + y)$ is defined to be $A + (x \times y)$. (Cf. Problem 8.) For a discussion of ideals the reader is referred to Albert [Al; 252ff] and van der Waerden [Wa; v. I, § 16].

2.4 Integral domains

A striking difference between the ring of integers and the ring of integers mod 4 is that in the latter case we may have—as in $2 \times 2 = 0$—a product of factors equal to zero although neither of the factors is zero! This happens nowhere in the arithmetic of integers, rational numbers, real numbers, etc. Indeed, in elementary algebra we *assume* that such can never be the case, as for example in the solution of equations when we assert that, if $(x - 1)(x - 2) = 0$, then the only possible solutions are obtained from $x - 1 = 0$ or $x - 2 = 0$ since the product cannot otherwise be zero.

Another way of looking at this is to notice that the cancellation law of multiplication breaks down in the case of the ring of integers mod 4. Thus, $0 \times 2 = 2 \times 2$, but it does *not* follow that the "2" on the right of each member can be cancelled.

Evidently, then, we obtain a more restricted system and more nearly approximate the operations of ordinary algebra by ruling out these possibilities. Actually, it is sufficient to rule out one of them (see Problem

22). The result is what is variously called an *integral domain*, *domain of integrity*, or *ring of integrity*:

2.4.1 Definition. A ring with a unit in which $\times$ is commutative and satisfies the cancellation law† is called an *integral domain*.

Thus the ring of integers mod 3 is an integral domain, but the ring of integers mod 4 is not. The ring of continuous real functions mentioned in 2.3.5.2 is not an integral domain (see Problem 27).

2.5 Fields

In the notion of integral domain we have not yet achieved the systems, such as the rational number system, which are used in ordinary algebra. Thus, the ring of integers is an integral domain, but it does not allow for *division*. The latter is attained in the notion of "field."

2.5.1 Definition. An integral domain in which every element $x \neq 0$ has an inverse x^{-1} with respect to $\times$ is called a *field*. If a system $(X, +, \times)$ has all the properties of a field except that $\times$ is not commutative, then it is called a *non-commutative field*.

The case $X = F$, and $+$, $\times$ with their ordinary meanings of addition and multiplication, exemplifies a *field*.

2.5.2 Some properties of fields. In a field, division (defined below), except by 0, is possible.

2.5.2.1 Lemma. *If* $(X, +, \times)$ *is a non-degenerate ring with a unit* 1, *then* $0 \neq 1$.

Proof. As X is non-degenerate, there exists $x \varepsilon X$ such that $x \neq 0$. Then $x \times 1 = x \neq 0$, whereas, if 1 and 0 were the same element, we would have $x \times 1 = 0$ (by Theorem 2.2.1).

2.5.2.2 Lemma. *If* $(X, +, \times)$ *is a non-degenerate ring with unit* 1, *then* 0 *has no inverse with respect to* $\times$.

Proof. For an inverse of 0, say 0^{-1}, would imply $0 \times 0^{-1} = 1$, whereas $0 \times 0^{-1} = 0$ by Theorem 2.2.1; and thus a contradiction of Lemma 2.5.2.1 would result.

2.5.2.3 Lemma. *In a field, the elements different from* 0 *form a group with respect to* $\times$.

† When $\times$ is commutative, the right-hand and left-hand cancellation laws, being equivalent, are called "the cancellation law for non-zero elements." The term "integral domain" is frequently used for systems satisfying Definition 2.4.1 without the requirement of a unit.

2.5.2.4 Definition. If $(X, +, \times)$ is a field, and $x,y \in X$, then by $x \div y$, or x/y, is meant a unique element z of X such that $x = y \times z$.†

2.5.2.5 Theorem. *If $(X, +, \times)$ is a field, and $x,y \in X$, $y \neq 0$, then for all $x \in X$ the element $x \div y$ exists, being identical with the element $x \times y^{-1}$, where y^{-1} is the inverse of y with respect to $\times$.*

The proof is immediate, following from Lemma 2.5.2.3 and Theorem 1.3.5, if $x \neq 0$, and from the fact that there are no zero divisors (see Problem 22).

2.5.2.6 Theorem. *If $(X, +, \times)$ is a non-degenerate field and $x \in X$, then $x \div 0$ does not exist.*

Proof. For $x \neq 0$, the theorem follows immediately from 2.2.1. For $x = 0$, both $z = 0$ and $z = 1$ satisfy the equation $0 = 0 \times z$, so that no *unique* z exists.

2.5.2.7 Although 2.5.2.4 is the usual definition of $x \div y$, an alternative definition would be that $x \div y$ is $x \times y^{-1}$, where y^{-1} is the inverse of y with respect to $\times$. Then Theorem 2.5.2.5 follows directly from Corollary 1.3.4.2, and Theorem 2.5.2.6 is an immediate corollary of Lemma 2.5.2.2.

2.5.2.8 Just as one goes about seeking the possible types of abstract groups of a given finite order by drawing up the operation tables (cf. 1.4.1), so may one investigate the possible fields of a given finite number of elements by constructing addition and multiplication tables. (See Problem 23.) It is easy, using the properties of rings and fields established above, to show that for the case of a field of only two elements only one type is possible, the latter being exemplified by the ring of integers mod 2 (cf. 2.3.5.4).

2.6 Vector spaces

The reader who has had some elementary physics or mechanics is familiar with vectors and their use. For example, he may recall that, in the plane, a vector **V** represents both a direction and a magnitude, and that, if $r \in R$, then $r\mathbf{V}$ is a vector of magnitude r times as great as that of **V** (the direction of $r\mathbf{V}$ being the reverse of that of **V** if r is negative). The number r is called a "scalar," and $r\mathbf{V}$ multiplication of **V** by a scalar. And, if he recalls the "parallelogram of forces," he knows that vectors are added.

† If the field were non-commutative, we could of course define "right-hand" and "left-hand" division. This would necessitate special symbols for the two types of division, however.

In modern algebra these ideas find their generalization in the notion of a vector space over a field $\mathfrak{F}$:†

2.6.1 Definition. A *vector space* $\mathfrak{V}$ *over a field* $\mathfrak{F}$ is a collection of elements called *vectors*, such that (1) $\mathfrak{V}$ is an abelian group relative to an operation $+$, called addition, (2) for $a \,\varepsilon\, \mathfrak{F}$, $V \,\varepsilon\, \mathfrak{V}$, there exists a unique element of $\mathfrak{V}$ denoted by αV; (3) for $\alpha,\beta \,\varepsilon\, \mathfrak{F}$, $V \,\varepsilon\, \mathfrak{V}$, $\alpha(\beta V) = (\alpha \times \beta)V$, and $(\alpha + \beta)V = \alpha V + \beta V$;‡ (4) for $\alpha \,\varepsilon\, \mathfrak{F}$, $V_1,V_2 \,\varepsilon\, \mathfrak{V}$, $\alpha(V_1 + V_2) = \alpha V_1 + \alpha V_2$; (5) if 1 is the unit of $\mathfrak{F}$, then $1V = V$ for all $V \,\varepsilon\, \mathfrak{V}$.

2.6.2 In addition to the example of plane vectors, in which $\mathfrak{F}$ is R, an example very useful in algebra, and its application in modern mathematics is that in which the vectors are the mappings (IV 3.2.3.1) of some set S into a field $\mathfrak{F}$. In this case, $+$ is defined by the relation $(f + g)(x) = f(x) + g(x)$, $x \,\varepsilon\, S$, for each pair of mappings f and g; and $(\alpha f)(x) = \alpha \times f(x)$ for $\alpha \,\varepsilon\, \mathfrak{F}$, $x \,\varepsilon\, S$.

2.6.3 Some properties of vector spaces. (Throughout Section 2.6.3 we use $\mathfrak{V}$ and $\mathfrak{F}$ as defined in 2.6.1.) The proofs of the following two theorems are left to the reader:

2.6.3.1 Theorem. *If* $\alpha_1,\alpha_2,\cdots,\alpha_k \,\varepsilon\, \mathfrak{F}$ *and* $V_1,V_2,\cdots,V_k \,\varepsilon\, \mathfrak{V}$, *then* $\alpha_1 V_1 + \alpha_2 V_2 + \cdots + \alpha_k V_k$ *is a unique element of* $\mathfrak{V}$.

2.6.3.2 Theorem. *There exists a unique element of* $\mathfrak{V}$ *called the zero element, which may be denoted by* $0'$, *such that* $0V = 0'$ *for all* $V \,\varepsilon\, \mathfrak{V}$ (the "0" in "$0V$" being the zero element of $\mathfrak{F}$).

2.6.3.3 Definition. The elements V_ν of a set $\{V_\nu\}$ of vectors are called *linearly independent* if there exists no relation of the form

$$\alpha_1 V_1 + \alpha_2 V_2 + \cdots + \alpha_k V_k = 0', \tag{2.6a}$$

where $\alpha_1,\alpha_2,\cdots,\alpha_k \,\varepsilon\, \mathfrak{F}$, $V_1,V_2,\cdots,V_k \,\varepsilon\, \mathfrak{V}$, and not all $\alpha_i = 0$. When vectors $V_1, V_2,\cdots, V_k$ satisfy a relation of the form (2.6a) in which not all $\alpha_i = 0$, then $V_1, V_2,\cdots, V_k$ are called *linearly dependent*.

2.6.3.4 Definition. A set $\{V_\nu\}$ is called a *base* for $\mathfrak{V}$ if (1) its elements V_ν are linearly independent, and (2) if $V \,\varepsilon\, \mathfrak{V}$, then there exists a relation $V = \alpha_1 V_1 + \alpha_2 V_2 + \cdots + \alpha_k V_k$ in which $\alpha_i \,\varepsilon\, \mathfrak{F}$ and $V_i \,\varepsilon\, \{V_\nu\}$, $i = 1, 2, \cdots, k$. That every vector space has a base can be proved by Zorn's Lemma.

† As a matter of fact, vector spaces over rings, called "modules," are also studied in modern algebra; see van der Waerden [Wa; 46]. Sometimes the term "linear space" is used to mean vector space.

‡ If V_1 and V_2 are vectors, then $V_1 = V_2$ means that V_1 and V_2 are identical.

2.6.3.5 Definition. The cardinal number of elements in a base for $\mathfrak{V}$ is called the *dimension* of $\mathfrak{V}$.

As stated, 2.6.3.5 implies that the dimension of a vector space is a unique cardinal number. Inasmuch as in general there are many bases for a given vector space, this implication is by no means obvious. It is not difficult to prove the uniqueness of dimension if there exists some base that is finite, but the proof that all bases of a given vector space have the same cardinal number is not elementary in the case where the dimension turns out to be infinite.†

A simple example of a vector space of infinite dimension is the collection of all finite linear forms in variables x_n, $n = 1, 2, 3, \cdots$, with rational coefficients; for example, $2x_3 + \frac{1}{2}x_7 - x_{16}$ would be an element of this vector space. The dimension in this case is $\aleph_0$. If linear forms in only three variables x_1, x_2, x_3 are used, then the dimension is 3 since x_1, x_2, x_3 themselves form a base. The latter vector space is a "subspace" of the former.

2.7 The real number system

Let us next consider the application of the above notions to the definition of the real number system. In Chapter VI we showed how the real number system may be considered either as a construction based on the natural numbers, or on the Peano axioms. In the present section we return to the former point of view.

By way of making connection with the algebraic ideas above, we notice that the real number system is a field; moreover it is an *ordered field*, in that its structure is that of a linear continuum (VI 1.4.1, Definition 1.4.1.1). We make these two properties part of a new definition of the real number system:

2.7.1 Definition. A *real number system* is a field R in which there is a binary order relation $<$ with respect to which R forms a linear continuum,‡ satisfying the monotonic law (VI 2.2.4) and the requirement that $0 < x$, $0 < y$ imply $0 < x \times y$.

2.7.2 This definition furnishes an axiomatic foundation which is categorical and from which, therefore, the properties of the real number system are derivable. As an example, let us consider the derivation of the

† For the case of finite dimension, see Birkhoff and MacLane [B-M; 169]; for the infinite dimensional case a proof based on Zorn's Lemma will be found in S. Lefschetz, *Algebraic Topology*, Amer. Math. Soc. Coll. Pub., vol. 27, New York, 1942, pp. 73–74.

‡ If we were to write out each axiom for a field, each axiom for a linear continuum, etc., we would find that the separability axiom [VI 1.4.1, Axiom (3)] is not independent. See, for example, Albert [Al; 110ff].

Laws of Signs. As these were derived in 2.2 for a system $(X, +, \times)$, they are not quite in the specialized form in which they are understood in arithmetic. The latter form is already partially stated in the above definition in the requirement that

(2.7.2a) $$0 < x, 0 < y \Rightarrow 0 < x \times y.$$

We should be able to *prove* that $x < 0$, $y < 0 \Rightarrow 0 < x \times y$ also. We may arrange the proof as follows (R denotes a real number system as defined in 2.7.1 and x, y, z denote elements of R):

2.7.2.1 Lemma. If $0 < x$, then $-x < 0$; and, if $x < 0$, then $0 < -x$.
[Here, as in a system $(X, +, \times)$, $-x$ denotes the inverse of x with respect to $+$.]

Proof. If $0 < x$, then by the monotonic law $-x + 0 < -x + x$ or $-x < 0$. (The other half of the lemma is proved in similar fashion.)

2.7.2.2 Theorem. *If* $0 < x$ *and* $y < 0$, *then* $x \times y < 0$.

Proof. By Lemma 2.7.2.1, $0 < -y$. Hence, by (2.7.2a), $0 < x \times (-y)$. By Theorem 2.2.3, $x \times (-y) = -(x \times y)$. Hence $0 < -(x \times y)$, which by Lemma 2.7.2.1 and Corollary 1.3.4.3 implies $x \times y < 0$.

2.7.2.3 Theorem. *If* $x < 0$ *and* $y < 0$, *then* $0 < x \times y$.

The proof is left to the reader.

2.7.3 The above theorems give a good example of how we would go about studying the real number system from an axiomatic point of view. In this system such properties as the Laws of Signs then assume a status like that of the theorems of plane geometry, when the latter is properly axiomatized.†

2.8 This completes our discussion of the role which the group notion plays in modern algebra. We have also exhibited, to some extent, the important role which the axiomatic method plays in algebra. Indeed, no branch of modern mathematics makes greater or more effective use of the method. For extensions of the above ideas the reader is referred to Birkhoff and MacLane [B-M]; A. A. Albert [Al]; van der Waerden [Wa].

3 The group notion in geometry

We have already noted the existence of groups in geometry, such as the group of rotations of the circle (1.2.1.3) and the affine group of plane

† See also the comment by Young [Y; 111]. Also see the geometric "proof" of the Laws of Signs cited by the same author [Y; 112ff].

There has been reported at least one case, in one of the teachers' "service journals," of a high school principal complaining that none of his mathematics teachers could "prove" the Laws of Signs; it being obvious that his idea of "proof" was impossible of realization.

transformations (1.2.1.4). The former may of course be extended to the group of rotations, about a fixed point, of the entire plane. Such a group is another example of a group of transformations. Let us define, in general fashion, what is meant by a group of transformations:

3.1 **Definition.** Let S be any set. Then a *transformation* of S is a mapping of S onto S (IV 3.2.3.1) which is (1-1); i.e., if $y = f(x)$ denotes the mapping, then every element of S is a y for some x, and for only *one* x.

It follows from the definition that a transformation always has an inverse; in the above symbols, the inverse may be denoted by $x = f^{-1}(y)$. This is of importance in the attainment of a *group* of transformations:

3.2 **Definition.** If $f = f(x)$ and $g = g(x)$ are transformations of a set S, then by fg is meant the transformation $f(g(x))$;† fg is called the *product* of f and g.

3.3 **Definition.** If $\mathfrak{T}$ is a set of transformations of a collection S, such that, in terms of the operation fg defined in 3.2 the set $\mathfrak{T}$ forms a group, then $\mathfrak{T}$ is called *a group of transformations*, or *transformation group*.

In particular, we have:

3.4 **Theorem.** *If $\mathfrak{T}$ is a set of transformations of a collection S such that* (1) *the product of every two elements of $\mathfrak{T}$ is an element of $\mathfrak{T}$,* (2) *the identity transformation, $x = f_0(x)$, is an element of $\mathfrak{T}$, and* (3) *the inverses of elements of $\mathfrak{T}$ are elements of $\mathfrak{T}$, then $\mathfrak{T}$ is a transformation group.*

[Since Axioms 1, 3, and 4 of 1.2 are provided for in the statement of the theorem, it is only necessary to show that the associative law holds.]

3.5 Geometry according to Klein

In the applications to geometry, the set S in the above definitions is some kind of *space*. In modern mathematics, the term "space" has extremely broad connotations, the difference between "set" and "space" often being very slight, a "space" being simply a set to which certain special properties have been added. The most common property is that of having a *metric*, or *distance, function*.

3.5.1 The space of analytic geometry

Consider, for instance, the plane of analytic geometry. As a *set*, it is the collection S of all ordered pairs (x, y) of real numbers x, y; i.e., a product set (see 1.1.1 and 1.1.2). As a *space*, it is the set S together with the Pythagorean distance function $d(p, q)$ which expresses the distance

† Since we are discussing a special type of group here, we drop the $\circ$ in $f \circ g$ in favor of the "multiplicative" form fg customarily used for transformations.

between p and q; if p is (x_1, y_1) and q is (x_2, y_2), then we have the usual expression:

$$d(p, q) = +\sqrt{(x_1 - x_2)^2 + (y_1 - y_2)^2}. \tag{3.5a}$$

In analytic geometry, when we draw the customary right triangle relative to rectangular axes and state that $d(x, y)$ is the length of the hypotenuse of this triangle, we really assume (3.5a) and thereby define our space. All the other properties of the analytic plane follow from this definition.

Similarly, to get the space of three-dimensional analytic geometry we start with the *set* of all ordered triples (x, y, z) of real numbers x, y, z; to define the *space* we add a distance function similar to (3.5a). Analogous remarks hold for the spaces of four-dimensional, five-dimensional, and generally n-dimensional analytic geometry.

3.5.2 The most general types of space

The most general type of space is the *topological space*. If we analyze the intrinsic effect of the addition of the distance function (3.5a) to the collection of pairs (x, y) of real numbers, we see that it amounts to the assignment of *position* to the pairs (x, y) relative to one another. Thus, if p, q, and r are points and $d(p, q) = 2$, $d(p, r) = 3$, then we consider q "nearer" to p than r. And, as the distance between points decreases, in terms of the usual ordering by magnitude of the real numbers $d(p, q)$, we consider that they come more nearly to occupy the same position in the plane of analytic geometry.

If we consider *position* of elements of a set as basic, then we arrive at the notion of a *topological space* (one of the earliest alternative names for topology was *Analysis Situs*—"position analysis"—due to Gauss). There are other ways of defining relative position than by the method of distance functions, however. One way is to assign a *neighborhood* to each element of the set, S, which we want to make into a space. Thus, in the ordered set R of real numbers, if $r \in R$, then for any two real numbers a and b such that $a < r < b$ we may call the set

$$\{x \mid a < x < b\} \tag{3.5b}$$

a neighborhood of r in R (sometimes called "open interval neighborhood"). Such notions as that of *limit* of a sequence, for example, are directly definable in terms of such neighborhoods as (3.5b). However, the most fundamental notion in a topological space T is that of *limit point*, a point p being a limit point of a set M of points in T if, for *every* neighborhood U of p in T, the set $M \cap (U - p)$ is not empty. Thus the number 0 is a limit point of the set of real numbers $\{x \mid 0 < x < 1\}$; also of $\{x \mid x = 1/n, n \in N\}$.

We notice, incidentally, that every simply ordered set can be considered a space, inasmuch as relative position of elements is given by the open interval neighborhoods. It may not always be possible to express this by defining a distance function like (3.5a), but we can define neighborhoods by (3.5b); the resulting topology is called the *order topology*. In the simply ordered set of ordinal numbers of the first and second classes, ω, $\omega \cdot 2$, ω^2, etc., are all *limit points* of sets of ordinal numbers; this is the reason for calling them "limit numbers."

3.5.3 Usually there is some motive for the manner in which we turn a set into a space. For example, the motive underlying (3.5a) is to obtain a *euclidean* geometry; and the motive underlying (3.5b) is to express the intuitive notion of "limit" in the collection (thus ω is the limit of $1, 2, \cdots, n, \cdots$ in a natural way). Any collection M can be turned into a topological space by defining each element to be a neighborhood of itself; but then the resulting space has no limit points and, unless there is a natural motive for this (such as, for example, where M is a finite collection, in which case we usually do not desire limit points), the resulting space is without significance.

3.5.4 Geometric properties and configurations

To sum up the foregoing remarks, to get a space from a set we add certain properties which in some way or other embody intuitive ideas of *position* in the set; the added properties or relations, and properties or relations definable from them, we may call *geometric*.

Thus, distance between points as defined by (3.5a) is a geometric property; it may also be considered a relation between p and q. Similarly, parallel, perpendicular, etc., are geometric relations between lines.

The elements of the space we usually call *points*, and special sets definable in terms of the geometric properties are called *geometric figures* or *configurations*. For example, the pairs (x, y) of the analytic geometry plane are called *points*, and the sets

$$\{(x, y) \mid (x - x_1)^2 + (y - y_1)^2 = r^2, r \, \varepsilon \, R^+\}$$

are called *circles* with *center* (x_1, y_1) and *radius* r. Properties definable in terms of sets alone, such as cardinal number, are not usually considered geometric (unless related particularly to a geometric property; see 3.5.5); and arbitrary collections of points, not defined in terms of geometric properties, are simply called *sets* of points, or *point sets*.

3.5.5 Geometric invariants

Now suppose that S is a space, P is a geometric property or relation of S, and T is a transformation of S. For example, S might be the space of plane analytic geometry defined in 3.5.1, P the distance function (3.5a), and T a rotation of S about the point (0, 0). Then $d(p, q) = d(T(p), T(q))$; i.e., distance between a pair of points is the same as the distance between their images under T. We express the latter fact by saying that P is an invariant under the transformation T. On the other hand, with the same space S, but with P denoting the relation of perpendicularity between lines and T an affine transformation (1.2.1.4), generally it is not true that if lines L_1 and L_2 are perpendicular then $T(L_1)$ and $T(L_2)$ are perpendicular, so that perpendicularity will generally fail to be invariant under an affine transformation.

Now, if a property P is invariant under *all* transformations of a *group* $\mathfrak{T}$ of transformations of a space S, then we call P an *invariant* of S under $\mathfrak{T}$, or simply a $\mathfrak{T}$-*invariant* of S. For example, if $\mathfrak{T}$ is the group of translations of the plane, S, of analytic geometry, then it is easy to show that parallelism of lines is a $\mathfrak{T}$-invariant of S. And, according to the point of view proposed by Klein in 1872,† one may speak of the $\mathfrak{T}$-*geometry of the space* S as the study of the properties of the space and its configurations that are invariant under $\mathfrak{T}$. That is, the $\mathfrak{T}$-*geometry of the space S is the study of the $\mathfrak{T}$-invariants of S.*

The importance of these ideas when they were first propounded cannot be fully appreciated without an understanding of the state of mathematics, particularly of geometry, at that time. Projective geometry had blossomed after the publication of the researches of such mathematicians as Poncelet in 1822 and von Staudt in 1847. Even some rudimentary but important beginnings had been recently made in topology by Riemann. But, with regard to the relations between the various geometries, their position in mathematics, and even the proper location of specific theorems, little had been done. Take for example the Euler polyhedral formula,‡ still to be found in high school textbooks on geometry; it is now recognized as a theorem of topology, but at the time of which we are speaking it turned up in all sorts of places.

Of more significance, however, than special theorems was the application of Klein's idea to the classification of geometries themselves. To take an example, let us consider the geometry taught in high schools. What

† See Klein [a]; this is a translation, with additional notes, of Klein's "Program on entering the philosophical faculty and the senate of the University of Erlangen in 1872," commonly called "Klein's Erlanger Program." It was customary, on appointment to a professorship in a German university, to deliver before the general faculty an "inaugural" lecture of this type.

‡ This is the theorem which states that, if the numbers of faces, edges, and vertices of a simple polyhedron are denoted by r, e, and v, respectively, then $r - e + v = 2$. The number 2 is here an invariant, although a cardinal number, since it is related to the geometric configuration (polyhedron) involved.

properties of figures do we study? They are such properties as length, area, congruence, parallelism, perpendicularity, similarity. All these are invariants under the *group of rigid notions*; this is the group whose elements are translations, rotations, and reflections through lines, as well as combinations of them of the sort defined in 3.2. The first part of high school geometry is devoted to the "rigid motion geometry of the plane and space." Later, attention is given to *similarity* or *equiform* geometry. Here the group is the group of *similarity transformations* which has the group of rigid motions as a subgroup, but contains also those transformations which preserve angles and reduce or increase distances between points in a certain ratio. Under this enlarged group such properties as area, volume, and congruence are no longer the subject of study, being no longer invariants. However, "similar" figures such as similar triangles and polygons are investigated.

3.5.6 Affine geometry

If one enlarges the group $\mathfrak{T}$ of transformations, then not only is a different geometry obtained, but also the objects, i.e., invariants, of interest change. In the first place, there are fewer of them since, if new transformations are added, one may expect that properties which were invariant before are no longer so. And, secondly, the decrease in the number of invariants leads naturally to the search for invariants not noticed before. For example, if one takes for $\mathfrak{T}$ the set of all affine transformations of the plane (1.2.1.4), then no longer are angles invariant; however, parallelism of lines is an invariant property, and hence a subject for study in "affine geometry" of the plane.

When a transformation group $\mathfrak{T}_1$ is a subgroup of a transformation group $\mathfrak{T}$, one may choose to study, in the $\mathfrak{T}_1$-geometry, only those $\mathfrak{T}_1$-invariants that are not $\mathfrak{T}$-invariants, reserving the latter for the $\mathfrak{T}$-geometry.

3.5.7 Projective geometry

To continue the process begun in 3.5.6, one may consider the case where $\mathfrak{T}$ is the group of projective transformations of the plane. A projective transformation, from the analytic standpoint, is by definition a transformation which replaces a point (x, y) by a point (x', y') according to the rule:

$$x' = \frac{ax + by + c}{dx + ey + f}$$

$$y' = \frac{gx + hy + i}{dx + ey + f}$$

where $a, b, c, \cdots$ are real numbers and satisfy the condition that the determinant

$$\begin{vmatrix} a & b & c \\ d & e & f \\ g & h & i \end{vmatrix} \neq 0.$$

The group of affine transformations of the plane is that subgroup of the group of projective transformations for which $d = e = 0$ ($f \neq 0$). Now even parallelism of lines is no longer an invariant of the larger group; however, *incidence* of points and lines (that a point p and a line L are incident means that p is on L), for example, is an invariant. For an excellent elementary discussion of projective geometry (including a description and invariance proof of one of its most important invariants, the "cross ratio" of four points), the reader is referred to Courant and Robbins [C-R; 165ff].

3.5.8 Although the Klein program formed an important landmark in the history of mathematics, serving to clear up many geometric misconceptions, arranging the mass of geometric material then existent in neat compartments, and pointing the way to new types of geometries,† what is today called "geometry" cannot be confined within its limits.‡ Nevertheless, the notion of a transformation group continues to be of great importance and to serve as a means of classification of geometries of the classic type.

4 Topology

As a branch of geometry, mostly developed during the present century, the field of *topology* is based on a group $\mathfrak{T}$ whose elements, called *topological transformations* or *homeomorphisms*, are extremely general; a topological transformation is applicable to any topological space (3.5.2), and is merely a (1-1)-mapping (see IV 3.2.3.1) such that both the mapping and its inverse preserve limit points. For the layman, a topological transformation is therefore often roughly described as any deformation of a configuration that does not "tear" or "fold" but is otherwise unrestricted; a circle may thus be deformed into an ellipse, a triangle, a polygon of any number of sides—but not into two non-intersecting circles (because of the "no tearing" condition) nor into a figure eight (because of the "no fold"

† To invent a new geometry it was necessary only to find a new type of transformation group $\mathfrak{T}$ and a space to which it was applicable; such a "synthetically" produced geometry might be of little significance to the progress of mathematics, of course.

‡ Regarding geometries outside the scope of transformation group classification, see Veblen and Whitehead [V-W; 31–33].

condition; more precisely, however, because of the "(1-1)" character of the transformation). Analogously, topology of the plane is sometimes facetiously called "rubber-sheet geometry." These descriptions are very unprecise, of course, although they may be made precise by suitable definition of terms (see Wilder [d]).

4.1 As for topological invariants, clearly distance is not an invariant, nor is any number which is a function of length, such as the cross ratio of projective geometry. However, to take a very simple case, the fact that two points, when deleted from a circle, separate it into two pieces is a topological invariant of the circle; for, no matter what the topological transformation T, the "transform" $T(C)$ of a circle C always has this property. Another, related topological property of the circle is that deleting *one* point from it does *not* "disconnect" it. The reader who has studied the theory of functions of a complex variable will recall the "simply connected regions" of the plane and the method of inducing simple connectedness by means of "cross cuts." Simple-connectedness and the minimum number of cross cuts needed to produce simple-connectedness are purely topological invariants.

4.2 It is hardly possible any longer to confine topology, as a branch of mathematics, to geometry in the Klein sense. Its early development by Riemann, Poincaré, Brouwer, Veblen, and others was geometric and quite classifiable within the Klein program. However, as it has most recently developed, topology has been adapted to all parts of mathematics, virtually any collection of mathematical objects being a "topological space" in some sense or other, with great advantages of simplification of method and breadth of perspective. Some have compared the rapidly growing role of topology in mathematics to the role which has been played by group theory.

Indeed, in much the same manner as the concept of group evolved and became a methodological (and unifying) tool for all of mathematics, so has topology grown from a strictly geometric study, concerned with the properties of polyhedrals, to a general body of concepts and methods which furnish tools and new perspectives to all branches of mathematics. As stated above, the notion of "topological space" is so general, any set whatsoever can be conceived of as a topological space. But this would be of little significance, of course, if it did not lead to fruitful results. We shall try to exemplify how this can happen.

4.3 Topological space defined by neighborhoods

As we pointed out in 3.5.2, the transition from "set" to "topological space" consists basically in the specification of "limit points." However, it would ordinarily be very difficult to specify, for each subset A of the set, just what points are to be limit points of A. Where a distance function (3.5.1) has been already assigned to the set, then limit points are determined as indicated in 3.5.2. Frequently, however, the "space" into which we wish to convert the given set will not be amenable to assignment of a distance function; spaces exist which cannot be assigned such functions. The more general approach is to assign neighborhoods of points, in terms of which limit points may be defined as described in 3.5.2. And with only this much to go on, a surprisingly large number of significant theorems can be proved about the "set" now turned "space." (Cf. Sierpinski [S_4; I].)

Most topological spaces are slightly more restricted, however, in that the neighborhoods assigned must satisfy certain axioms. Of particular interest, from a historical standpoint, are the Hausdorff axioms which were stated in Hausdorff's classic book on set theory of 1914 (see Hausdorff [H; 213]): One supposes given a set S in which to each element x is assigned a non-empty collection $\{U_x\}$ of subsets of S called *neighborhoods* of x, such that

(a) For each U_x, $x \varepsilon U_x$.
(b) If U_x and V_x are neighborhoods of x, then there exists a neighborhood W_x of x such that $W_x \subset U_x \cap V_x$.
(c) If $y \varepsilon U_x$, then there exists a neighborhood U_y of y such that $U_y \subset U_x$.

In terms of neighborhoods, limit points are defined just as in 3.5.2, namely, x is a limit point of a subset A of S if for every neighborhood U_x of x the set $A \cap (U_x - x) \neq \emptyset$. Intuitively, this means that there are points of A arbitrarily near to x (and different from x).

For example, any simply ordered set, such as the real number system R, forms a topological space in terms of the open interval neighborhoods (3.5.2). And in this space, the number 0 is a limit point of the set $A = \{x \mid x = 1/n, n \varepsilon N\}$; but 0 is not a limit point of the set $\{x \mid x = 1 + 1/n, n \varepsilon N\}$.

4.4 Closed and open sets

In terms of limit points, many basic topological notions are defined. For instance, a set F is called *closed* if it contains all its limit points. Thus, in R, the set A defined in the previous paragraph is not closed; but if we add the number 0 to it, the augmented set is closed. The complements of closed sets are called *open*; i.e., if F is closed, $S - F$ is open (and conversely). Thus the collections of closed and open sets, being complementary, are of equal cardinality. It should be noted at this point that the effect of axiom (c) above is to make every neighborhood an open set. For if $y \varepsilon U_x$, then y cannot be a limit point of $S - U_x$ since by (c), there exists a neighborhood $U_y \subset U_x$ and *a fortiori* $(U_y - y) \cap (S - U_x) = \emptyset$. Consequently, $S - U_x$ is closed and U_x is open.

Now it may seem just as difficult a task to assign neighborhoods to each point as it would be to assign limit points to each subset, and that therefore nothing is gained by resorting to the definition in terms of neighborhoods. However, it turns out that (1) only the open sets need be defined, and (2) in a majority of cases only a certain subcollection of the collection of all open sets need be specified; in particular, while S may have cardinality c, for instance, often only $\aleph_0$ open sets need be defined in order to get *all* open sets.

To see how this may be, recall that it is the limit points that are basic. Now limit points may be defined either by the neighborhoods, as above, or by the open sets. Another way of putting this is to say that the open sets are themselves an "admissible" collection of neighborhoods if we make the convention that if U is open and $x \varepsilon U$, then U is a neighborhood of x; i.e., an open set is a neighborhood of *every* point in it, and neighborhoods need then not be specified for each and every point. And with this convention it is easy to see that with the open set neighborhoods, we get precisely the same limit points for a given set as before. (See Problem 32.)

In view of these considerations, the definition of topological space is now usually given in terms of open sets. That is, if S is a set, then a collection $\mathfrak{T}$ of subsets of S, to be called *open sets*, constitutes a *topology of* S if the following axioms are satisfied:

(1) Both S and $\emptyset$ are elements of $\mathfrak{T}$.
(2) Every union of elements of $\mathfrak{T}$ is an element of $\mathfrak{T}$.
(3) The finite intersections of elements of $\mathfrak{T}$ are elements of $\mathfrak{T}$.

Limit points are defined as before, but substituting "open set" for "neighborhood." Of course, the same set S can be made into different topological spaces by varying the assignments of neighborhoods; we show an example of this below. Now, by virtue of (2), we can assign a topology by giving only a *base* for the topology; i.e., a collection $\mathfrak{B}$ of open sets such that every open set is a union of elements of $\mathfrak{B}$. For example, a base for the usual topology of R is the denumerable collection of open interval neighborhoods having rational endpoints. Because of (3), the union of a finite number of closed sets is closed (see Problem III 5).

4.5 Examples. A classic theorem, developed by Euclid, states that if M is any finite collection of prime numbers, then there exists a prime number not in M. The modern mathematician usually states, "The collection of all prime numbers is infinite." An amusing proof of this can be given by using only the meager topological apparatus already described (see Furstenberg [a]). Let S be the collection of all integers, and let a base of open sets in S be the collection of all arithmetic progressions (from $-\infty$ to $+\infty$). Now not only is each arithmetic progression open, but is also closed (since the complement is the union of a finite number of arithmetic progressions and hence open by (2)). Consider the set $A = \bigcup_p A_p$, where for every prime $p \geqq 2$, A_p is the set of all integral multiples of p. Then $S - A = \{-1, 1\}$. The set $S - A$ cannot be open, hence A cannot be closed. But if the collection of all primes p were finite, A would be closed since it would be the union of a finite number of closed sets.

Notice, in the above example, that the topology assigned to the integers is not the "natural" topology; the latter makes each integer an open set, hence a neighborhood of itself, and thus no point is a limit point of any set in the natural topology (the so-called "discrete" type of space). However, the topology assigned above accomplishes the purpose for which it was chosen, namely, to prove the number of primes infinite. Another, quite different type of topology, may be assigned the integers. It is necessary only to consider the set N of natural numbers: Let the arithmetic progressions of type $\{an + b\}$, where a and b are prime to one another, be taken as a base for open sets. Again it will follow that the number of primes is infinite, in a manner similar to that used above. But this time the space, N, has many interesting topology properties.† In particular, it is *connected*; i.e., it is impossible to express it as the union of two disjoint, non-empty, closed sets ("open" may be substituted for "closed" here). This property of "connectedness" is a characteristic property of the space R of real numbers; indeed, using the order topology, Axiom (2) of VI 1.4.1 for the linear continuum may be replaced by "C is connected." Another interesting aspect of the topology just assigned to N is related to Dirichlet's theorem which states that every progression $\{an + b\}$, where a and b are prime to one another, must contain an infinity of primes. This theorem is equivalent to the assertion that the primes form a "dense" subset of N in that every natural number n is a limit point of the set of all primes.

This is as far as we shall go with the notion of a topology. For further study of the subject, the reader is referred to the numerous books now available in "general topology" (e.g., Hall and Spencer [H-S]).

5 Concluding remarks

With the discussion of these fundamental algebraic and geometric ideas we bring to a close our résumé of what may be called the foundations

† See Golomb [a].

of classical mathematics. The ideas that form the content of this and the foregoing chapters are derived from the end result of centuries of mathematical thought and evolution, and constitute what we may consider the basis of the type of mathematics—algebra, geometry, analysis, etc.—that is commonly taught in universities at the present time.

In the following chapters we shall inquire more closely, from various points of view, into the source, validity, and acceptability of what has gone before.

SUGGESTED READING

Albert [Al; I, II, XI]
Baumgartner [Ba]
Bell [B_1; III–VI] [B_4; III–IX]
Birkhoff and MacLane [B–M]
Campbell [Ca; II–V]
Kerschner and Wilcox [K-W; VII, XX]
McCoy [Mc]
Northrop [N; VIII]
van der Waerden [Wa]
Weyl [b]

PROBLEMS

1. Do the natural numbers and their reciprocals, with × as operation, form a group?

2. Let G be a set with three distinct elements $1, x, y$, and an operation whose operation table is:

	1	x	y
1	1	x	y
x	y	1	x
y	x	y	1

Does G form a group with respect to this operation? If not, which of the axioms for a group fail to hold?

3. Form a group of order 6 analogous to the group of 1.2.1.3. Relate it to the regular hexagon.

4. Prove Theorem 1.3.4 without using the commutativity of i.

5. Show that, if in Axioms 3 and 4 of 1.2 the expressions "$x \circ i = x$," "$x \circ x^{-1} = i$" are replaced by "$i \circ x = x$," "$x^{-1} \circ x = i$," respectively, then the new axiom system is equivalent to the old (cf. Problems 11 and 12 of Chapter II).

6. Show that, if only one of the two changes in Axioms 3 and 4 indicated in Problem 5 is made, then the new axiom system is not equivalent to the old.

7. Show that, if Axioms 3 and 4 of 1.2 are replaced by the statement of Theorem 1.3.5, then the new axiom system is equivalent to the old.

8. Show that, if G is of finite order n, then Axioms 3 and 4 of 1.2 may be replaced by the right- and left-hand cancellation laws (1.3.2, 1.3.4.1).

9. Show that the transformations of the coordinate plane of type (1.2.1.4a), with $ad \neq bc$, form a group.

10. Show that parallelism of lines is invariant, but that angles are not invariant, under the group cited in Problem 9.

11. Show that parallelism is not an invariant of the projective group of transformations of the plane (3.5.7).

12. Show that all groups of order 3 are isomorphic.

13. Exhibit two operation tables such that every group of order 4 has one of them for its operation table.

14. In any group we may represent $x \circ x$ by x^2, $x \circ x \circ x$ by x^3, and so on. Also we let $x^0 = i$ and $(x^{-1})^k = x^{-k}$ for each $k \varepsilon N$. If a group G has an element g such that, for every $x \varepsilon G$, there is an integer k such that $x = g^k$, then G is called *cyclic*. Show that for every $n \varepsilon N$ there exists a cyclic group of order n.

15. Show that if a group G is an infinite cyclic group, then G is isomorphic to the group of integers with operation addition.

16. Show that every infinite cyclic group has infinitely many subgroups.

17. Show that a group which has only finitely many subgroups must be finite.†

18. Show that an integral domain having only finitely many subsets forming integral domains need not be finite. Also show that a field having no proper subfield may be infinite.†

19. Show that the real numbers mod 1 do not form a system $(X, +, \times)$ as defined in 2.1.

20. Do the real numbers mod 1 satisfy the Laws of Signs as embodied in Theorems 2.2.2 to 2.2.4?

21. Show that the ring of integers mod 4 is not a field. What can we say in this respect about the integers mod p, where p is a prime?

22. If r is an element, not 0, of a commutative ring R with a unit, and there exists $x \varepsilon R$, $x \neq 0$ such that $r \times x = 0$, then r is called a *zero-divisor* of R. Show that a necessary and sufficient condition that a ring R be an integral domain is that it have no zero-divisors.

23. Construct addition and multiplication tables for a field having four elements, in which $1 + 1 = 0$.

24. Show that the set of all real numbers of the form $a + b\sqrt{2}$, where $a,b \varepsilon F$, with $+$ and $\times$ in the ordinary sense, is a field.

25. Prove that no non-degenerate proper subset of F is a field (if $+$ and $\times$ have their usual meaning in F).

26. Show that, for any two elements x and y of a system $(X, +, \times)$, the number $[(-x) \times y] + [x \times y] + [-(x \times y)]$ is both $(-x) \times y$ and $-(x \times y)$ and hence $(-x) \times y = -(x \times y)$.

27. Show that the ring of all continuous, single-valued real functions is not an integral domain.

28. Using the notation of 2.3.6, show that, if A is an ideal of X, then the operation $\times$ is uniquely defined for the cosets of X/A; i.e., if $x_1 \varepsilon A + x$, $y_1 \varepsilon A + y$, then $x_1 \times y_1 \varepsilon A + (x \times y)$.

29. Show that the integers form a subring, but not an ideal, of the ring of all polynomials in one variable x with integer coefficients (with $+$, $\times$ as usually understood).

30. Show that, if X is the polynomial ring of Problem 29 and I the subring of integers, then the cosets of X mod I (rel. $+$) do not satisfy uniquely an operation $\times$ as defined for the case if I were an ideal of X.

31. In Definition 2.7.1, replace the requirement that "$0 < x$, $0 < y$ imply $0 < x \times y$" by "$0 < x$, $0 < y$ imply $x \times y < 0$." What do the "Laws of

† Cf. *Amer. Math. Mo.*, **70** (1963), p. 332, Problem E1522.

Signs" become in the resulting system? Is this new "real number system" consistent?

32. Prove that in a topological space defined by neighborhoods (as in 4.2), a point x is a limit point of a point set M if and only if for every open set U containing x, $M \cap (U - x) \neq \emptyset$.

33. Show that if the set R of real numbers is assigned the order topology (3.5.2), then every element of R is a limit point of R. More generally, if S is a non-degenerate connected topological space in which each point forms a closed point set, then every point of S is a limit point of S.

PART TWO

Development of Various Viewpoints on Foundations

Up to this point our chief concern has been with what might be called foundation *material*; i.e., the ideas and methods which form the basis of the mathematics taught in most modern universities. Except for incidental remarks, there has been little formal criticism or philosophical speculation on the *nature* of the material.

In this second part, however, the discussion is no longer confined to the "orthodox," although much of what is included may justly be so labeled, or at least is of such a character that it is rapidly becoming so. Although the preceding chapters may have seemed, with few exceptions such as the Russell and Burali-Forti contradictions, to consist of what the "non-cognoscenti" would expect of mathematics—precise, logical ideas and methods of deduction—and to depict a situation which appears fairly "austere and serene," the fact is that in modern mathematics the austerity and serenity are mainly on the surface. At the heart of mathematical creation there is sharp disagreement as to the origin, meaning, and validity of mathematical notions. The remark in II 6.3 that it would be almost a miracle to get as many as five mathematicians to agree upon a definition of *mathematics*, was not an exaggeration; certainly with few exceptions no mathematician would seriously propose such a definition.† This is due to the nature of the subject; it is not a dead compendium of laws regarding "quantity" and "space," but a live, growing aspect of human culture that would quickly break the bonds of any confining definition.

During periods which have been conducive to abstract thought and intellectual freedom, such as those of ancient Greece and the European Renaissance, mathematicians and mathematical philosophers have speculated on the nature of their activities. In modern times, such speculation has been particularly active and has led to sharply differing conclusions. It will be the function of this second part to describe some

† By "definition" is meant the type of definition used in mathematics; see the second paragraph of Chapter III.

of the chief aspects of this development and its present status; not exhaustively (which would require volumes), but enough to draw a general picture whose details may be filled in, wherever desired, by reference to the bibliographical material cited. The descriptions of the differing views of various individual mathematicians or "schools" will be as objective and impartial as possible. At times it may seem that "sides" are being taken; such an impression, however, will be due to the desire to present the advocate's point of view as advantageously as possible.

It will be impossible, as a matter of fact, to do justice to all points of view in a work of this size and nature. The available space is too limited, and probably many readers would not want to labor with all the details. The best that can be done is to include certain general ideas to the extent of *provocation*; that is, that the reader will be provoked into going to original sources (furnished in the citations to the bibliography) and forming his own conclusions on the merits of the case in hand. If I succeed in doing this, I shall have fulfilled my function of reporter and expositor.

VIII

The Early Developments

Some description of the development of the present-day situation has already been included in the earlier chapters. For instance, in Chapter I we remarked on the background and development of the modern axiomatic method. And in III 1 and III 2 the circumstances which led up to the invention of the set-theoretic contradictions and some of the resulting impact on mathematical thought were described. In the present chapter these matters will find their place in a brief systematic presentation, in approximately chronological order, of the origin and sources of the various opinions now in existence.

1 The eighteenth-century beginnings of analysis

Any student of the history of mathematics is familiar with the discussions that followed the work of Leibniz and Newton on the calculus. The literature of the time shows that many different concepts of the meaning of differentials were then current; and that the methods of proof then in vogue would for the most part be wholly unacceptable today. One prominent mathematician† of the period postulated that "A quantity, which is increased or decreased infinitely little, is neither increased nor decreased." We may wonder just what "infinitely little" meant; to the modern mathematician the phrase would be indicative of a lack of comprehension of the fundamental ideas of the calculus. It is probably not unfair to state that for a long time the calculus had no foundation. Each writer chose what seemed to him a "foundation" and built thereon. The period can be characterized as one of experimentation. Mathematicians were aware of the importance of the new theory (calculus), and in general showed remarkably good sense in developing it. As a matter of fact, we should not be unjustly critical of the pioneering work done at this time; rather, our admiration should be evoked by the manner in which sound mathematical theory was erected on unsound bases.

For example, Lagrange tried to avoid the difficulties in the meaning of the differential by using the derivative exclusively. Instead of proceeding, like Johann Bernoulli and Euler, to set up expressions in dx and treating

† Johann Bernoulli.

some of the occurrences of dx as zero, and others as "infinitely small" quantities, he calculated his derivatives by expanding functions in Taylor's series and then selecting the respective coefficients as (by definition) the derivatives. He attempted to show that any function can be so handled, which is of course incorrect; but he may be considered to have incidentally made a beginning on the theory of analytic functions.

When any attention at all was paid to divergence of series, it was as a rule of a desultory and unrigorous character. Even after Cauchy had given a rigorous foundation for the theory of infinite series, old habits prevailed. The notion of *limit*, so simple from the modern standpoint, remained fairly clouded in mystery throughout the eighteenth century. D'Alembert gave a good approximation to the modern definition (his definition gave Cauchy the starting point for his "Cours d'analyse"), but unfortunately the idea of "approach"—a quantity approaching another "nearer than by any given quantity"—led to such futile questions as "Can a variable *reach* its limit?"

As we shall see, modern "rigor" is not beyond reproach, although it would hardly be possible for a "layman" to criticize, with justification, the present standards of rigor, such as happened in the eighteenth century when Bishop Berkeley attacked Newton's "fluxions."†

2 The nineteenth-century foundation of analysis

Early in the nineteenth century Cauchy placed the calculus on an essentially modern basis; Abel, Gauss, and Cauchy developed a rigorous treatment of infinite series; and later in the century Weierstrass worked at the "arithmetization" of analysis. A characteristic of the latter was the freeing of analysis from the intuitive geometric type of proof so prevalent at the time. For example, the geometric approach to continuous real functions leads naturally to the belief that such a function, representable by a "smooth" graph, certainly must have derivatives at *some* points at least. Weierstrass' famous counter example‡ of a continuous function defined over the reals and having a derivative at *no* point dispelled this illusion. We must here keep in mind, however, that to "free" a theory of certain notions does not necessarily imply that those notions forever thereafter cease to be of use. Just as in the case of the undefined terms of an axiomatic system we find it extremely useful, and in practice necessary, to keep certain interpretations in mind, so in the development of

† See the entertaining account in Struik's remarkable little history of mathematics [St; 178]. The reader may find it profitable to read the entire account of the beginnings of analysis in this book.

‡ Riemann also possessed such an example (unpublished).

calculus and function theory we find it advantageous to continue to use geometric interpretations. It is generally conceded, for instance, that the teaching of calculus is facilitated by the use of graphs. Strictly speaking, however, no analytic geometry is needed for either calculus or function theory.

We might recall here that we make similar concessions to pedagogy in the teaching of high school geometry, not striving to place it on a truly axiomatic basis but allowing the student to use "preconceived notions" such as "straight," "distance," "measure." In the teaching of calculus, one may accept the geometric foundation of the sine function which the student has studied previously in his trigonometry course, and give an unrigorous geometric proof that $\lim_{x \to 0} (\sin x)/x = 1$, before establishing the derivatives of the trigonometric functions. (In analysis, the latter are *analytically* defined, not geometrically.) Such pedagogic devices are entirely justifiable, of course, except that it is questionable just how long the student should be kept insulated from the underlying "concessions." Certainly he should not be allowed to believe that he is receiving *dogma.*

The work of Weierstrass was paralleled by that of other researchers, especially Dedekind and Cantor. Dedekind's approach to the foundation of the real numbers has already been exemplified in the "Dedekind cut" (see VI 1.4 and 3.3 for instance), and contact has been made with that of Cantor in the characterization of the order type of the real number system by means of the separability principle.† Both Dedekind's methods and those of Cantor were based on an acceptance of the "actually infinite." We have seen, for example, how the Dedekind cut defines a single real number in terms of infinite classes of numbers.

Vigorous opposition to these ideas, especially to the ideas of Cantor (which we discussed in some detail in Chapters IV and V), were expressed by Kronecker.

2.1 Kronecker's "intuitionism"

Kronecker was a contemporary and (ultimately) a colleague‡ of Weierstrass at the University of Berlin, but their different notions of what constitutes mathematical existence did not make for harmony

† Cantor designated the order type of $\{x \mid 0 \leqq x \leqq 1\}$ as "order type θ." Instead of the Dedekind cut axiom which we used in VI 1.4, however, Cantor employed the notion of "perfect set"; to define the latter would have necessitated employing the notion of limit point.

‡ Weierstrass was professor at the University of Berlin for thirty years, commencing with the year 1856. Kronecker (whose private resources freed him of the necessity of seeking a salaried position) commenced his sojourn in Berlin in 1855, teaching informally at the university until finally accepting a professorship made vacant by Kummer's retirement in 1883.

between them.† Because of their "prophetical" significance and relation to the present-day discussions on foundations, we shall outline briefly Kronecker's chief views:

2.1.1 In the first place, he *objected strenuously to "clouding" mathematics by basing the notions of finite set and of real numbers on the actually infinite.* Although he agreed that "arithmetization" was the correct approach, not only to analysis but also to all mathematics, his ideas of arithmetization ruled out all use of infinite sets (as in the Dedekind cut) in both definitions and number constructions. Much quoted is his remark,‡ "The integers were made by God, but everything else is the work of man." Hence:

2.1.2 He asserted that *the natural numbers and operations with them are "intuitively founded,"* and that *algebraic numbers and operations with them can be based on the natural numbers and their properties*; but that *the real numbers are not capable of such a foundation.*

Here he was striking at the heart of the question of *what constitutes mathematical existence.* Previously there had been considerable reluctance, because of fear of ridicule or of ecclesiastical denunciation, to introduce certain concepts (as, for example, negative and imaginary numbers, and non-euclidean geometries), but evidently Kronecker's objections were based on a serious and deeper-lying philosophy. He seemed to fear that there was a danger of mathematics drifting into fantasy and mysticism. To him, the Cantor theory of transfinite numbers was not mathematics but mysticism;§ it started with the assumption that infinite sets exist in mathematics—an assumption that to Kronecker was untenable.

The reader may be puzzled by the conjunction of this statement with the preceding statement that Kronecker accepted natural numbers and operations with them. But notice that we did not say "*the set* of natural numbers . . ." The latter might be taken to imply an *already existing* "totality." In the latter sense, we recognize, for example, that

$$2,867,944,782,612,713,942$$

is the symbol for a natural number, although it is quite likely that no one has ever seen it before it appeared here in print! That is, it represents an element of the "set of all natural numbers" which existed *prior* to our selecting this particular one of its elements for exhibition. Quite different is the conception of the natural numbers as

† In a letter to Sonja Kowalewski, Weierstrass complained bitterly of the fate that compelled him to endure the "interesting, but not mathematics" type of comment which Kronecker frequently used in appraising Weierstrass' work and that of others.

‡ Bell [B_1; 34] facetiously remarks that inasmuch as "he said this in an after dinner speech perhaps he should not be held to it too strictly."

§ Cantor was aware that his work was in a sense (to cite Struik [St; 243]) "a continuation of the ancient scholastic speculations on the nature of the infinite" and "defended St. Augustine's full acceptance of the actually infinite."

a kind of constructible or "growing" collection which consists of a *beginning* set of numbers—1, 2, 3, 4, 5, 6, 7, 8, 9, 10, etc.—and a rule for deriving new elements of the "collection" (thus $10 + 1 = 11$; $11 + 1 = 12$; going as *far as we like or are able*). In the latter case no actual infinite is ever attained. And the above natural number is not a natural number because it is a member of some *already existing* totality, but because it can be exhibited as a result of the construction just indicated. If one objects that each and every natural number is so attainable and hence the existence of the "set of all natural numbers" again follows, it can be pointed out that this assumes an *infinity* of operations in order to construct each and *every* natural number. For the "finitist," only the "constructible" concept of the natural numbers is acceptable. By "intuitively founded" Kronecker evidently meant that we have an intuition of the natural numbers founded on our experience of time and the succession of events.

When informed of Lindemann's proof (1882) that π is a transcendental number, Kronecker is reported to have remarked that this was of interest, excepting for the fact that π does not exist!

This is perhaps an exaggeration, the evidence pointing rather to a rejection of the general notions of irrationals as set up by his contemporaries. In order specifically to avoid these and, indeed, avoid the use of all numbers except natural numbers (even negative integers!), Kronecker developed a modular arithmetic, employing congruences mod $x + 1$, or mod $x^2 + 1$ (to avoid use of the complex unit i), etc. (For a simple discussion of these matters, see Couturat [Co_2; 603–616].)

2.1.3 *All definitions and proofs should be "constructive,"* that is, a definition of a mathematical entity should permit one to construct the entity; for example, by giving a *rule* for its construction from mathematical elements already known to exist. (In particular, the "set of all real numbers" is not definable in this sense.) And, if a proof is given of the existence of a mathematical entity, then the proof should be "constructive"; i.e., one following the steps of the proof should thereby commence constructing the entity.

For example, most proofs of the Fundamental Theorem of Algebra are non-constructive, in that they show that the assumption of the existence of an algebraic equation having no root leads to contradiction. Such proofs give no inkling as to how to find a root. Generally, *reductio ad absurdum* existence proofs are of this nature, hence, from Kronecker's point of view, unacceptable.

Another type of existence proof to which Kronecker would have objected is that which relies on the Choice Axiom (III 6.3). An outstanding example of such a proof is that of the Well-Ordering Theorem (V 3.1.2). In this proof (V 5.1.4) it is assumed, on the basis of the Choice Axiom, that, for each subset S' of a given set S, there exists a representative element $x(S')$. But no rule is given for specifying $x(S')$, so that one is unable actually to well-order a given set by following the steps of the proof.

Like comments hold regarding the proof (V 3.3.1) that the set R (assuming its existence for the moment) has a basis. Before giving the rule for constructing such a basis, a well-ordering of R or equivalent thereof has to be provided. Yet from the existence of a basis one can prove many beautiful mathematical theorems.

2.1.4 *Arguments of a purely logical nature do not necessarily yield legitimate mathematical theorems.* This is a natural corollary of the stipulation (2.1.3) that existence proofs must be constructive. Thus, to use again the example of the *reductio ad absurdum* proofs of the Fundamental Theorem of Algebra, such existence proofs are logically correct perhaps, but they are of no value as *mathematics.* For mathematics the proofs must provide a way of *finding* the root asserted to exist.

We note here the first inkling of a possible limitation on the use of logic in mathematics. From a position of utter dependence on logic, one turns to a new criterion: *construction* on the basis of the natural numbers. One may still use logic—for example, one will prove *non-existence* of an entity by the usual demonstration that assumption of existence leads to contradiction. But to squeeze *existence* out of such logical manipulation is not valid *mathematically.*

2.1.5 Thus Kronecker's meaning of the term "arithmetization" is quite literal! From the arithmetic of the natural numbers one constructs, by finite methods, the rest of mathematics. As we shall see, in "the rest of mathematics" we do not reach all that is commonly taught today, not even all the real numbers which are at the basis of present-day analysis.

Kronecker found, in his day, no supporters for his opinions. Otherwise, our discussion of his ideas would not have been relegated to a few paragraphs of a summary nature. No doubt this was due in considerable measure to the severe limitations these notions placed on both method and content, and the radical proposal which they contained regarding the purely *logical* as opposed to the *mathematical.* But still his ideas might have found more favor, and certainly would have been accorded more sympathetic discussion among his colleagues, if it were not for the advent of Cantor's theory of the transfinite.

2.2 Cantor's theory of the transfinite

During the period 1879–1897 Cantor published his investigations on the theory of sets, including the theory of cardinal and ordinal numbers. It is interesting to observe that he was led to his ideas by his researches in analysis, particularly in the theory of trigonometric series. We have already referred (footnote in connection with V 3.5.2.2) to his observation of the well-ordered sequence of derivatives of a given set of real numbers, for instance. Cantor's ideas, as applied to the foundations of analysis, form an extensive continuation of the work of Weierstrass.

Their fruitfulness in analysis, however, was almost eclipsed by their influence on mathematics in general. In geometry, for example, they made possible a new perspective. And in the new field of topology they became indispensable.

On all sides, then, Cantor's ideas bore fruit. Is it any wonder that the objections of Kronecker† were of no avail and that the Cantorian theory prevailed? It is true that some of the "fruit" was sour—Cantor himself was probably aware of the fact that care and discretion had to be used in "storming the infinite"—else one might transcend consistency entirely. But the good seemed by far to outweigh the bad. Consequently today we find the ideas of Cantor, especially his extensions of Weierstrass' treatment of the real number system, almost universally accepted and taught in the universities and colleges of the world. This, incidentally, partially explains the prominence which we have accorded them in the earlier chapters.

Before continuing in this vein, however, we should note another development that was taking place during the latter part of the nineteenth century, one which was to have considerable influence on present-day thought.

3 The symbolizing of logic

We have already had occasion to comment, in III 2.3, on the rise of "symbolic logic." In order that the nature of this development may be clear, we recall that logic, as developed by the Greeks (particularly Aristotle), and by the scholastic philosophers of the Middle Ages, made little use of symbols. It is true that not infrequently symbols were used as a kind of shorthand. For example, the syllogism might be stated in the form, "If all A's are B's, and x is an A, then x is a B." Here A and B stand for classes of perfectly general character; the classical logicians understood well the generality of their logical laws. But such use of symbols is a far cry from a *calculus* of symbols, such as was introduced in mathematics in very early times.

3.1 Meaning and purpose of a symbolic logic

When we speak of *symbolic logic*, then, we mean something analogous to the use of symbols in mathematics. When we apply algebra to the solution of some concrete problem, we are employing a *symbolic calculus* in which the steps involved in solving the particular problem have all been provided for; little further thought need be expended, once the

† Cantor's ideas also met criticism, but when their fruitfulness became evident, opposition dwindled (although it survived in notable instances—see Section 7).

symbols of the algebra ($x, y, \cdots$) have been properly interpreted or adapted to the problem. In a simple case where only one symbol, x, receives a meaning, and, say, a quadratic equation results, the solution of the latter by the "quadratic formula" gives the desired result "ready made." So, in logic, we desire not only symbols to designate the entities (such as propositions and classes) with which we deal, but symbolic representation of *relations* (such as $\supset$, ε, $<$) and *rules for operating with them*. The reasoning involved in special problems should be provided for in the resulting calculus.

It was undoubtedly the example of what a good calculus of symbols had accomplished in mathematics that stimulated the search for a similar apparatus to employ for the *reasoning process*. The purpose of such an apparatus would be not simply a way of making reasoning—more specifically, the drawing of conclusions by the deductive process—an easier feat, but it was realized, in view of the example of mathematics, that (1) greater rigor and (2) broader perspectives would be attainable. Both of these aspects of the symbolic method are involved in the facility with which the mathematician sees new properties and relations in known theories, and the opening up of new avenues of research.

3.2 Influence of Leibniz

Probably no one realized these facts better than Leibniz, who, though neither he nor anyone else can be credited with *beginning* symbolic logic, was one of the prime movers in its direction. With Leibniz's somewhat grandiose schemes for systematizing, and reducing to a few simple elements, all human thought we are not concerned here; but his recognition of the importance of a good symbolism, and in particular of instituting a calculus of reasoning (*calculus ratiocinator*), together with his subsequent propagandizing for it, were unquestionably a factor in stimulating the logicians of the eighteenth and nineteenth centuries.†

3.3 Boole's algebra of logic

What can perhaps properly be called a "foundation" for symbolic logic was laid in England in the first half of the nineteenth century, especially by Boole and De Morgan. In Boole's work we obtain for the first time a calculus of logic, specifically a calculus of sets, complete with rules of operation. It is properly called an algebra—not quite the modern form of "Boolean algebra," however—since all four elementary operations of algebra are defined. Subtraction had the same meaning as in III 3.3.2,

† The reader might consult the section on Leibniz in Struik [St; 154ff].

and multiplication the same meaning as in III 3.3.3; but addition was somewhat different in that† $x + y$ was defined only for x and y disjoint. Rules such as the commutative laws of addition and multiplication and distributive laws were given, as well as such peculiar laws as $x^2 = x$. The numbers 1 and 0 occupied special positions in that 1 designated the entire universe of elements (of which all sets are subsets); 0 designated the null set (our $\emptyset$ of III 3.4). The difficulties arising in the interpretation of such expressions as $x + 1$, x/y, $2x$, were circumvented by noticing that, if one treated the algebra as though the variables were restricted to the values 0 and 1, then all ordinary laws and operations were applicable. The result was, then, that if one carried through the algebra without worrying about interpreting the formulas derived,‡ and used the devices worked out by Boole to reduce the resulting expressions, the end product would be both valid and capable of interpretation.§

3.4 Subsequent development of the algebra

One of the principal changes subsequently made in Boole's algebra of logic was the interpretation (Jevons) of $x + y$ as defined as III 3.3. This enabled one to write $x + x = x$ and thus eliminate the $2x$ which had no interpretation. And, to get rid of other non-interpretable formulas, division was altogether eliminated (Peirce). The inclusion symbol, $\subset$, of III 3.1.2 was introduced by C. S. Peirce. The monumental treatise of Schröder [Sch], published in 1890, more or less brought this development to a final satisfactory form.

It would take us too far afield to go into the details of the Boole–Schröder algebra. But more recent developments in the applications of Boolean algebra should not go without mention at this point, since lack of space will not permit consideration of them later. These have extended to applications in computer and switching circuit design (see Hohn [Hoh], for instance). Relations to set theory and logic, in both of which occur important models of Boolean algebra, are more noteworthy from our point of view, however.

† To denote sets we shall use here the usual symbols $x, y, \cdots$ of algebra rather than the capital letters $(S, A, B, \cdots)$ of Chapter III.

‡ These were usually incapable of interpretation.

§ The student familiar with Boole's *Calculus of Finite Differences*, still used as a reference in present-day instruction (particularly of actuarial students), should understand this process perfectly. The algebraic operations with $\sum$, $\int$, etc., performed with meaningless formulas, but ultimately reduced to a form where they can be affixed to the functions under consideration, form a nice analogue of the logical algebra.

In the modern forms of Boolean algebra, the "$\cup$" and "$\cap$" of III 3.3 are commonly employed instead of "$+$" and "$\cdot$," with "$+$" reserved for the "symmetric difference"; specifically, $x + y = (x - y) \cup (y - x)$. The relation to the calculus of sets found concrete expression in the Stone representation theorem to the effect that every Boolean algebra is isomorphic with a certain algebra of sets; also, every Boolean algebra is isomorphic to a certain class of rings with unit. (See Stone [St].) These results remind us of the relation between abstract groups and permutation groups (cf. Theorem VII 1.4.2.1).

Applications to logic are based on the fact that the calculus of propositions (to be discussed in IX 3) forms a model for Boolean algebra, and recent investigations in this theory have been made using Boolean algebra as principal tool. Regarding these matters, the reader is referred to such works as those of Rosenbloom [Ros; I] and Stoll [St; VI].

4 The reduction of mathematics to logical form

By the latter half of the nineteenth century, mathematical thought had matured to the point where it was recognized that mathematics could be developed along "pure" or abstract lines, with no reference to material reality. Undoubtedly the acceptance of conflicting (euclidean and non-euclidean) geometries and the consequent denial of one of the tenets of Kantian philosophy had a good deal to do with this evolution. And, too, the ideas of Leibniz and the subsequent development of logic along formal lines must have exerted an influence. In addition, both Peirce and Schröder had begun to apply their logical calculus to arithmetic.

4.1 Frege's arithmetic

The first work of a definitive character, so far as basing mathematics on logic is concerned, was that of Frege. In his *Begriffschrift* [Fr_1] he introduced the notion of "propositional function"—called by him "Function," the term "propositional function" being due to Russell (see Chapter IX)—and a complete system of basic formulas for propositions in terms of implication and negation (compare the comment of Hilbert-Bernays [H-B; 64f]). Frege's *Die Grundlagen der Arithmetik*, published in 1884 (now available in parallel paged German and English translation [Fr]), is a landmark in the development of the modern logical foundations of mathematics, being the first attempt to base mathematics on pure logic; specifically, to found, on the notions of primitive logic, both definitions of number and laws of arithmetic. To quote Frege's own appraisal of the work†

† By permission of Basil Blackwell.

[Fr; 99[e]], "Arithmetic thus becomes simply a development of logic; and every proposition of arithmetic a law of logic, albeit a derivative one." This work (in which, incidentally, the "Frege-Russell" definition of number given in IV 4.1.1 first appeared) was not accorded proper credit until Russell called attention to its true worth and significance. We shall postpone to Chapter IX the continuation of Frege's line of thought in the work of Russell and Whitehead.

4.2 The Peano school

Of a somewhat different type was the work of Peano and his colleagues. Instead of trying to *base* mathematics on logic, they analyzed the methods of mathematics and tried to *express* them in a form similar to that of the logical calculi. It could be characterized as a union of mathematics and a calculus of logic. Instead of trying to give "natural number" a *meaning* such as is embodied in the Frege-Russell definition, i.e., reducing it to the logical notion of set, Peano treated the notion as undefined and used essentially the axiom system which we gave in VI 3.1.

However, as Peano formulated the axioms, they were stated in the logical symbols which constitute his "pasigraphy." For "number" the symbol "No" was used, $a+$ designated the successor of a, etc., and the analogue of Axiom (2) of VI 3.1, for example, was stated in the form

$$a \, \varepsilon \, \mathrm{No} \, . \supset . \, a+ \, \varepsilon \, \mathrm{No},\dagger$$

where ε corresponds to the symbol denoting "is an element of" which we introduced in III 3.1. (The introduction of this symbol ε, implying a recognition of the necessity for making a distinction between ε and $\subset$, was an important innovation, incidentally.) The derivation of theorems becomes an algebraic process, only symbols and formulas being employed; each theorem is stated as a symbolic formula. The reader can consult Peano's *Formulaire* [P_3] for details, especially Volume V.‡

This method of deriving mathematical theorems as symbolic formulas from a given set of axioms stated as formulas, using what corresponds to a logical calculus for the derivation, finds its counterpart in the so-called "formal systems" of modern mathematical logic, to which we come later. It reduces mathematics to a strictly "formal" process, with no direct reference to any "real" interpretation of the symbols involved. It has the advantages that accrue from avoiding errors due to varied interpretations

† Literally translated: "If a is a number, then $a+$ is a number"; the symbol $\supset$ denoted implication.

‡ The volumes of the *Formulaire* were published between 1894 and 1908, and essentially represent a series of reports by Peano and his collaborators.

of terms or unsuitable connotations, such as are frequently made in the use of ordinary language, as well as from rendering the steps of a proof more precise and less vaguely dependent upon logical rules. It has its opponents among those who dislike such formality and divorcement from "reality." It does, however, represent an important development in the relations between mathematics and the methods of symbolic logic, as well as in the evolution of mathematics itself, and has had an influence on both mathematics and logic which continues to the present day.

5 Introduction of antinomies and paradoxes

Another development of great importance, which occurred at the end of the nineteenth century and the beginning of the twentieth, was the introduction of the contradictions to which the "unrestricted" theory of sets leads (cf. III 1.1, III 2, IV Problem 28, V 3.5.3). That this new theory, so beautiful and fruitful, should lead to such logical consequences came as a profound shock to many mathematicians. Frege (who lived until 1925), for example, considered that all his work, based on the theory of sets, was jeopardized. Many mathematicians, as a result of the antinomies, as well as of later developments in the foundations, ceased to work on aspects of mathematics which depend upon an unqualified acceptance of set theory. Poincaré characterized the theory of sets as "a disease from which mathematics will some day recover." Others, more courageous perhaps, or more convinced of the ultimate validity of the theory of cardinal and ordinal numbers, set out to correct the errors into which mathematics had drifted. Prominent among these was Russell, whose use of the "logistic" method we shall discuss presently; also Zermelo, who attacked the problem of providing a set theory free of contradiction by trying to furnish a consistent axiom system which would avoid the "too large" sets.

6 Zermelo's Well-Ordering Theorem

The publication of Zermelo's proof of the Well-Ordering Theorem (Zermelo [a]) in 1904 added fuel to the discussion, since even those mathematicians who would grant the existence of the set R were, in many cases, loath to admit the possibility of its being well-ordered. There was no question of any contradiction here; it was a question of admissible methods and existence. Already the question of what constitutes mathematical existence had arisen in connection with Kronecker's views, and the appearance of contradiction in set theory had revived some of the ideas expressed by Kronecker. This was particularly to be noted in Poincaré's writings.

7 Poincaré's views

Poincaré's reaction to the Zermelo theorem was one of complete rejection. His basis for this was the *lack of definition* of representative elements involved in the use of the Choice Axiom, which we have already stated (in 2.1.3) would have been opposed by Kronecker. In a way, Poincaré can be considered a follower of Kronecker in his philosophy of mathematics. Of Poincaré's views we should mention specifically the following:

7.1 *Recognition of the basic character of the natural number system, and especially of the mathematical induction principle, as incapable of reduction to logic;* i.e., *a "synthetic a priori judgment" in the Kantian sense.* (See Poincaré [Po; I, 64, 452].)

7.2 *Every mathematical entity should be definable in a finite number of words; concepts should be built up by proceeding from the particular to the general, not conversely.* (Poincaré [Po; 382].) This reminds us of Kronecker's objections to "basing the finite on the infinite" (2.1.1).

7.3 *Most of the concepts and conclusions of the Cantor theory of sets should be excluded from mathematics* [Po; 483–484].

In 7.1 to 7.3 Poincaré was clearly in agreement with Kronecker. Kronecker did not have the antinomies before him as a guide in attacking the problem of definition, however, as Poincaré did. Poincaré's reaction to them was:

7.4 *Non-predicative definitions should not be employed* [Po; 480–481]. By a *non-predicative definition* is meant a definition of an entity E which defines E in terms of a class of which E is an element. It not infrequently happens that, in order to specify a certain real number, for example, we determine a class to which we know it belongs, and then we define the number in terms of this class. Thus we define the maximum of a continuous function, which has been defined over a closed interval, as the greatest of all its values. Such a definition is non-predicative.†

But it seems impossible to give up this type of definition entirely, especially in the theory of sets. To reject sets defined non-predicatively would, to be sure, bar such self-contradictory and non-predicatively defined sets as the set of all sets and the well-ordered set of all ordinal numbers. But the cost appears too great unless some modification can be introduced to retain the apparently legitimate uses of this type of definition.

† It is amusing to note that, soon after Poincaré made known his opposition to the non-predicative type of definition, one of his colleagues pointed out to him a situation in one of his (Poincaré's) own articles, where he had used this objectionable type of definition. (Unfortunately, Poincaré was able to point out how it could be avoided in this case!)

7.5 *Poincaré rejected and ridiculed attempts to base mathematics on logic* [Po; 448–485]. He asked, "If . . . all the propositions [which mathematics] enunciates can be deduced one from another by the rules of formal logic, why is not mathematics reduced to an immense tautology? The syllogism can teach us nothing essentially new, and, if everything is to spring from the principle of identity, everything should be capable of being reduced to it. Shall we then admit that the enunciations of all those theorems which fill so many volumes are nothing but devious ways of saying *A* is *A*?" [Po; 31.†] Someone has written "Logic is barren, whereas mathematics is the most prolific of mothers." Poincaré exclaims "Logic is no longer barren; she has brought forth a contradiction!" (in reference to the Russell antinomy).

He believed that attempts to place the method of mathematical induction on a contradictionless basis were doomed to run into a vicious circle, in that the induction procedure would of necessity be used in the proof of freedom from contradiction. For mathematical induction, he felt, is forced on us by our intuition. Nevertheless, he stated repeatedly—and in this he agrees with the Formalists (Chapter XI) rather than the Intuitionists—that, in general, proof of freedom from contradiction, as in the case of a geometry, is adequate basis for assertion of mathematical existence [Po; 61, 439–440, 454, 474].

During the first decade of the present century a discussion took place between Poincaré and Russell which fortunately took the form of a series of articles in the French periodical *Revue de metaphysique et de morale.*‡ These are available for the reader familiar with French. Poincaré's *Foundations of Science* is available in English translation [Po].

7.6 Many of Poincaré's French colleagues were equally vehement in voicing their opinions concerning "le grand débat." Their views make interesting reading; some will be found in Note IV of the appendix to Borel [Bo; 150–160]. In particular there are included there five letters interchanged by Borel, Hadamard, Lebesgue, and Baire (*loc. cit.*, pp. 150ff) which were prompted by the appearance of Zermelo's Well-Ordering Theorem.§ Also see Menger [a].

8 Zermelo's set theory

In Section 5 we mentioned the reactions of Russell and Zermelo to the introduction of the antinomies. Because of the scope of Russell's work

† Quoted with permission of Science Press.

‡ Probably many of Russell's early ideas on the relations between mathematics and logic were formed during this exchange.

§ These appeared originally in the *Bulletin de la société mathématique de France*, 1904.

we defer its exposition to the next chapter; we close this chapter with some details concerning Zermelo's axioms for set theory. Zermelo's system is of importance in that it formed a point of departure and a model for many later systems which place set theory on an axiomatic foundation.

8.1 Zermelo conceived the idea of providing an axiom system, similar to those used in geometry, in which the undefined terms would be *set* and ε. (Compare *point* and $<$ as used in the simple order axioms, II 7). The aim would be to provide a consistent system of axioms in these terms, so that, if the only sets used in practice actually constituted *models* of the axiom system, no contradictions would result; and which, while not permitting the "too large" sets in the models, would allow large enough sets for all ordinary purposes of mathematics.

8.2 The manner in which Zermelo carried out this idea (see Zermelo [c]) was substantially as follows:†

With the undefined term *set* and binary relation ε between sets, the terms *subset* ($A \subset B$), *disjoint* and relation $=$ between sets may be defined essentially as in III 3.1.2, 3.1.3, and 3.6. ("$A = B$" would be defined as "$A \subset B$ and $B \subset A$"; some authors make an axiom of this, and call it the "Axiom of Extensionality)." If a relation $x \varepsilon M$ holds between sets x and M, we call x an *element* of M. The first five axioms are:

I. If a, b, and A are sets such that $a \varepsilon A$ and $a = b$, then $b \varepsilon A$.‡

II. If a and b are distinct sets, then there exists a set, denoted by $\{a, b\}$, such that $a \varepsilon \{a, b\}$, $b \varepsilon \{a, b\}$, and, if $x \varepsilon \{a, b\}$, then x is either a or b.

III. If $\mathfrak{M}$ is a set and $M \varepsilon \mathfrak{M}$ for at least one set M, then there exists a set $\mathfrak{S}\mathfrak{M}$ whose elements are the elements of all the elements of $\mathfrak{M}$, and only these. (In symbols, $x \varepsilon \mathfrak{S}\mathfrak{M}$ implies $x \varepsilon M$ for some M such that $M \varepsilon \mathfrak{M}$; and, if $x \varepsilon M$ such that $M \varepsilon \mathfrak{M}$, then $x \varepsilon \mathfrak{S}\mathfrak{M}$.)

IV. If M is a set, then there exists a set $\mathfrak{U}M$, called the *power set* of M, whose elements are the subsets of M, and only these. [Thus $\mathfrak{U}M$ is the analogue of the set 2^M defined in IV 3.2.3.4.]

V. If M is a set and P a property having significance for the elements of M, then there exists a subset $M(P)$ of M whose elements are all those elements of M that have property P, and only these.

By the term "significant property" in Axiom V is meant a property such that, if $x \varepsilon M$, then either x has property P or it does not. Thus, if M is the set of all natural numbers, P may be the property of being odd, or it

† For our elementary exposition we are following the first formulation of Fraenkel [F_2; 58ff], which differs in minor respects from Zermelo's. Fraenkel's axioms, as translated by R. L. W., are used with permission of B. G. Teubner.

‡ The necessity for this axiom rests on the undefined character of ε.

may be the property of being prime. If M is the set of all real numbers, P could be the property of being algebraic, even though there exist numbers about which we do not at present know whether they are algebraic or not. If M is the set of natural numbers and P is the property of being red, then P is not a significant property for the elements of M. (That the use of this notion in the axioms is objectionable was recognized by Fraenkel and others. For further comment on this axiom see Section 9.)

Since the next Axiom (VI) involves the null set and product set of a given set, we prove the following theorems which introduce and establish their existence:

8.2.1 Theorem.† *If there exists any set at all, then there exists a unique set* $\emptyset$ *called the null set, which has no elements and is a subset of every set.*

Theorem 8.2.1 follows from Axiom V if, for given set M, we let P be the property of *not being an element* of M; and from the definition of subset (cf. III 3.1.3–3.1.4).

8.2.2 Theorem. *If* M *is a set, then there exists a set* $\{M\}$ *whose only element is* M.

Proof. Given a set M, there exists the set $\emptyset$ by 8.2.1. And by Axiom IV there exists the set $\mathfrak{U}\emptyset$, which evidently is a set $\{\emptyset\}$ having exactly one element, namely $\emptyset$.‡

Thus, no matter whether M is $\emptyset$ or not, there exist two distinct sets M and N, and hence by Axiom II the set $\{M, N\}$. If we let P denote the property of being "equal to $(=)$ M," then from Axiom V we get $\{M, N\}(P)$ to be $\{M\}$.

8.2.3 Theorem. *If* M *is a set different from* $\emptyset$, *whose elements are disjoint sets, then there exists a set* $\mathfrak{P}M$ *whose elements are those subsets of* $\mathfrak{S}M$ *that have exactly one element in common with each element of* M. *If* $\emptyset \varepsilon M$, *then* $\mathfrak{P}M = \emptyset$.

Proof. The set $\mathfrak{S}M$ exists by Axiom III, and the set $\mathfrak{U}\mathfrak{S}M$ by Axiom IV. Let P be the property of having exactly one element in common with each element of M. Then by Axiom V the set $\mathfrak{U}\mathfrak{S}M(P)$ exists, and is the desired set $\mathfrak{P}M$. If $\emptyset \varepsilon M$, no element of $\mathfrak{U}\mathfrak{S}M$ has property P and in this case $\mathfrak{P}M$ is $\emptyset$.

† In Zermelo's formulation (see Zermelo [c]), the existence of the sets $\emptyset$ and $\{M\}$ (cf. 8.2.2) was postulated along with $\{a, b\}$ in Axiom II.

‡ We have to distinguish between the set $\emptyset$, which has *no* elements, and the set $\{\emptyset\}$, which has *one* element. Similarly, we have to distinguish, for any set M whatsoever, between M and $\{M\}$; the former may have many elements, whereas the latter has only one.

8.2.4 We can now state the next axiom which, it will be noted, is another form of the Choice Axiom:

VI. If M is a set different from $\emptyset$, whose elements are disjoint sets, and $\emptyset \notin M$, then the set $\mathfrak{P}M$ is different from $\emptyset$.

8.2.5 It will be noted that all the Axioms I to VI, as well as Theorems 8.2.1 to 8.2.3, are of the "If . . ., then . . ." form. None of them is an assertion of *existence*; in other words, none of them gives us permission to assume that any sort of set exists, *a priori*.

One way to remedy this defect is to assert the existence of some set; for example, we might add the axiom, "There exists at least one set." Or we might be more explicit and state the existence of the set $\emptyset$ (cf. footnote to Theorem 8.2.1). In either case, we would then have postulated the existence of sets with any finite cardinal number. For, if M is a set, then $\emptyset$ exists by Theorem 8.2.1, and the set $\{M\}$ with exactly one element exists by Theorem 8.2.2. Axiom II provides the set $\{\emptyset, \{\emptyset\}\}$ with two elements. And so on.

However, the existence of an infinite set would not follow from the axioms unless a stronger existence axiom is added. For this purpose we might use:

VII. There exists a set Z having the following properties: The set $\emptyset$ is an element of Z, and, if $x \varepsilon Z$, then $\{x\} \varepsilon Z$.†

With this axiom it is provable‡ that there exists a set Z_0 of the type postulated in this axiom which is a subset of *every* set Z of the type postulated here. The set Z_0 is a "smallest" set of this type; in other words, its elements are $\emptyset$, $\{\emptyset\}$, $\{\{\emptyset\}\}$, $\{\{\{\emptyset\}\}\}$, etc. Its similarity to the set of natural numbers (which has the smallest transfinite cardinal number $\aleph_0$) is evident. (Compare IV 4.1.1.) Indeed, many prefer, even when a non-axiomatic approach is undertaken (the so-called "naive" approach), to let these sets or combinations thereof serve either as the standards or norms (IV 4.1.1) for the natural numbers or as the natural numbers themselves. When we say that "a set M has 3 elements," we can take the position that this means that there is a (1-1)-correspondence between the elements of M and the elements of the set composed of $\emptyset$, $\{\emptyset\}$, and $\{\{\emptyset\}\}$.§

8.2.6 Even with Axiom VII, the existence of sets with arbitrarily large cardinal number is not assured. Although from the set Z_0 (defined in 8.2.5) we obtain, by virtue of Axiom IV, the set $\mathfrak{U}Z_0$ of cardinal number

† This is VII b of Fraenkel [F_2; 99].

‡ See Fraenkel, *loc. cit.*

§ See, for instance, Halmos [Hal].

$2^{\aleph_0} = c$; and (denoting $\mathfrak{U}Z_0$ by Z_1) a set $Z_2 = \mathfrak{U}Z_1$ of still larger cardinal number—in fact, then, a sequence of sets $Z_0, Z_1, \cdots, Z_n, \cdots$ such that, for each n, $\overline{\overline{Z}}_n < \overline{\overline{Z}}_{n+1}$; no set of cardinal number greater than *all* the numbers $\overline{\overline{Z}}_n$ is assured. In the "unrestricted" set theory (see Problem 26 of Chapter IV) we obtain such a number by first forming the set $Z = \{Z_0, Z_1, \cdots, Z_n, \cdots\}$, then the set $T = \mathfrak{S}Z$. Then $\overline{\overline{T}} > \overline{\overline{Z}}_n$ for all n. But, with Axioms I to VII alone, the "set" Z is not postulated as a *set* by these axioms.

From the standpoint of consistency, perhaps this is desirable. The axioms seem to provide all sets of the type ordinarily used, without going so far as to permit "too large" sets of the self-contradictory type such as the "set of all sets" or the "well-ordered set of all ordinal numbers." However, we may go even further, providing for the existence of such sets as the set T of the preceding paragraph by strengthening the assertions of the axioms. The details of this we do not go into here. The reader will find them, together with extended commentary, in Fraenkel [F_2; 114–115].

8.2.7 Relations and functions. In order to introduce the notion of cardinal number on the basis of the above axioms,† it is necessary to set up the relation of (1-1)-correspondence. It is possible to go further, however, since by introducing the notion of *ordered pair* we may set up general binary relations between sets and hence the concept of function as in II 4.4.1. This may be done by the Wiener-Kuratowski method‡ as follows:

8.2.7.1 **Definition.** If x and y are distinct sets, then the *ordered pair* (x, y) is the set $\{\{x\}, \{x, y\}\}$.

Note that, if x and y exist, then $\{x\}$ exists by Theorem 8.2.2, $\{x, y\}$ exists by Axiom II, and consequently the set $\{\{x\}, \{x, y\}\}$ exists by Axiom II; thus the ordered pair (x, y) exists. The symbol (x, y) can be extended to the case where $x = y$ by making the convention that (x, x) is the set $\{x\}$.

On the basis of Definition 8.2.7.1 it is easy to set up definitions of binary relations, correspondences, and functions, the latter two in substantially the same manner as was done in II 4.4.1 and IV 3.2.3.1. A *binary relation* in a set S can be defined as a collection $\mathfrak{R}$ of pairs $\{\{x\}, \{x, y\}\}$, where $x, y \in S$. For each such pair we may write, for example, $x < y$, or $x\mathfrak{R}y$, to indicate that the pair $\{\{x\}, \{x, y\}\}$ is an element of $\mathfrak{R}$.§

† The discussion in 8.2.5 and 8.2.6, in which sets of various cardinalities were mentioned, was of course "extra-axiomatic" and not *within* the system.

‡ See Wiener [a] and Kuratowski [a].

§ For a systematic development of set theory based on the Zermelo-Fraenkel axioms, see Suppes [Su].

9 Amendments to the Zermelo system

We have already referred (8.2) to the objectionable use of the notion "significant property" in Axiom V. As we pointed out in III 3.2.1, set and property are interchangeable notions. Hence to introduce into the axioms for sets the idea of "property" as though it were an intuitively known concept is clearly unsatisfactory. In the first place, it is tantamount to favoring one of two equivalent notions by accepting it as a kind of "universal" while subjecting the other to axiomatic treatment. And, in the second place, to sneak an equivalent notion into the axioms under the guise of a "universal" smacks of a vicious circle procedure [Fraenkel calls it a "jellyfish" ("quallenhaft") device].

9.1 The problem of how to eliminate this defect from the axiom system was attacked by both Skolem (see Skolem [a]) and Fraenkel ([F_1; 285ff], [F_2; 103ff]). As they reformulated the offending axiom, the "properties" admissible for determining subsets are defined by formulas built up from formulas of the type $x \varepsilon M$. Thus Fraenkel commences with elementary "set-functions"; a "set function" of a set x may be (1) any fixed given set, (2) the (variable) set x, (3) the set $\mathfrak{S}x$ (Axiom III), and (4) the set $\mathfrak{U}x$ (Axiom IV); and (5) if $\varphi(x)$ and $\psi(x)$ are set-functions, then so are $\{\varphi(x), \psi(x)\}$ and $\varphi(\psi(x))$. Axiom V is then restated as follows:

V′. Given a set M and two set-functions $\varphi(x)$ and $\psi(x)$; then there exists a subset M' of M which has as its elements exactly those elements x of M such that $\varphi(x) \varepsilon \psi(x)$.

For example, if M and A are given sets, then the assignment of meanings $\varphi(x) = x$, $\psi(x) = A$, with M as in V′, proves the existence of the set $A \cap M$.

Further details will be found in Fraenkel (*loc. cit.*). Also see Stoll [Sto; VII].

For axiom systems for set theory containing further improvements, one may consult papers of von Neumann [a, b], Bernays [a], and Gödel [G]. (See also Ackermann [a].)

ADDITIONAL BIBLIOGRAHY

Bell [B_3; 553ff]
Bernays [Ber]
Fraenkel [F_1; V], [F_2; 58ff], [F_3]
Fraenkel and Bar-Hillel [F-B]
Kelley [Ke; 6–17]
Lewis [Le; I–III, VI]
Menger [a]
Skolem [b]
Struik [St; V 2, VII, VIII]
Suppes [Su]

PROBLEMS

1. Prove on the basis of the Axioms I to VI of Section 8.2, augmented by the axiom "There exists at least one set," that there exists for every natural number n a set having n elements.

2. What is the relation between Axiom VI of 8.2.4 and the existence of product sets (VII 1.1.1)? How would you define products of sets S_ν, where the index ν has for domain an arbitrary set A (i.e., A could be N, or R, for instance)?

3. How would you define *ternary relation* in analogy with the definition of "binary relation" in 8.2.7.1; and, in general *n-ary relation*?

4. What would you think of a consistency proof of the system of Axioms I to VII of Section 8 which is based on the satisfiability criterion of Chapter II, using as model the set theory of Chapters III to V?

5. Now that we have discussed both the non-axiomatic (sometimes called "natural" or "naive") set theory (III–V), and the axiomatic set theory (Sections 8, 9), which would you prefer as a basis for your work if you were going to do a type of mathematics which requires set theory tools?

6. On the basis of Axioms I to VII of Section 8, can we give the definitions of "ordinary infinite" and "Dedekind infinite" as in III 4 and prove their equivalence as in III 5?

7. In view of the set theory contradictions and the alternative set theories which have been proposed (Zermelo-Fraenkel, von Neumann, Bernays, etc.) do you think it is feasible to speak of "*the* theory of sets" as though it were a unique thing?

IX

The Frege-Russell Thesis: Mathematics an Extension of Logic

In this chapter we again pick up the logical thread which had its origin in ancient Greece, and was strengthened by the introduction of symbolic methods during the last two centuries by Leibniz, De Morgan, Boole, and others. As we saw in the last chapter, despite the criticism of Kronecker, the set-theoretic approach became more and more popular and was freely used by Weierstrass, Cantor, and Dedekind in the founding of analysis. Frege, and Peano and his colleagues, attempted to make the relations of logic and mathematics even more intimate, using symbolic methods to construct what seemed to them a firmer and more rigorous foundation of mathematics.

1 The Frege-Russell thesis

Early in the present century a new shift of emphasis occurred; from being a method or a *tool* for the construction of mathematical theory, logic was advanced as a progenitor of mathematics. What is variously called the "logistic thesis," or the "Frege–Russell thesis," viz., that mathematics is derivable from, or an extension of, logic, was presented in extensive detail in the *Principia Mathematica* of Whitehead and Russell [P.M.].†

This thesis was not entirely new; its germ can be found in the writings of earlier logicians and mathematicians. Thus Jevons remarks:‡ "I hold that algebra is a highly developed logic, and number but logical discrimination." Frege held§ that "... inferences which on the face of it are peculiar

† Hereafter we shall denote the *Principia Mathematica*, 2nd ed., by P.M. (See Whitehead and Russell [P.M.].) This was preceded by Russell's book, *The Principles of Mathematics* [R_2], which, as explained in the Preface to P.M., was to have been the first volume of P.M. As the writing of P.M. progressed, however, it became clear that revisions, etc., of material contained in *The Principles of Mathematics* necessitated the writing of a completely new work.

‡ As quoted by Frege; cf. Frege [Fr; 22^e]; used here with permission of Basil Blackwell.

§ *Loc. cit.*, IVe; quoted by permission of Basil Blackwell.

to mathematics, such as that from n to $n + 1$, are based on the general laws of logic"; and† "... essential for mathematics ... is the recognition of its close connection with logic. I go so far as to agree with those who hold that it is impossible to effect any sharp separation of the two."

To Peano and his followers, however, the use of symbolic logic would seem to have been purely a means to an end. Logic, to them, was the *servant* of mathematics; a servant that mathematics admittedly could not dispense with. It was probably inevitable that a school of thought should arise that would conceive of this dependence on logic as the outward evidence of a relationship that was more like that of parent to child than servant to master. (See VIII 4.1.)

1.1 Background of the thesis

We can hardly do better here than let the chief protagonists of this theory speak for themselves: "[This work, i.e., P.M.] has arisen from the conjunction of two different studies, ... On the one hand we have the work of analysts and geometers, in the way of formulating and systematizing their axioms, and the work of Cantor and others on such matters as the theory of aggregates. On the other hand we have symbolic logic, which, after a necessary period of growth, has now, thanks to Peano and his followers, acquired the technical adaptability and the logical comprehensiveness that are essential to a mathematical instrument for dealing with what have hitherto been the beginnings of mathematics. From the combination of these two studies two results emerge, namely (1) that what were formerly taken, tacitly, or explicitly, as axioms, are either unnecessary or demonstrable; (2) that the same methods by which supposed axioms are demonstrated will give valuable results in regions, such as infinite number, which had formerly been regarded as inaccessible to human knowledge. Hence the scope of mathematics is enlarged both by the addition of new subjects and by *a backward extension into provinces hitherto abandoned to philosophy*" (P.M., Preface).‡

1.2 Axioms of logic as a basis

The ideal goal of this "backward extension" was a set of "primitive" axioms of logic, statements that the logician would call "true," on which one could base the whole of mathematics. In this way the main thesis might be considered established. Numbers, for instance, would be so defined as to have unique meanings; "1" would emerge as what we

† *Loc. cit.*, IX[e]; quoted by permission of Basil Blackwell.

‡ The italics are mine; quoted with permission of Cambridge University Press.

ordinarily think of as "unity," and not merely as a candidate for the initial element of a sequence as postulated in the Peano axioms (which might equally well be "0," "2," "3," or . . . , as "1," so far as the Peano axioms are concerned). At the same time, one would attempt so to construct the theory as to eliminate contradictions such as had appeared in the theory of sets and which had threatened the foundations of Frege's† work.

In what follows we shall give a sketch of the manner in which this program was carried out in P.M., together with some subsequent developments. Only enough detail will be given (1) to illustrate the general procedure which was followed and (2) to furnish a basis for the later discussion of intuitionist logic as well as of the modern "formalistic" approach to logic and mathematics.

2 Basic symbols; propositions and propositional functions

Much as in the axiomatic method, one starts with certain ideas that are left undefined, and with certain assumptions analogous to axioms. However, while recognizing this, Russell and Whitehead provide an auxiliary explanation of the "meanings" of these basic ideas. This is necessary, particularly since they are not dealing with such matters as euclidean geometry, for example, where one can usually assume that a reader is already familiar with the fundamental ideas in some form or other, as well as with just what the ultimate purposes are. The explanations are accompanied by the proviso that they "do not constitute definitions, because they really involve the ideas they explain."

2.1 Primitive ideas

The most elementary notion employed in P.M. is that of *proposition*; as used therein, it appears to be a *statement* that involves only definite or constant notions: thus "the sun is bright" or "this ink is black" are propositions. They are similar to the constants of elementary algebra or the points of geometry. But statements such as "x is red" or "x is a

† Whitehead and Russell make clear their debt to Frege: "In all questions of logical analysis, our chief debt is to Frege. Where we differ from him, it is largely because the contradictions showed that he, in common with all other logicians ancient and modern, had allowed some error to creep into his premises; but apart from the contradictions, it would have been almost impossible to detect this error." And to Cantor: "In Arithmetic and the theory of series, our whole work is based on that of Georg Cantor." And, although the geometric portion of P.M. was never completed, the geometric (as well as axiomatic) work of "v. Staudt, Pasch, Peano, Pieri, and Veblen" is acknowledged. (Quotations from P.M., Preface, by permission of Cambridge University Press.)

father of y" are not propositions since they contain *variables*. "Variable" is used much as in mathematics, every variable having associated with it a valid "domain of values." Statements that contain variables and which become propositions when specific values, i.e., constants, are substituted for all variables, are called *propositional functions* (see Section 4).

To propositions are assigned "truth-values," the truth-value of a proposition being *truth* if it is true, and *falsehood* if it is false.

These notions (cf. VIII 4.1) embody concepts whose nature is not generally agreed upon even today by logicians and philosophers, who point out that in P.M. they seem not to be used consistently. With good reason, one may insist that to confine the notion of "proposition" and "propositional function" to actual sentences is reminiscent of the early status of the concept of "function" in mathematical analysis as an expression of a certain kind containing variables. In P.M. it is stated (p. 92),† for instance, that by "elementary propositional function" is meant "an expression containing an undetermined constituent, i.e., a variable, or several such constituents, and such that, when the undetermined constituent or constituents are determined, i.e., when values are assigned to the variable or variables, the resulting value of the expression in question is an elementary proposition."

Probably the majority of logicians would hold that a proposition is the "meaning" expressed by the sentence embodying it, so that "This table is small" and "Dieser Tisch ist klein" are the same proposition although their forms are different. And it is quite possible that this is the view intended in P.M.‡

2.2 Basic symbols

Propositions are denoted by small Latin letters, usually p, q, r, etc. Propositional functions are denoted variously: Thus φx denotes a function with a single variable x.§ By φp one might indicate a propositional function whose variable p is replaceable by propositions. The assertion of the truth of a proposition or propositional function is indicated by the symbol $\vdash$. Thus

$$\vdash . p$$

asserts the truth of the proposition p. All axioms are assertions, and hence preceded by the symbol $\vdash$. Without this symbol, no assumption is

† Quoted with permission of Cambridge University Press.

‡ Compare the discussion in IV 4.1.1 regarding the concept of number and the symbols used to denote numbers.

§ Parentheses are avoided to a large extent in P.M.

being made regarding the truth of a proposition p; in particular, p, q, $r, \cdots$ are symbols for propositions which may be either true or false.

Most of the parentheses, brackets, etc., are replaced by dots on the line; e.g., "$.\, p \vee q\, .$" instead of "$(p \vee q)$." For brackets enclosing parentheses, a "double dot" : is used. And generally a larger number of dots indicates an "outside" parenthesis, a smaller number an "inside" parenthesis. Hence, if one encounters : in a formula, it indicates a bracket enclosing everything to the next double dot : or to the end of the expression. (Cf. pp. 9ff, 16ff of P.M.)

Negation is represented by the $\sim$ already used in II 3.1; thus $\sim p$ represents denial of a proposition p. The and/or symbol $\vee$ is used as in III 3.3.1. The result of the application of $\sim$ or $\vee$ to propositions is again a proposition (if p is a proposition, then $\sim p$ is a proposition, etc.).

2.2.1 Implication. P.M. uses $\supset$ for implication, which is *defined* in terms of the basic symbols as follows:

(2.2a) $$p \supset q \,.=.\, \sim p \vee q \qquad \text{Df}$$

The Df stands for "definition" and in conjunction with = denotes "is defined to mean." Both = and, ultimately, $\supset$ are usually set off by dots. Expression (2.2a) can be read "p implies q is defined to mean that either p is false or q is true."

This definition of implication may at first seem strange, since in ordinary affairs of life we so commonly assume that in the statement "p implies q" or, as more commonly put, "If p, then q," the "antecedent" p is necessarily connected in some way with the "consequent" q. Thus, it would hardly seem natural to assert that "if water flows downhill, then light is bright." For in this case there is no apparent connection between antecedent and consequent. Nevertheless, if p is the proposition "Water flows downhill" and q the proposition "Light is bright," then $p \supset q$ (see 3.5—especially 3.5.3). For in view of the meaning of $\vee$, the expression "$\sim p \vee q$" is considered true whenever (1) p is true and q is true, (2) p is false and q is either (i) true or (ii) false; the only case ruled out by (2.2a) is that where p is true and q is not true. What one is interested in, here, is truth-value; whether there is some necessary connection between p and q is not of importance. The truth-value of $p \supset q$, like the truth-values of $p \vee q$ and $\sim p$, depends on the truth-values of the constituent propositions p, q; in the case of $p \vee q$, the truth-value is truth if the truth-value of either p or q is truth. In order to distinguish it from other types of implication, (2.2a) is called *material implication* (cf. 3.8).

If we examine the kind of implication that we use in mathematics, we find that it actually corresponds to the "material" type. Consider, for example, the axiom system O″ obtained by adjoining to the simple order axioms (II 7) the axiom, "There exist two and only two distinct points in C," and changing the "If" part of Axiom 3 to "If x, y, z are distinct." The new Axiom 3 is not independent in the resulting system. For suppose we let p denote the statement "There exist three distinct points x, y, and z such that $x < y$ and $y < z$," and q the statement "$x < z$"; then the new Axiom 3 states "If p, then q," or "p implies q." Now, in view of the new axiom, p is false, or, as we sometimes put it, "p cannot be satisfied"; and whatever the statement "p implies q" may be, under these conditions we take it as valid in the system. In other words, if p is false, we accept that p implies q, no matter whether q is true or false. We might recall here the discussion in II 1.2 of vacuous satisfiability.

2.3 The primitive propositions

The axioms of what may be called the *propositional calculus*, which formed the initial portion of P.M., and which P.M. called "primitive propositions" (see P.M., p. 96), were as follows:

I	$\vdash : p \vee p \,.\supset.\, p$
II	$\vdash : q \,.\supset.\, p \vee q$
III	$\vdash : p \vee q \,.\supset.\, q \vee p$
IV	$\vdash : p \vee (q \vee r) \,.\supset.\, q \vee (p \vee r)$
V	$\vdash :. q \supset r \,.\supset:\, p \vee q \,.\supset.\, p \vee r$

(The dots after $\vdash$ always indicate range, which includes everything following the dots until either an equal number of dots is encountered preceding a symbol $\supset$, or the end of the expression.) These axioms may be read as follows:

I. If p is true or p is true, then p is true.

In P.M., I is called the "principle of tautology."

II. If q is true, then p or q is true.

P.M. calls II the "principle of addition"—to a true proposition any alternative may be added and the resulting proposition is true.

III. If either p or q is true, then either q or p is true.

This is obviously the commutativity of $\vee$; in P.M. it is called the "principle of permutation."

IV. If p is true, or either q or r is true, then q is true, or either p or r is true.

This is clearly a type of associative law.

V. If q implies r, then "p or q" implies "p or r." Notice the similarity to the monotonic law of real number arithmetic (VI 2.2.4). P.M. calls V the "principle of summation."

These "principles," from the logistic standpoint, are "primitive logical truths," and constitute a system sufficient for the theory of deduction.† They are to be considered true regardless of the truth values of the propositions p, q, r. The usual name for such expressions is "tautology." They must, however, be accompanied by rules for determining *their own consequences.*

3 Calculus of propositions

Axioms I to V constitute, in their symbolic form, what would today be called a set of basic *formulas.* Formulas of a similar character proved from them would correspond to "theorems."

3.1 Manner of proving theorems

Inasmuch as the system so constituted is supposed to embody a "theory of deduction," it raises what seems at first sight to be a paradoxical question, viz., What kind of deduction shall we use to deduce the properties of deduction? The answer is that we give two simple rules for calculation of formulas from formulas.

They are:

(1) Substitution of a formula A for *all* occurrences of a proposition p in a given formula. (For example, substitution of $p \vee q$ for p in I gives $(p \vee q) \vee (p \vee q) \, . \supset . \, (p \vee q)$.)‡

(2) *Modus ponens:* "Anything implied by a true proposition is true." (That is, no matter what the formulas A and B, if we have "$\vdash . A$" and "$\vdash . A \supset B$," then we may infer "$\vdash . B$.")

3.2 Some sample proofs

We insert here some examples of how formulas are derived from the "primitive propositions." In particular, we show through the following sequence of lemmas how to derive the Law of the Excluded Middle.

† As a matter of fact, I to III and V constitute such a system, since Bernays [b] showed that IV is derivable from them. Regarding the independence of these postulates, see Henle [a].

‡ This rule is not explicitly stated in P.M., but is employed throughout. Cf. Russell [R_1; 151f].

3.2.1 $\vdash :. q \supset r . \supset : p \supset q . \supset . p \supset r$

Proof. Using rule (1), substitute $\sim p$ for p in V:

(3.2.1a) $\vdash :. q \supset r . \supset : \sim p \vee q . \supset . \sim p \vee r$

By (2.2a), $\sim p \vee q$ is the same as $p \supset q$. Hence (3.2.1a) becomes

$$\vdash :. q \supset r . \supset : p \supset q . \supset . p \supset r$$

3.2.1.1 It follows easily from 3.2.1 that, whenever we have propositions of the form $a \supset b$, $b \supset c$ asserted, and $a \supset c$ is to be proved, then the proof may proceed as follows: Replacing p, q, r in 3.2.1 by b, c, a respectively gives

$$\vdash :. b \supset c . \supset : a \supset b . \supset . a \supset c$$

$$\vdash . b \supset c$$

Then by rule (2) we get

$$\vdash : a \supset b . \supset . a \supset c$$

$$\vdash . a \supset b$$

Again by rule (2),

$$\vdash . a \supset c$$

3.2.2 $\vdash : p . \supset . p \vee p$

Proof. Substitute p for q in II.

3.2.3 $\vdash . p \supset p$

Proof. In 3.2.1, substitute $p \vee p$ for q, and p for r:

$$\vdash :: p \vee p . \supset . p : \supset :. p . \supset . p \vee p : \supset . p \supset p$$

By virtue of I, $p \vee p . \supset . p$ is a true proposition. Hence, by rule (2), we get

$$\vdash :. p . \supset . p \vee p : \supset . p \supset p$$

which in turn gives, by 3.2.2 and *modus ponens*,

$$\vdash . p \supset p$$

3.2.4 $\vdash . \sim p \vee p$

Proof. This is merely a restatement of 3.2.3, by virtue of Definition (2.2a).

3.2.5 $\vdash . p \vee \sim p$ (Law of the Excluded Middle)

Proof. By III, with $\sim p$ and p substituted for p and q respectively, we get

(3.2.5a) $\vdash : \sim p \vee p . \supset . p \vee \sim p$

But $\sim p \vee p$ is true by 3.2.4; hence by *modus ponens* (3.2.5a) and 3.2.4 give

$$\vdash . p \vee \sim p$$

3.3 The *reductio ad absurdum*

The type of proof which involves showing a proposition p false, by showing that p leads to absurdity, is embodied in the assertion:

3.3.1 $$\vdash : p \supset \sim p . \supset . \sim p$$

Proof. In I substitute $\sim p$ for p to get

$$\vdash : \sim p \vee \sim p . \supset . \sim p$$

Then by (2.2a) this becomes the desired assertion.

3.3.2 As a matter of fact, there is a more general form for *reductio ad absurdum*, of which 3.3.1 is a special case, namely:

$$\vdash : p \supset q . \supset : p \supset \sim q . \supset . \sim p$$

(That 3.3.1 is obtainable from 3.3.2 may be seen by substituting p for q in 3.3.2 and applying 3.2.3 and *modus ponens.*)

3.4 The double negative

An important principle of classical logic is that the double negative of a proposition is equivalent to the proposition. As soon as equivalence is defined, this may be derived from the primitive propositions and the Law of the Excluded Middle as follows:

3.4.1 $$\vdash . p \supset \sim(\sim p)$$

Proof. In 3.2.5 substitute $\sim p$ for p:

$$\vdash . \sim p \vee \sim(\sim p)$$

Then by Definition (2.2a) this becomes the desired assertion.

3.4.2 $$\vdash . p \vee \sim\{\sim(\sim p)\}$$

Proof. In V replace q and r by $\sim p$ and $\sim\{\sim(\sim p)\}$ respectively:

(3.4.2a) $\vdash :. \sim p . \supset . \sim\{\sim(\sim p)\} . \supset : p \vee \sim p . \supset . p \vee \sim\{\sim(\sim p)\}$

In 3.4.1 replace p by $\sim p$:

(3.4.2b) $$\vdash : \sim p . \supset . \sim\{\sim(\sim p)\}$$

Relations (3.4.2a) and (3.4.2b), together with *modus ponens*, give

$$\vdash : p \vee \sim p . \supset . p \vee \sim\{\sim(\sim p)\},$$

which, with 3.2.5 and *modus ponens*, gives the desired assertion.

3.4.3 $$\vdash . \sim(\sim p) \supset p$$

Proof. In III replace q by $\sim\{\sim(\sim p)\}$:

$$\vdash : p \vee \sim\{\sim(\sim p)\} . \supset . \sim\{\sim(\sim p)\} \vee p$$

This, together with 3.4.2 and *modus ponens*, gives

$$\vdash . \sim\{\sim(\sim p)\} \vee p$$

Definition (2.2a) yields the desired assertion.

3.5 Miscellaneous theorems

The following exemplify other theorems important in the applications of the theory of deduction.

Frequently used is the type of argument which proves a proposition p by showing that the assumption of its falsity implies p. This is embodied in the P.M. calculus of propositions by:

3.5.1 $$\vdash : \sim p \supset p . \supset . p$$

Proof. From 3.4.1 and 3.2.1 (in the latter we substitute p for q, $\sim(\sim p)$ for r, and $\sim p$ for p), we get

$$\vdash : \sim p \supset p . \supset . \sim p \supset \sim(\sim p)$$

Substituting $\sim p$ for p in 3.3.1 gives

$$\vdash : \sim p \supset \sim(\sim p) . \supset . \sim(\sim p)$$

Hence by 3.2.1.1 we have

$$\vdash : \sim p \supset p . \supset . \sim(\sim p)$$

Application of 3.4.3 and 3.2.1.1 completes the proof.

A useful formula, also, is the following:

3.5.2 $$\vdash : p . \supset . p \vee q$$

Proof. Axioms II and III (with II in the form $p . \supset . q \vee p$, etc.) together with 3.2.1.1 give the desired relation.

A feature of material implication, often termed "paradoxical," is the theorem that a false proposition implies every proposition: This derives from:

3.5.3 $$\vdash : \sim p . \supset . p \supset q$$

Proof. Substitute $\sim p$ for p in 3.5.2 and apply (2.2a).

Thus if A is a false statement—"$\vdash . \sim A$"—and B is any statement whatsoever, substitution of "A" for "p" and "B" for "q" in 3.5.3 gives "$\vdash : \sim A . \supset . A \supset B$"; whereupon *modus ponens* gives "$\vdash . A \supset B$."

Another favorite device in mathematics is to prove that a proposition p implies a proposition q by showing, instead, that the denial of q implies the denial of p. The basis for this is the theorem:

3.5.4 $$\vdash : p \supset \sim q . \supset . q \supset \sim p$$

Proof. In Axiom III replace p by $\sim p$ and q by $\sim q$, giving

$$\vdash : \sim p \vee \sim q . \supset . \sim q \vee \sim p$$

Application of (2.2a) gives the desired assertion.

Note that 3.5.4 takes the form stated in words prior to 3.5.4, if we change p to $\sim q$ and q to p (and apply 3.4.3).

Another concomitant of material implication is the assertion that a true proposition is implied by *any* proposition. This follows from

3.5.5 $$\vdash : q . \supset . p \supset q$$

Proof. In Axiom II replace p by $\sim p$ and apply the definition of implication (2.2a).

Thus if B is a true statement—"$\vdash . B$"—and A is any statement, 3.5.5 and *modus ponens* give "$\vdash . A \supset B$."

3.6 The logical product

An important detail of the Russell-Whitehead propositional calculus is the non-independence of the logical product, there indicated by a dot ., and the symbols $\sim$, v. In P.M. (p. 109), the symbol . is defined as follows:

(3.6a) $$p . q . = . \sim(\sim p \vee \sim q) \qquad \text{Df}$$

The intuitive idea behind (3.6a) is that assertion of p and q is equivalent to asserting the denial of the logical disjunction of the denials of p and q.

3.6.1 A consequence of Definition (3.6a) is the equivalence, in the P.M. calculus of propositions, of the Law of the Excluded Middle and the Law of Contradiction. The latter has the form:

3.6.2 $$\vdash . \sim(p . \sim p)$$

In words, 3.6.2 states that it is false that a proposition is both true and false. To derive 3.6.2 from (3.6a) and the Law of the Excluded Middle (3.2.5), note that, by 3.2.3, Definition (3.6a) gives

$$\vdash : p . q . \supset . \sim(\sim p \vee \sim q)$$

and hence, by 3.5.4 and *modus ponens*,

(3.6.2a) $$\vdash : \sim p \vee \sim q . \supset . \sim(p . q)$$

Since, by 3.2.5, $\vdash . \sim p \vee \sim(\sim p)$, we get the desired relation by replacing q by $\sim p$ in (3.6.2a).

3.6.3 Note that combination of 3.5.3 and 3.6.2 furnishes the basis of the much-quoted assertion that, *if a logical system contains a contradiction —i.e., an assertion* $\vdash . A . \sim A$, *where A is a formula—then all propositions are provable in the system.*

3.7 Sheffer's "stroke" symbol

In closing this section, we might point out some developments which should prove of great interest to the student of mathematics. Since P.M. was first published, with its two undefined symbols (or "logical constants") $\sim$ and $\vee$ and five axioms (later reduced to four axioms by Bernays—cf. 2.3), it has been shown that *one* undefined symbol and *one* axiom are sufficient. (Cf. problems 29 and 30 of Chapter II.)

The symbol referred to is / and is called "stroke." It was introduced by H. M. Sheffer [a], and its intuitive meaning can correspond to either (1) total rejection (employed by Sheffer, *loc. cit.*), or (2) incompatibility. To be more specific, p/q would mean, in case (1), that *both* p and q are false—in the previous symbols, $\sim p . \sim q$; and, in case (2), that *at least one* of the propositions p, q is false—in the previous symbols, $\sim p \vee \sim q$. In either case, $\sim p$ would be defined as p/p, whereas (1) would require that we define $p \vee q$ to mean $(p/q)/(p/q)$, and (2) that $p \vee q$ mean $(p/p)(q/q)$. The intuitive meaning (2) is generally preferred, however (as leading to somewhat simpler forms).†

Thus, treating the new symbol / as undefined, we would define the P.M. symbols $\sim$ and $\vee$ as follows:

(3.7a) $$\sim p . = . p/p \quad \text{Df}$$

(3.7b) $$p \vee q . = . (p/p)/(q/q) \quad \text{Df}$$

Since $\supset$ and . are defined in terms of $\sim$ and $\vee$, it is clear that by means of (3.7a) and (3.7b) they are definable in terms of / alone. For example,

(3.7c) $$p \supset q . = . p/(q/q)$$

If meaning (1) had been selected for /, then $p \supset q$ would have had to be defined as $\{(p/p)/q\}/\{(p/p)/q\}$. The simpler form (3.7c) is one of the reasons for preferring meaning (2).

That Axioms I to V may be replaced by a single axiom in terms of / was shown by J. Nicod [a]. This axiom is a single formula in terms of propositional symbols $p, q, \cdots$ and the stroke /, and is not a "fictitious"

† This is the sense in which Quine uses the symbol; see his [Q_1; V].

combination of separate axioms of the sort discussed in II 3.6. It does, however, have the disadvantage of being rather complicated, so that it is probably better to found the system of P.M. on the basis outlined previously. For further commentary, the reader is referred to Quine [Q; 45–49]. Also see the comment of Hilbert-Bernays [H-B; 64–65] and the work of Tarski referred to therein.

3.8 Other types of "implication"

In dealing solely with propositions whose "truth values" are of concern to us, Definition (2.2a) seems to embody the most generally accepted type of implication, and is certainly the one commonly employed in mathematics. It is not, however, the only kind definable, nor, when only the forms or structure of propositions are under consideration (with no attention paid to their content or "truth"), is it necessarily the most desirable.

The first attempt at an alternative type of implication was made by C. I. Lewis. This resulted in a theory of deduction which he called *strict implication*. A new undefined concept, denoted by $\Diamond p$, which may be read "p is possible," was introduced. The relation of strict implication was then defined thus:

(3.8a) $$p \prec q \,.=.\, \sim\Diamond(p \,.\, \sim q) \qquad \text{Df}$$

The formula $p \prec q$ may be read "p strictly implies q." Between this type of implication, and $p \supset q$, as defined in (2.2a), which may now be read "p materially implies q," the following relation holds:

(3.8b) $$p \prec q \,.\prec.\, p \supset q$$

(The "converse" implication does not hold.) Using the interpretation of 2.2.1, if p is "Water flows downhill," and q is "Light is bright," then, although $p \supset q$ holds, $p \prec q$ does not hold.

The paradoxical properties of material implication proved in 3.5.3 ("A false proposition implies every proposition") and 3.5.5 ("If a proposition is true, then it is implied by every proposition") do not hold for strict implication. (See, however, Halldén [a].)

For further details regarding strict implication, the reader is referred to Lewis and Langford [L-L; 122ff]. In all, Lewis studied five different systems of deduction (see Lewis and Langford [L-L; 492ff]), remarking (*loc. cit.*, pp. 501–2), "Prevailing good use in logical inference—the practice in mathematical deductions, for example—is not sufficiently precise and self-conscious to determine clearly which of these five systems expresses the acceptable principles of deduction."

Regarding the general problem, the reader is also referred to Tarski [T; 23ff], Emch [a], Lewis [a], Curry [Cu].

3.9 Truth-table methods

An important alternative to the methods of proof employed in the propositional calculus of P.M. is that of matrix or "truth-table" methods. Although implicit in the work of some earlier authors, they were first systematically exploited by E. L. Post [a] and J. Lukasiewicz in 1921. Their importance, from the present point of view, derives from the fact that they provide an *algorithm* or rule for determining whether a given formula is a tautology (i.e., true independently of the truth values of the constituent propositions) or not.

Instead of starting with primitive propositions, tables defining the truth values of the basic constants ., v, ~, and ⊃ are provided. Denoting "true" and "false" by "t" and "f" respectively, these are given in the following form:

I

p	q	$p \cdot q$	$p \vee q$	$\sim p$	$p \supset q$
t	t	t	t	f	t
t	f	f	t	f	f
f	t	f	t	t	t
f	f	f	f	t	t

In each of the last four columns are given the truth values corresponding to the possible combinations of truth values of p and q to the left.

On the basis of the tabulation I, tests of arbitrary formulae may be made. For example, consider the formula constituting Axiom I of 2.3: $p \vee p \,.\supset .\, p$. The following table illustrates the method:

II

p	q	$p \vee p$	$p \vee p \,.\supset .\, p$
t	t	t	t
t	f	t	t
f	t	f	t
f	f	f	t

In Table II, the column under "$p \vee p$" is formed by observing that "$p \vee q$" is true in Table I whenever p or q has a "t" value; then the column under "$p \vee p \,.\supset .\, p$" is formed by observing that in Table I, $p \supset q$ is true whenever both p and q have value "t" or when p has value "f." That the last column contains only "t" indicates that the formula is valid irrespective of the truth values of p and q; i.e., the formula is a tautology.

In case more than two primitive propositions are involved in a formula, the possible combinations of truth values necessitates using a more extended matrix, of course; for k propositions, the number of rows will be 2^k. For example, the table validating Axiom V is:

p	q	r	$q \supset r$	$p \vee q$	$p \vee r$	$p \vee q . \supset . p \vee r$	$q \supset r . \supset : p \vee q \supset p \vee r$
t	t	t	t	t	t	t	t
t	t	f	f	t	t	t	t
t	f	t	t	t	t	t	t
f	t	t	t	t	t	t	t
t	f	f	t	t	t	t	t
f	f	t	t	f	t	t	t
f	t	f	f	t	f	f	t
f	f	f	t	f	f	t	t

The first three columns contain the eight possible combinations of truth values of p, q, and r. Then the next three columns contain the corresponding truth values of the innermost parentheses $q \supset r$, $p \vee q$, and $p \vee r$. The next inner parenthesis is "$p \vee q . \supset . p \vee r$", whose truth values are obtained from the values of $p \vee q$ and $p \vee r$, using the table for "$\supset$" in I. The final column is made up from the fourth and seventh columns of the table, again using the table for "$\supset$." Since the final column contains only "t's," the formula being tested is a tautology.

As already observed, the virtue of this method is that it forms an algorithm; like that for finding a square root of an integer, or the greatest common divisor of two integers ("euclidean algorithm"), it always works. We have a method of deciding, for every formula, whether it is a tautology or not, and from this standpoint—the "truth-table" standpoint—the propositional calculus is a "decidable" theory (see XI 5). Moreover it can be shown, although we do not go into this, that the two methods, (1) provability on the basis of the axioms as in P.M. and (2) validation by truth table, are *equivalent*; if a formula is provable from the axioms, then it will turn out to be a tautology by the truth-table method, and conversely. As a consequence, the propositional calculus is *complete* in the sense that every tautology can be proved. As a matter of fact it can be shown that the system is complete in the sense that if any formula not provable from the axioms is added to them, then contradiction results (this is the logical analogue of completeness in the sense of II 4.3.1; see Problem 21 at the end of Chapter II). The distinction will be made clearer when we consider the predicate calculus below.

4 Forms of general propositions; the predicate calculus

Although the propositional calculus is a complete system in the senses described above, it is not sufficient for the needs of mathematics. In the propositional calculus no attention is paid to the *form* or *structure* of a proposition; e.g., whether it may be a statement such as "For every two numbers x and y, $x + y = y + x$," or "There exists a number x such that $x^2 - 4x + 4 = 0$." It is necessary to analyze the various forms of propositions that may arise, particularly if we are to achieve a foundation for arithmetic.

For this purpose, we may first symbolize the subject-predicate form of a proposition by a so-called *predicate variable* or *propositional form* $P(x)$, in which P denotes the predicate and x the subject; or, more generally, a functional symbol $P(x_1, x_2, \cdots, x_n)$ in case more than one variable enters into the predicate. And as we may expect, attention must be given to the domains of these functions, just as in the case of the set-functions introduced in II 4.4.1; the range, however, will generally be two-valued—truth or falsity—and for some values of the individual variables in P, P may be expected to be true, and for others false.

4.1 Free and bound variables

For the study of the "for all" and "there exists" type of propositions exemplified above, the customary symbols are $\forall$ and $\exists$. Thus, for a propositional function of one variable x, we have the forms:

(4.1a) $$\forall x\, P(x)$$

(4.1b) $$\exists x\, P(x).$$

We may read (4.1a): "For every x (in its domain), $P(x)$"; and (4.1b): "There exists x such that $P(x)$," or "For some x, $P(x)$." The occurrences of x in these expressions may be compared to that of x in $\int_0^1 x^2\,dx$; as the latter is a constant, the appearance of x in it was called "apparent" in P.M., x being termed an *apparent variable* in both (4.1a) and (4.1b). Today the more common term for such an x is *bound variable*; while the x in $P(x)$ itself is called a *free variable*. [In P.M. the symbol (x) was used instead of $\forall x$).]

The symbols $\forall$ and $\exists$ are called *quantifiers*; $\forall$ is often called the "universal quantifier" and $\exists$ the "existential quantifier." For functions of more than one variable, various combinations may occur: "$\forall x \forall y\, P(x, y)$," "$\forall x \exists y\, P(x, y)$," and "$\exists x \forall y \forall z\, P(x, y, z)$," for example.

4.2 Predicate calculus

We may now extend the propositional calculus to a more general theory variously called the *predicate calculus*, *calculus of propositional functions*, or *functional calculus*, of which the propositional calculus forms only a part. We shall describe as much of it as will serve our purposes, referring the reader to treatises on logic for details and extensions. (See, for instance, Hilbert and Ackermann [H-A].) In particular, we need consider only the *first-order* predicate calculus, in which the predicate variables are not themselves quantified (i.e., no forms such as $\forall P\, P(x)$ occur).†

4.2.1 Formulas

First we must specify what may be called a *formula*. As in the propositional calculus, we have propositional variables $p, q, r, \cdots$, as well as the logical constants $\vee$, $\sim$, $\cdot$, and $\supset$. In addition, we have individual ("subject" or "object") variables $x, y, z, \cdots$ as well as predicate variables (functions) $P(x)$, $P(x, y)$, $Q(x)$, etc., in which occur individual variables; and, finally, the quantifiers $\forall$ and $\exists$. As formulas we have, in addition to propositional and predicate variables, the various combinations that we can build up from these by using both the logical constants and quantifiers. Thus, if A is any formula, then $\sim A$ is a formula. If A and B are formulas, then $A \cdot B$, $A \vee B$, $A \supset B$ are formulas, provided that A and B do not contain any variable that is free in one and bound in the other. If A is a formula in which x occurs as a free variable, then $\forall x\, A$ and $\exists x\, A$ are formulas.

4.2.2 Axioms and methods of proof

In the case of the propositional calculus, we gave a set of axioms and two proof methods (substitution and *modus ponens*) in order to establish theorems concerning the tautological character of formulas. We also pointed out, however, that one could alternatively use truth tables to arrive at the same results, thereby achieving not only greater simplicity in many cases, but establishing an algorithm for deciding in every case whether a formula is a tautology or not. Can we proceed analogously in the case of the predicate calculus?

Consider the formula $\forall x.\ P(x) \supset Q(x)$, P and Q being predicate variables and x the sole individual variable involved. What would we mean by specifying that it is a "tautology" or, to use the more commonly preferred term, "universally valid"? Aside from the possible interpretations of P and Q, consider the question of the domain of the variable x.

† Quantification of predicate variables leads to higher-order calculi. In following the now common procedure of making a separation of the first-order calculus from those of higher order, we are emulating Hilbert and Ackermann [H-A] as well as later authors.

This domain might be of any cardinality, finite or infinite; this feature alone might be enough to discourage us from attempting truth-table methods, since *a priori* an infinite, even uncountably infinite, variety of truth tables would be needed. Despite this, we can develop truth-table methods and a mode of reasoning about them which will yield results. However, we shall avoid further consideration of this possibility and turn, instead, to an axiomatic presentation.

As axioms we select those axioms of the propositional calculus given in 2.3 and the following:

VI $\vdash . A(s) \supset \exists x\, A(x)$

VII $\vdash . \forall x\, A(x) \supset A(s)$

We may read VI: "If A holds for a particular s, then there exists an x for which A holds"; and VII: "If A holds for all x, then it holds for any particular individual s."

As rules for proof, in addition to *modus ponens* we have:

(1) From $A \supset P(x)$ we may infer $A \supset \forall x\, P(x)$,

(2) From $P(x) \supset A$ we may infer $\exists x\, P(x) \supset A$,

where in both (1) and (2), A is any proposition not containing x as a free variable and $P(x)$ has x as a free variable. We must also extend the substitution rule so as to allow, for example, substitution of a formula A for an individual variable x in a formula F so long as (i) A contains no individual variable which is bound in F, and (ii) all occurrences of x in F are replaced by A, and (iii) the result of the substitution is a formula as defined above.

Since we have proved the Law of the Excluded Middle—$p \vee \sim p$—in 3.2.5, we may substitute $A(x)$ for p and obtain: $A(x) \vee \sim A(x)$. By 3.5.5, $A(x) \vee \sim A(x) . \supset : (p \vee \sim p) . \supset . A(x) \vee \sim A(x)$. Hence, by *modus ponens*, $p \vee \sim p . \supset . A(x) \vee \sim A(x)$. Then by rule (1) above and *modus ponens*, we may infer $\forall x . A(x) \vee \sim A(x)$. Such a formula, we would conclude, is true for every domain and, hence, valid in the sense of the predicate calculus.

For a more extended presentation of the calculus, we refer the reader to treatises on logic (e.g., Hilbert and Ackermann [H-A]).

4.2.3 Completeness

We found that the propositional calculus was not only complete in the sense that every tautology is provable, but also in the "logical" sense that addition of an unprovable proposition would result in contradiction (cf. II, Problems 21 and 22). Does the like hold for the predicate calculus?

That the predicate calculus is not complete in the logical sense can be seen from a consideration of the formula $\exists x \,.\, A(x) \supset \forall x\, A(x)$. That this is not provable from the axioms (i.e., is not universally valid), can be inferred from a consideration of any domain for x containing at least two individuals a and b; but in a domain containing a single element the formula would be true, so it is not logically inconsistent with the axioms of the predicate calculus. (A formal proof of the impossibility of proving it may be found in Hilbert–Ackermann [H-A], pp. 92–95.)

However, the predicate calculus is complete in the sense that every universally valid formula is provable. This was first shown by K. Gödel in 1930 (Gödel [c]). Gödel's theorem† may be stated as follows:

For every formula A of the first-order predicate calculus, either $\sim A$ *is provable, or A is satisfiable in the domain of the natural numbers.* (A formula is called "satisfiable" in the domain N of natural numbers if it holds for some assignment of natural numbers to its individual variables and truth values for the predicate variables.)

It follows from this theorem that *if a formula A is universally valid it is provable*; for if it is universally valid, it certainly is valid in N, and hence $\sim A$ is not satisfiable in N. Thus, by the theorem, $\sim(\sim A)$ is provable; and by the propositional calculus, $\sim(\sim A)$ and A are equivalent. By the same line of reasoning, we have a famous theorem of Löwenheim (1915) to the effect that *if a formula is valid in the domain N, then it is valid in every domain* (since provability implies universal validity).

Moreover, we can now get another famous theorem, the so-called Löwenheim–Skolem theorem: *If a formula A is satisfiable in some non-empty domain D, then it is satisfiable in N.* For if A is satisfiable, then $\sim A$ is not universally valid and therefore not provable. It follows from Gödel's theorem that A is satisfiable in the domain N.

4.2.4 Skolem's Paradox

From the result just stated has come a famous paradox called "Skolem's Paradox" which may be interpolated here. We recall that mention was made, in the discussion of the axiomatic method in Chapters I and II, of the way in which the logical apparatus was taken for granted; only the "technical" terms were regarded as undefined, while the logical terms were placed in a different—"universal"—category (see II 4.10, for instance). We are now in a position to alter this situation by "imbedding" the technical theory in the predicate calculus, the logical apparatus thereby

† This theorem is not to be confused with the more famous theorem to be discussed later in the next chapter.

becoming also axiomatized. Instances of this will be seen in the discussion of formal systems in Chapter XI.

But how will the set theory apparatus be handled in such a treatment? We can, if we wish, employ a development of classes similar to that exemplified in P.M., which will be described in the next section. However, it is now more common to use an axiomatic treatment such as that of Zermelo–Fraenkel (cf. VIII, 8,9), in which the axioms are formulated in the logic of the first-order predicate calculus (with addition of suitable set-theoretic predicate variables such as a predicate symbol for "$x = y$").†

Let us now consider such an axiomatized set theory. Since it is formulated in the logic of the first-order predicate calculus, all formulas of the theory must be satisfiable in N. But how is this compatible with the fact that such theorems as that expressing the uncountability of the set of all subsets of N are provable in the theory?

One way of looking at this is to observe that the collection of ordered pairs constituting the (1-1)-correspondence between the new domain D_1 and the set N is not one of the "sets" of the axiom system. What is a set in the usual "naive" sense is not, in other words, a set in the axiomatic sense of the term. And, according to the Löwenheim–Skolem theorem, it is necessary to conclude that no set theory, fully formalized within the predicate calculus, can force the usual interpretation of the "uncountable." Of course, it may be that no reasonable axiom system for set theory is satisfiable, thus implying that the conjunction A of all the axioms is not satisfiable and that $\sim A$ is provable (by Gödel's theorem); that is, all such systems are inconsistent. Even if this unpleasant alternative is the case, the "Skolem paradox" still applies to any consistent theory which seems to require an uncountable set (such as a theory of the real numbers—if there is such a consistent theory).

5 Classes and relations as treated in P.M.

It was remarked in III 3.2.1 that the notion of set could be replaced by that of property, a set being considered to consist of all things which have some given property. A similar notion is at the basis of the treatment in P.M. of sets, or "classes" as they are called in P.M.

Consider a propositional function φx. Then by $\hat{x}(\varphi x)$ we may denote the class of all things x such that φx is true. In the symbolism of III 3.2, $\{x \mid \varphi x\}$ is evidently the same as $\hat{x}(\varphi x)$, and we could obviously use the former symbol here. But as it is more commonly used by logicians, we shall in the present discussion employ the latter symbol.

† It is in such formulations of set theory that the proof of independence of the Choice Axiom and continuum hypothesis, discussed in V 3.6, may be carried out.

5.1 Avoidance of contradiction

Although, from the standpoint of mathematics, the main purpose of developing the theory of classes is to define number and set up arithmetic, it is necessary also to take care that the theory is so constituted as to avoid the contradictions to which the "unrestricted" theory of sets leads (cf. III 2, V 3.5.3).

Let us suppose that C denotes the class determined by the function φx, and let us express this by the relation

$$C = \hat{x}(\varphi x). \tag{5.1a}$$

That y is an element of C we may denote by the usual ε : $y \, \varepsilon \, C$. That y is not an element of C we may denote by $y \sim\varepsilon \, C$.

Now the x in the function φ defining C should have a definite range. Thus, if φx is the propositional function "x is a professor of mathematics," we would not substitute "New York" for x. A value of x which we should particularly avoid in assertion of either truth or falsity is the class C itself; for C is only a symbol which was defined (5.1a) in terms of φ (compare VIII 7.4). It is violation of this principle that gives rise to such contradictions as that of the "set of all sets that are not elements of themselves." For, if $\sim\varphi C$, and hence $C \sim\varepsilon \, C$, is a permissible assertion, then we may define

$$S = \hat{C}(C \sim\varepsilon \, C).$$

Then either (1) $S \, \varepsilon \, S$ or (2) $S \sim\varepsilon \, S$. But (1) $S \, \varepsilon \, \hat{C}(C \sim\varepsilon \, C)$ implies $S \sim\varepsilon \, S$; and (2) $S \sim\varepsilon \, \hat{C}(C \sim\varepsilon \, C)$ implies $\sim(S \sim\varepsilon \, S)$; i.e., $S \, \varepsilon \, S$.

5.2 Theory of types

To avoid such contradictions, P.M. introduced a "theory of types" for the handling of propositional functions. As it was presented in P.M., this theory involved the use of an axiom called the "Reducibility Axiom" (to be discussed in 5.2.3), at whose non-primitive and arbitrary character much criticism was directed. Some critics pointed out that the axiom, from a logical standpoint, is probably not true, and in any case renders void the contention that mathematics has been shown to be derivable from postulates of primitive logic. Whitehead and Russell were not unaware of this; thus in the Introduction to the second edition of P.M. they stated: "This axiom has a purely pragmatic justification: it leads to the desired results, and to no others. But clearly it is not the sort of axiom with which we can rest content." Subsequent attempts by Chwistek [a], Wittgenstein [Wt], Ramsey [Ra], and later logicians to eliminate or modify the use of the axiom have nullified this objection to a great extent, and the "simple theory of types" such as pioneered by Chwistek and Ramsey indeed renders pointless the axiom of reducibility. (See especially Church [e].)

5.2.1 Simple theory of types

Because of the parallel between (1) classes and (2) propositional functions, a type theory may be set up for either classes or functions.

A simple type theory for classes would consist of a hierarchy of levels, each level above the first encompassing a class type; specifically, on the first or 0-th level would be placed the elements of lowest order, what would be called "individuals," from which classes of the first level are formed. The classes of level 1 may be used as elements to form classes of level 2; and so on. In applying such a type theory, one follows the rule that the elements of a class are all to belong to a given level; in an expression $A \varepsilon B$, the symbol B may stand for a class of any level above the 0-th, but then A must be the symbol for a class (or an individual) of next lower level. Adherence to such a rule would, it will be noted, avoid such contradictions as that of Russell and "the set of all sets."

The parallel simple type theory for functions is set up analogously. To classes of the first level correspond propositional functions of the first level; to classes of the second level, functions of the second level (whose variables are, then, functions of the first level); etc.

5.2.2 Relations

From propositional functions of two or more variables may be formed relations: Thus, if φxy is a propositional function of two variables x, y, there corresponds a relation

$$\hat{x}\hat{y}(\varphi xy). \tag{5.2.2a}$$

[Obviously one can think of this as the class of all (ordered) pairs (x, y) satisfying the propositional function φxy.]

For example, if φxy is the propositional function "x is the father of y," then $\hat{x}\hat{y}(\varphi xy)$ would denote the relationship of fatherhood; in particular, if x_1 and y_1 are particular (constant) individuals bearing this relationship, one could write

$$x_1 \quad \hat{x}\hat{y}(\varphi xy) \quad y_1$$

in which formula (5.2.2a) is used as the symbol for the relationship.

5.2.3 Extension of the type theory

With the introduction of relations, it becomes necessary to extend the theory of types in order to accommodate functions of several variables. For the theory of classes, it is sufficient to designate the primary "individuals," and thence the classes of first, second, . . ., levels, as in 5.2.1.

But, where two or more variables are involved, these may not all be of the same type.

In P.M., a so-called "ramified" type theory was introduced which would enable one to avoid not only the set-theoretic contradictions, but also the "semantical" contradictions such as that of Grelling (see Problem 13 of Chapter III) and Richard (see Problem 31 of Chapter IV). In this ramified theory, propositional functions of the same type were divided into "orders" according to the occurrence of functions as bound variables. Thus, on the first level, propositional functions, φx, where x is an individual (0-th level), would be designated as of first order, but $(\varphi)\psi(\varphi, y)$, in which a first-order function occurs as bound variable, would be designated as a second-order function. Unfortunately the resulting theory necessitated at this point the introduction of the "Reducibility Axiom" mentioned above, which asserted that every propositional function is formally equivalent† to a "predicative function" (i.e., a propositional function whose order is next above the highest order of its variables). And, as remarked above, this axiom was criticized on various grounds. The work of later logicians makes it apparent, however, that the axiom is by no means necessary. Ramsey [Ra] made the distinction between the set-theoretic contradictions and the contradictions of semantic type, and gave an alternative theory sufficient to avoid the former while maintaining the non-essentiality of the latter so far as mathematics is concerned.

The current use of the theory of types probably stems from the formulation of Carnap [Car; 19ff]. He denotes class types by symbols $t0$, $t1$, $t2, \cdots$, where 0 is the type of an individual. If in a relation xRy we call x the "first member" and y the "second member," then a relation whose first member is of type tm and second member of type tn is designated as of type $t(mn)$. Here, the m and n may denote not merely numbers, but combinations of numbers, etc. Thus, if m denotes the type of an individual and n a relation of type $t(02)$, then the relation itself would be designated as of type $t(0(02))$. Analogous conventions are made for relations involving three variables, four variables, etc. For classes, if the elements of a class have type tm, then the type symbol of the class is $t(m)$. Thus, if the elements of a class are relations of type $t(02)$, then the class is of type $t((02))$. For $t(0)$ we may write $t1$, for $t(1)$ write $t2$, as before.

In employing this theory of types, it is stipulated that all values of a given variable must have the same type. In particular, all elements of a given class must have the same type; if the class is of type $t(m)$, then its elements must be of type tm, no "objects" of any other type than those of type tm being even considered as allowable candidates for elements of the

† Functions are called "formally equivalent" if they are true or false for the same values of their variables.

class. (Thus no class C can stand in either relation $C \varepsilon C$ or $C \sim\varepsilon C$, each of these possibilities being ruled out.) Likewise all first elements of a relation must be of the same type, viz., type tm if the relation is of type $t(mn)$. [Exceptions are made in the case of class inclusion, such as $A \subset B$, class intersections, $A \cap B$, etc., where it leads to no difficulty if A, B are allowed to vary in type; strictly speaking, a special symbol, such as $\supset_{(mn)}$ should be used if A and B are of type $t(mn)$, for example.]

For further details, Carnap's work (*loc. cit.*) may be consulted. See also Hilbert–Ackermann [H-A; 152ff], and Church [e]. For a critique of the classical methods of constructing the continuum of real numbers, and alternative methods, see the work of H. Weyl [We]; also see the elementary expositions of Weyl [a] and Quine [b], [Q_1; V].

5.2.4 In addition to the Reducibility Axiom, it was found necessary to include other axioms whose "logical" character is at least debatable. Thus an Axiom of Infinity, providing for the infinite classes of mathematics, and "multiplicative axioms" (choice axioms) were introduced in P.M.

5.3 When the concepts of class and relation have been adequately developed, it is possible to proceed to the definition of cardinal number and to ordinary arithmetic. From our earlier discussions of number it can easily be surmised how we would go about doing this, without entering into the details.

Also, since mathematics contains geometry, it would be necessary, after arithmetic has been set up, to introduce geometry by the usual methods of coordinates. The geometric part of P.M. never appeared, however. Shortly after Whitehead's death, Russell [b] wrote: "Whitehead was to have written a fourth volume on geometry, which would have been entirely his work. A good deal of this was done, and I hope still exists. But his increasing interest in philosophy led him to think other work more important. He proposed to treat a space as the field of a single triadic, tetradic, or pentadic relation, a treatment to which, he said, he had been led by reading Veblen."

By way of an interesting historical footnote, we might add that in the same connection Russell pointed out the importance of Whitehead's contribution to P.M., attributing to him entire responsibility for the treatment of apparent (bound) variables, for instance; and commenting "Neither of us alone could have written the book; even together, and with the alleviation brought by mutual discussion, the effort was so severe that at the end we both turned aside from mathematical logic with a kind of nausea."

5.4 Before leaving these matters we should emphasize something that becomes apparent in the above discussion, namely, that in dealing with a symbolic logic we operate at different language levels. Thus, in discussing P.M., we have the language symbolized in terms of p, q, v, $\sim$, etc., as well as the language ("metalanguage") in which we discuss the former language, set up our rules of syntax, etc. For example, the symbols 0, 1, etc., used in discussing the theory of types, are in the metalanguage; later they appear in the arithmetic as it evolves from the theory of classes and relations, but now in the language of our formal logic.

As a matter of fact, the use of a metalanguage is also essentially involved in any discussion of non-formalized axiom systems, for instance, in Chapters I and II. A striking example of the contrast between language of a system Σ and the metalanguage occurs frequently in practice concerning the use of the Law of the Excluded Middle. If an axiom system Σ is not complete, and if A is a Σ-statement such that both $\Sigma + A$ and $\Sigma + \sim A$ are consistent, the statement "Either A holds or A does not hold" would not be valid in the metalanguage. It would, however, be correct to state within the system Σ, "Either A holds or A does not hold," since this is a valid use of the Law of the Excluded Middle. And we use this principle in the proof of theorems. For example, in the system Γ of Chapter I we may wish to prove some theorem T, and are able to show that (1) if every line contains at most a finite number of points, then T holds; and (2) if there exists a line which contains infinitely many points, then T holds. Since either every line contains at most a finite number of points or not every line contains at most a finite number of points, we would consider T proved.

6 Concluding remarks

Whether the Frege–Russell thesis—that mathematics is an extension of logic—has been established or not is a matter of opinion. Obviously, an affirmative opinion would seem to assume that one is satisfied in his mind as to what mathematics is; indeed, the affirmation of the thesis does, in a way, assert a definition of mathematics, inasmuch as one points to P.M. or one of its more modern counterparts, and says "What you can thus and so derive from this system, using the rules assigned for such derivation, is mathematics."

However, as will be seen in Chapter XI, no system such as P.M. can be complete in the sense that it will answer all questions in mathematics (even of the arithmetic of integers); and no matter how the axioms are augmented, there will always remain theorems that can neither be proved nor disproved. This objection can, of course, be countered by the observation that this is inherent in the nature of mathematics; that mathematics is ever capable of expansion.

There still remains, nevertheless, the debatable question as to whether the theory of sets, which necessitates axioms peculiar to itself, can be considered as part of logic. The fact that we must introduce axioms of infinity, Choice Axiom, etc., is now augmented by the proof recently given that both the Choice Axiom and the continuum hypothesis are independent axioms in the theory of sets. And this necessity of introducing into set theory axioms whose only purpose is to achieve technical mathematical ends certainly weakens the case for the "logistic thesis."

We shall return in Chapter XI to the subject of Mathematical Logic. But to appreciate the manner in which later developments evolved, it seems desirable to take account of a point of view which stems from Kronecker's "intuitionism" (VIII 2.1) and emphasizes both *the role of the natural numbers* and *constructivity* in the Foundations of Mathematics. This new Intuitionism, although opposed as a body of doctrine by the majority of mathematicians, has nevertheless exerted a subtle influence on the more recent developments in mathematical logic.

ADDITIONAL BIBLIOGRAPHY

Black [Bl]
Eaton [E; Part III]
Journal of Symbolic Logic
MacLane [a]
Rosenbloom [Ros]
Russell [R_1]

PROBLEMS

1. Use a truth table to show that the following formula is a tautology: $\sim p \,.\vee.\, \sim q \vee p \,.\, q$ [the last dot is "and"; the others denote parentheses, thus: $(\sim p) \vee (\sim q \vee p \,.\, q)$].

2. Use a truth table to verify that the formula given in Axiom V is a tautology.

3. Prove the formula given in Problem 1 as a theorem from Axioms I to V (formulas already proved in the text may be used in the proof). As a corollary, obtain $\vdash :.\, p \,.\supset:\, q \,.\supset.\, p \,.\, q$.

4. Let us define:

$$p \equiv q \,.=.\, p \supset q \,.\, q \supset p \qquad \text{Df}$$

Then prove as a theorem $\vdash : p \equiv p$. [*Hint:* In the result of Problem 1, substitute "$p \supset p$" for "p"; use 3.2.3 and *modus ponens.*]

5. Prove as a theorem $\vdash : p \supset q \,.\supset.\, \sim q \supset \sim p$. [*Hint:* Substitute "$\sim(\sim q)$" for "$r$" in 3.2.1 and use *modus ponens*; then substitute "$\sim q$" for "q" in 3.5.4 and refer to 3.2.1.1.]

6. Prove as a theorem $\vdash : \sim q \supset \sim p \,.\supset.\, p \supset q$. [*Hint:* Substitute "$\sim q$" for "$p$" and "$p$" for "$q$" in 3.5.4; use 3.4.3 and 3.2.1.]

7. Prove as a theorem $\vdash : p \supset q \,.\equiv.\, \sim q \supset \sim p$. (To what type of proof in mathematics does this formula correspond?)

8. Prove as a theorem $\vdash : \sim(p \,.\, q) \,.\supset.\, \sim p \vee \sim q$.

9. Prove as a theorem $\vdash : \sim p \vee \sim q \,.\supset .\, \sim(p \,.\, q)$.
10. Prove as a theorem $\vdash : (p \,.\, q) \,.\supset .\, (q \,.\, p)$.
11. Prove as a theorem $\vdash : (p \equiv q) \,.\equiv .\, (q \equiv p)$.
12. The following is given as a definition in P.M.:

$$\sim(\forall x P(x)) \,.=.\, \exists x(\sim P(x)) \qquad \text{Df}$$

Justify it on intuitive grounds.
13. Treat the following similarly (also given as a definition in P.M.):

$$\sim(\exists x P(x)) \,.=.\, \forall x(\sim P(x)) \qquad \text{Df}$$

14. Would you judge on intuitive grounds that the following formula is universally valid?

$$\forall x(P(x) \,.\, Q(x)) \,.\equiv .\, \forall x P(x) \,.\, \forall x Q(x)$$

15. The following implication is universally valid:

$$\forall x P(x) \vee \forall x Q(x) \,.\supset .\, \forall x(P(x) \vee Q(x))$$

Show that the converse implication is not universally valid.
16. Would you judge on intuitive grounds that the following formula is universally valid?

$$\exists x(P(x) \vee Q(x)) \,.\equiv .\, \exists x P(x) \vee \exists x Q(x)$$

17. The following implication is universally valid:

$$\exists x(P(x) \,.\, Q(x)) \,.\supset :\, \exists x P(x) \,.\, \exists x Q(x)$$

Is the converse implication universally valid?
18. Is the converse of the following formula universally valid?

$$\exists y \forall x P(x, y) \,.\supset .\, \forall x \exists y P(x, y)$$

19. Would you judge the following formula to be universally valid?

$$\sim[\exists x\, \forall y P(x, y)] \,.\supset .\, \forall x\, \exists y[\sim P(x, y)]$$

20. What is your judgment regarding the validity of the following implication?

$$\sim[\forall x\, \exists y\, \forall z P(x, y, z)] \,.\supset .\, \exists x\, \forall y\, \forall z[\sim P(x, y, z)]$$

X

Intuitionism

During the first half of the present century it was frequently stated that there were three "schools of thought" regarding the origin and nature of mathematics: Logisticism, Intuitionism, and Formalism. This did not mean that the class of all mathematicians could be subjected to a class decomposition, relative to an equivalence relation denoting membership in the same "school" (as in II 8.5)! Probably the great majority of mathematicians have spent little, if any, time speculating on the question of possible membership in a "school of thought." They have been either too busy doing research at the higher levels of their fields, or disdainful of such a question.

Nor was this so-called division into "schools" a partition of the relatively small body of mathematicians and philosophers who make research in the foundations their main occupation. Rather, it represented an attempt at classification of *thought tendencies.* Some, for example, L. E. J. Brouwer and his students, could justifiably be classified "Intuitionists." Brouwer, a Dutch mathematician and founder of modern Intuitionism, wrote his thesis (1908) on the Law of the Excluded Middle and its limitations as a mathematical tool. Thereafter he continued to espouse a doctrine unyielding in its philosophical aspects (which stemmed from Kronecker) while developing, along with his students, the type of mathematics to which this philosophy leads. Simultaneously, the German mathematician D. Hilbert (cf. I 1.5–1.6, for instance) and his students undertook to use symbolic logic as a tool for giving a contradiction-free development of classical mathematics including the theory of the infinite. The sequel brought about a recasting of mathematical logic which no longer needs to make any appeal to a philosophy such as the "logistic thesis." Indeed, stripped of the surrounding philosophical detail, P.M. is easily conceived as a purely axiomatic, or "formal," building up of mathematics. Consequently, so far as the symbolic framework is concerned, distinction between the Logistic and Formalistic "schools" has disappeared.

As a philosophy, however, Intuitionism is of such a distinctive character that it still deserves separate attention. Moreover, although violently opposed to Intuitionist tenets, Hilbert was eventually compelled to adopt principles in his "metamathematics" that savored strongly of the constructive character of Intuitionism—a tendency which became even more marked in the subsequent evolution of Mathematical Logic.

1 Basic philosophy of Intuitionism

In our discussion of Kronecker's views in VIII 2.1, we emphasized his insistence on the notion of mathematics as a construction on the basis of the "intuitively given" natural numbers. This notion is the kernel of Intuitionism as it is generally understood† today. And the consequences of this thesis, particularly in set theory and logic, will be shown below so that the results can be contrasted with the "classical" concepts treated in Part I.

Before going into such details, however, let us review briefly the underlying philosophy of present-day mathematical Intuitionism. The most striking aspect of it is probably what we might call its *self-sufficiency*. It is *self-generating*, relying in no way on other philosophies or logic. Its basic ideas are to be found in the *intuition*, which seems to be similar to the time (not the spatial) intuition of Kant. Specifically, it recognizes the ability of the individual person to perform a series of mental acts consisting of a first act, then another, then another, and so on endlessly. In this way one attains "fundamental series," the best known of which is the series of natural numbers. (Compare the viewpoint of Poincaré, VIII 7.1.)

This operation is not dependent upon the use of a language. To quote Brouwer (f; 86) in this regard: ". . . neither the ordinary language nor any symbolic language can have any other role than that of serving as a nonmathematical auxiliary, to assist the mathematical memory or to enable different individuals to build up the same set." As a consequence of this principle, mathematics is basically *independent of language*. In Heyting's words [He_1; 8], "Intuitionist mathematics consists in mental constructions; a mathematical theorem expresses a purely empirical fact, namely the success of a certain mental construction." Thus, "$2 + 2 = 3 + 1$" means "I have effected the mental constructions indicated by

† As a philosophy, Intuitionism has undoubtedly been misinterpreted in many of its details. This is quite probably not due to carelessness on the part of interpreters in general, however, but rather to the complicated nature of the basic definitions of Intuitionism as presented by Brouwer. Of recent years much of this unfortunate misunderstanding has been cleared up, owing in large measure to the investigations and expositions of Brouwer's students and colleagues (such as, for instance, A. Heyting; see especially Heyting [He], [He_1]. Also see Beth [Be] and Kleene [Kle]). The present author is quite aware, let it be made plain, of his own weaknesses of understanding and susceptibility to error; so the reader is admonished to take the following exposition "with a grain of salt" and, if sufficiently provoked, to consult the bibliography cited herewith.

'2 + 2' and by '3 + 1' and I have found that they lead to the same result" (Heyting, *loc. cit.*). For the *communication* of mathematics the usual symbolic devices, including ordinary language, are necessary, but this is their only function. This seems to make of mathematics virtually an *individual* affair rather than an organized or *cultural* phenomenon, and is perhaps the tenet of Intuitionism that it is most difficult to accept. However, it is possible that the Brouwer thesis only requires that one set apart the basic ideas—the intuition of the natural numbers and the ideas regarding set-formation—from linguistic influences, while acknowledging the subsequent role of language in the development, on this basis, of mathematics as a cultural phenomenon. Whether this is a tenable doctrine or not is still debatable. (Compare Chapter XII.)

For the related insistence of Intuitionism that *mathematics is not dependent upon classical logic*, a stronger case can be made. Brouwer goes to considerable lengths to build his system without use of any logic other than what he can justify and found on his mathematics. Here the argument that mathematics is not dependent upon logic is supported by *doing*. There emerge general rules, supposedly intuitively arrived at, for the derivation of new theorems from old. And, as Heyting has shown, these can be presented in the form of a symbolic logic which he calls a *mathematical logic*. This mathematical logic is thus a subdomain of mathematics whose use outside of mathematics would be "senseless" (cf. Heyting [He; 13ff]). A brief description of it is given below (Section 7) in order that the reader may compare it with the classical type of logic as described in Chapters VIII and IX.

To summarize, the basic philosophy of mathematical Intuitionism is that, although historically mathematics was derived from the world of experience by means of the senses, in its abstract formation it is purely intuitive and not dependent upon logic or science. On the contrary, the latter depend upon and use the methods of mathematics. All mathematics may be derived from the fundamental series of natural numbers by "intuitively clear" constructive methods. Language and other symbolic apparatus are not mathematical tools, but (imperfect) means of communicating mathematical ideas.

2 The natural numbers and the definition of set; spreads

Kronecker was a strict finitist. To him, the "constructible" concept of the natural numbers, as described in VIII 2.1.2, embodied the only acceptable manner in which the natural numbers form a "set." Such a "set" is not a *set* in the Cantorian or generally accepted sense, which is that of an "already formed" or "already existing" totality. The reader may refer to the material in fine print in VIII 2.1.2, and compare with the following, which forms the opening paragraph of what may be taken as the "official" presentation of Brouwer's views [a]:

"At the basis of mathematics lies an unlimited sequence of symbols or finite symbol arrays, which is determined by a first symbol and the law which derives from each of these rows of symbols the next. For this purpose† the sequence ζ ,of (natural) numbers 1, 2, 3, 4, 5, $\cdots$ is especially useful."

Thus the intuitionist does not base the natural numbers on an axiom system, such as that of Peano. Except for the mathematical induction axiom, the properties embodied in the Peano system are obvious from the law of generation of the natural numbers; while the induction property can be proved therefrom. Thus we can argue as follows: Let $P(x)$ be a property of natural numbers such that $P(1)$ holds, and $P(n)$ implies $P(n + 1)$. Then given a natural number k, the Intuitionist observes that in generating k by starting with 1 and passing over to k by the generation process, the property P is preserved at each step and hence holds for k.

The distinction between this manner of conceiving the natural numbers and that of classical mathematics, which considers them as forming a completed (infinite) totality, is brought out in the following example [He_1; 2]. Define two natural numbers m and n as follows: (1) m is the greatest prime such that $m - 1$ is also prime; if no such prime exists, let $m = 1$. (2) n is the greatest prime such that $n - 2$ is also prime; if no such prime exists let $n = 1$. Now certainly m is 3. As for n, the classical concepts of the completed totality N of natural numbers would permit us to argue that either there exists an infinite series of twin primes (in which case n is 1), or there does not exist such a series (in which case n is the larger of the greatest pair of twin primes). Thus (2) defines a number n. But the Intuitionist would not admit this argument, since the manner of generating natural numbers has not, so far, yielded definitive information regarding the existence of an infinity of twin primes. To use the Law of the Excluded Middle and assert "either the sequence of twin primes is finite or it is not finite" makes no sense to the Intuitionist.

The intuitionist conception of "set" must, of course, be based on the natural numbers, and be of a constructive nature. The definition as formulated by Brouwer in [a; 245] was quite complicated, although it does bring out the central role played by the natural numbers:

"A *set* is a *law*, on the basis of which if an arbitrary natural number is repeatedly chosen, then each of these choices either generates a definite

† Meaning, presumably, "for the purposes of such a sequence."

symbol array with or without termination of the process,† or brings about stoppage of the process and rejection of its result; for every $n > 1$ following a non-terminating and non-stopped sequence of $n - 1$ choices, at least one natural number can be given which, if it is chosen as the nth natural number, does not bring about stoppage of the process. Every sequence of symbol arrays generated in this manner by an unlimited sequence of choices (which is then generally not presentable in finished form) is called an *element of the set*. The common origin of the elements of a set M we shall usually designate briefly as *the set* M."

It is to be observed that a set, according to this definition, is not a finished totality; rather it is a *law*, according to which *elements* ("of the set") can be constructed. Involved in the construction of an element is the sequence of natural numbers. Not even an element is necessarily presentable in finished form—the array of symbols constituting it is to be generated by an unlimited sequence of choices. Thus a real number, π for instance, is determined only to the extent to which the construction of its successive digits is carried out.

The laws determining finite sequences as unlimited sequences of symbol arrays of the kind found in the sequence ζ form special cases of sets whose elements are formed by the single symbols. The set of natural numbers, that is, the symbol arrays of ζ, will be designated by A.

As to the distinction between "finite" and "infinite" sets, Brouwer defines: "If for each n in ζ there is determined a natural number k_n, such that every time a natural number greater than k_n is selected as the nth choice, the process is stopped, then the set is called *finite*." As an example, he points out that the "unlimited" sequences of numbers of *one digit* form a finite set.

2.1 Later Brouwer avoided the term "set," replacing it by "spread" as being more graphic and less likely of confusion with the common denotation. The definition given by Heyting in [He$_1$; 34ff] is quite lucid and utilizes two laws—a "spread-law" and a "complementary law"—a device which separates out the difficulties and helps bring out more clearly and precisely the role played by the natural number and construction therewith. Without going into technicalities, the spread-law may be described as a rule for specifying "admissible" finite sequences of natural numbers; and the complementary law assigns a specific mathematical entity to each such admissible sequence. The sequence generated by the complementary law is called an *element* of the spread.

† Brouwer notes here that the possibility of the "termination of the process" can obviously be replaced by the possibility "that after a certain choice every further choice generates nothing."

In particular, in order to get the real numbers as conceived by the Intuitionist, the mathematical entities assigned are rational numbers, the real numbers then emerging in the form of Cauchy sequences. However, the assignment made according to the complementary law† may be quite arbitrary within, perhaps, certain limitations. For example, in the definition of a Cauchy sequence $\{a_n\}$, defining a real number, the assignments may be quite arbitrary so long as for every n $|a_n - a_{n+r}| < 1/n$ for all r. It will be observed that this makes the intuitionist real number "system" something in the nature of a spectrum where the "points" are not well marked out but are in a process of appearing. In particular, like the natural numbers, it is not a completed totality. Owing to the arbitrary nature of the sequential elements, however, the real numbers generate a cardinal number (in the intuitionist sense described below) different from that assigned to the natural numbers.

Two elements are called *equal* if it is certain that for every n the nth choice for each element generates the same symbol array. Of just what this "certainty" consists, and its character, is presumably to be decided for the particular case. For example, if we were to present any two of the many well-known definitions of π, we could give a proof that the decimal developments of the two numbers so defined would agree in the nth digit for every n; and so the two numbers are equal. But suppose that N is defined as follows: N is a number such that $0 \leqq N \leqq 1$; in decimal form its nth digit after the decimal point, a_n, is 0, unless the nth digit of the decimal part of π is the first of a sequence of seven 7's—7777777—in which case the digit a_n is 1. We cannot say that N and 0 are equal, since we have no way of knowing that there does not exist a sequence of seven 7's in the decimal part of π. (Possibly there is such a sequence. Incidentally, π is now known, by computation on modern computers, to thousands of digits.)

Two spreads are called *equal* if for each element of either "can be given" an equal element of the other.

3 Species

On the basis of spreads and elements, which he calls *mathematical entities*, Brouwer erects a hierarchy of concepts which he calls *species* (*Spezies*; compare the type theory of P.M., IX 5.2). A *species of the first order* is a *property* which only a mathematical entity can possess, in which case the entity is called an *element of the species of first order*. Thus the property of being prime is a species of first order, since it is a property only

† A like freedom is allowed in regard to the spread-law, so long as the resulting sequences are admissible.

of natural numbers (which as elements of A are mathematical entities). And spreads are themselves species of first order (since, if M is a spread, then the property of being an element of M is possessed only by the elements of M—hence the latter are elements of a species, M, of the first order). Equality of species is defined in a manner similar to that for spreads. A *species of second order* is a property which only a mathematical entity or species of first order can possess, in which case it is called an *element of the species of second order*. And similarly *species of the nth order*, as well as their equality, are defined for arbitrary number n. *Subspecies* are defined in a manner similar to that for subsets.

4 Relations between species

Equality was defined above, and the notion of subspecies introduced. Two elements are called *different* if it is *certain* that their equality is an impossibility. For example, the number N defined in Section 2 is not *different* from zero; thus we can say neither that N *equals* zero nor that N is *different* from zero. Similarly, two species are called different if it is certain that their equality is an impossibility. A species S is called *discrete* if it is possible to determine of each two elements of S that they are either equal or different. Thus any set which has both zero and the number N of Section 2 as elements cannot be called discrete.

4.1 Union and intersection of species

Union (addition) of sets is defined as in ordinary set theory (III 3.3), and union of species is defined similarly. A like remark holds regarding *intersection* (product); see III 3. However, although the union of two sets is again a set, the intersection of two sets is not necessarily a set; as, for example (Brouwer, *loc. cit.*) if S is a set whose single element is the number 0, and S' a set whose single element is the binary fraction N defined in Section 2.

5 Theory of cardinal numbers

If there can be established a (1-1)-correspondence between two species S and S', that is, a law which makes correspond to each element of S an element of S' in such a way that equal and only equal elements of S correspond to equal elements of S', and every element of S' to an element of S, then we write $S \sim S'$, and say that S and S' *have the same cardinal number*, or are *cardinally equivalent*. (Brouwer cites the example of the set of all natural numbers and the set of all natural numbers except 1.) Two species are not necessarily comparable, however (compare V 3.6.2).

5.1 Finite and infinite species and their cardinal numbers

A species is called *finite* if it is cardinally equivalent to the set of natural numbers of a certain initial segment† of the sequence ζ. A species is called *infinite* if it has a subspecies cardinally equivalent to the species A of natural numbers. It should be noted that there is no basis for ascertaining that every set or species is either finite or infinite; hence, a species that is not finite is not necessarily infinite. On the other hand it is certain that a species cannot be both finite and infinite. For one can prove:

5.1.1 Theorem.‡ Principal property of finite species. *For every manner of representation of the* (1-1)*-correspondence between a finite species S and the set of numbers in an initial segment of ζ, the same initial segment is used.*

Proof. Let S_1 and N denote two "enumerations" of S, such that S_1 uses up the initial segment $1, 2, \cdots, k$ of ζ, and N uses up an initial segment that includes $1, 2, \cdots, k$. Let S_2 denote the enumeration derived from S_1 by exchanging in S_1 the elements which correspond to 1 in S_1 and N. Then S_2 uses up the same initial segment $1, 2, \cdots, k$ of ζ as S_1, while in both S_2 and N the same element of S now corresponds to 1.

Next let S_3 be the enumeration derived from S_2 by exchanging in S_2 the elements which correspond to 2 in S_2 and N. Then S_1 and S_3 use the same initial segment $1, 2, \cdots, k$ of ζ, while S_3 and N have the same elements corresponding to 1 and 2.

Proceeding in this manner, we obtain finally a last enumeration, S_ι, which uses the same initial segment of ζ as S_1, and at the same time has the same elements corresponding to $1, 2, \cdots, k$, respectively, as N. But S_ι contains all elements of S, and consequently the elements of N corresponding respectively to $1, 2, \cdots, k$ contain all elements of S; in other words, N has no elements corresponding to an element of ζ greater than k.

5.1.2 Cardinal numbers of finite sets. As a consequence of this theorem, Brouwer states that we can denote the *cardinal number* of a finite species by the last number of the sequence ζ which is used in an enumeration of the species. The cardinal number of a species having no elements is called *null* and denoted by 0. These are evidently to be taken as definitions of "cardinal number" for finite species, since Brouwer gives no general definition such as that of Frege-Russell.

5.1.3 *A proper subspecies of a finite species S cannot be cardinally equivalent to S*; for the supposition of such an equivalence would lead to a

† That is, section; see V 3.1.1. Note that Brouwer's "finite" corresponds to the "ordinary finite" defined in III 4.1.1.

‡ Compare with Theorem III 5.2.1.

contradiction of Theorem 5.1.1. And in particular *a finite species cannot also be infinite.*

Of course species exist about which one cannot say whether they are finite or infinite.

5.2 Denumerable sets

As in the classical theory (as given in Chapters III to V), the simplest example of an infinite species is A. Brouwer states (*loc. cit.*, p. 249, § 3) that he will denote the "cardinal number" of A by a (again no explicit definition of "cardinal number" is given). Species which have the same cardinal number as A are called *countably infinite*; since no confusion would result, we may use the same term *denumerable* as we used in the classical case. Hence, *every infinite species has a denumerable subspecies.*

Again, as in the classical theory, the species of integers, rational numbers, and algebraic numbers are each denumerable. The proof of the denumerability of the species of the algebraic numbers is worth noting, but will not be repeated here for lack of space (see Brouwer, *loc. cit.*, pp. 249–250). Also, a species which can be separated into a denumerable species of denumerable species is itself denumerable.

5.3 The cardinal number c

As another example of an infinite set, Brouwer offers the species C of unrestricted sequences of natural numbers,† whose cardinal number he denotes by c. The species C_n each of whose elements consists of n unrestricted sequences of natural numbers has the cardinal number c. And as in the classical theory, this fact permits a relation of cardinality between points of a line and points of n-dimensional euclidean space (for details see Brouwer, *loc. cit.*, pp. 251–252).

5.4 Other types of equivalence between species

The notion of cardinal equivalence does not exhaust the possible types of equivalence relations between sets. In the classical theory, it was shown that, if A and B are sets such that there exist (1-1)-correspondences between A and a subset of B, and between B and a subset of A, then there exists a (1-1)-correspondence between A and B (Bernstein Equivalence Theorem, IV 4.2.3). Brouwer defines:

5.4.1 Definition. Two species S and S' (and likewise the corresponding cardinal numbers s and s') are called *equivalent* if by some law L_1 there

† See Heyting [He$_1$].

is made to correspond to every element of S an element of S' so that to equal and only to equal elements of S equal elements of S' correspond; and if by some law L_2 there is made to correspond to every element of S' an element of S so that to equal and only equal elements of S' equal elements of S correspond. This equivalence is expressed by the formula $s = s'$.

5.4.2 Definition. If for species S and S' such a law as L_1, defined in 5.4.1, exists, but no such law as L_2 can possibly exist, then we write $s < s'$ (or $s' > s$) and say that S (s) is *smaller than* S' (s'), and S' (s') is *greater than* S (s).

5.4.3 Definition. If we know that to every element of S there corresponds an element of S' such that to equal elements and only equal elements of S do equal elements of S' correspond, then we write $s \leqq s'$ (or $s' \geqq s$), although in this case *there need not necessarily one of the relations* $s < s'$, $s = s'$ *hold.*

It will be noted that we have here a denial of the Law of the Excluded Middle; since, if such a law as L_1 of 5.4.1 exists, then, according to the Law of the Excluded Middle, either such a law as L_2 exists (and hence $s = s'$) or such a law as L_2 does not exist (and hence *cannot* exist in the classical sense, so that $s < s'$).

In particular, the species C is greater than the species A; $a < c$. If K is the species of *lawfully defined* unlimited sequences of natural numbers (i.e., each element of K is defined by some law, as opposed to the unrestricted nature of the elements of C), then C is also greater than K.

The "equivalence" possibilities are not yet exhausted, but we refer the reader to works cited for such further details.

5.5 Types of countability

Corresponding to the above types of equivalence (and non-equivalence), Brouwer defines various types of countability (see Heyting [a]). With regard to the possibility of cardinal numbers "higher" than c, nothing seems to have been done in the intuitionist mathematics. The possibility of a "species of all subspecies of a given species" seems in general excluded by the difficulty of finding suitable defining properties, for instance. And, similarly, the diagonal procedure (IV 3) is in general not definable.

This does not mean, of course, that higher cardinal numbers are not definable. In any case, however, one obviously must distinguish between the cardinal numbers of intuitionist mathematics and those of the classical type as expounded in Chapter IV.

6 Order and ordinal numbers

As might be expected, in view of the above description of Brouwer's set theory, the treatment of order and ordinal numbers is quite different from that expounded in Chapter V. For a description of it, the reader is referred to Brouwer [b]. In that work, incidentally, Brouwer gives a definition of "continuum" which is very similar to the one given in VI 1.4 above, where use is made of the "Dedekind cut."

The theory of well-ordered species is given in Brouwer [c]. It is probably unnecessary to remark that for Intuitionism the theorem of Zermelo to the effect that "every set has a well-ordering" (Theorem V. 3.1.2) is devoid of sense. To form well-ordered species two "generating operations" are permissible; addition of a finite number of known well-ordered species, and addition of a fundamental series of known well-ordered species. Consequently every well-ordered species is countable. The introduction of "null" elements renders the theory even less analogous to the classical theory. (Cf. Heyting [He; 27].)

For work of Brouwer concerning "partial order," and related types of order with applications to the continuum, one may consult Brouwer [d, e].

7 The intuitionist logic

In mathematics as it is usually taught, logic of the classical type is assumed as *a priori*, and plays an important role in proofs of theorems. Of special importance is the Law of the Excluded Middle and its consequence that the falsity of the falsity of a proposition p implies p (see IX 3.4.3). Thus, to prove a "Theorem T," if one can demonstrate that the assumption of the falsity of T leads to contradiction, then T is true. For any proposition which leads to contradiction must be false; hence, if the falsity of T is false, then T is true by the principle cited above. In particular, Theorem T may be an assertion of the type, "There exists a natural number having property P." If property P is "evenness," one naturally proves the theorem by exhibiting an even number. But, in the event that property P is of such a character that no number having this property has been found, one may resort to the device of showing that the assumption of non-existence of such a number leads to contradiction and concluding that one has thereby proved Theorem T.

The intuitionist would not accept such a "proof" as this. The former proof—by *exhibition* of a number with the desired property P—is of the constructive character demanded by the intuitionist. But the latter proof introduces into mathematics a new principle that is not capable of being founded on the ideas set forth above. If one examines the proofs given in Section 5, one sees that 5.1.1 involves only a constructive procedure; however, 5.1.3 employs the argument that a certain two species cannot be cardinally equivalent since the supposition of such an equivalence would lead to contradiction of 5.1.1. Thus the intuitionist accepts the principle that any proposition which implies contradiction must be false. This does not involve a use of the Law of the Excluded Middle, however, nor its corollary that the falsity of the falsity of a proposition p implies p.

It appears, then, that the intuitionist accepts the Law of Contradiction, but not the Law of the Excluded Middle. Of course, he accepts the Law of the Excluded Middle where it is "intuitively" clear; if G were a *finite* set of natural numbers, then he would say that it is intuitively clear that either G contained a prime number or that no element of G was prime (no matter how great the number of elements in G, so long as it is finite). But, if G were infinite, the intuitionist would not admit the assertion, unless, of course, a prime number is exhibited in G or it is shown that the assumption that G contains a prime number leads to absurdity (the latter being accepted as proving that no element of G is prime). To show that the assumption that G contains no prime number leads to contradiction would not be accepted as a proof that G contains a prime number, since this would constitute a replacement of the falsity of the falsity of a proposition by the proposition itself.

We saw in IX 3.6 that, as the P.M. propositional calculus is set up, the Law of the Excluded Middle and the Law of Contradiction are equivalent; a consequence of the interdependence of the logical constants $\sim$, ., and $\vee$. And from the material in IX 3.4 it follows that p and $\sim(\sim p)$ are equivalent. Since the intuitionist accepts none of these principles, it becomes an interesting problem to analyze, if possible, what can be described as the logical apparatus to which the intuitionist mathematics leads.

7.1 An intuitionistic symbolic logic

Such an analysis was first made by Heyting in 1930. We shall give sufficient indication of Heyting's system to enable the reader to compare it with the logistic calculus as described in Chapter IX. (We base our description on Heyting [He] and Kolmogoroff [a].)

Instead of commencing with *two* "logical constants" as in Chapter IX, namely $\vee$ and $\sim$, and defining $\supset$ and the logical product (as in IX 3.6), *four* symbols, $\wedge$, $\vee$, $\neg$, and $\supset$, are immediately introduced, standing respectively for conjunction ("and"), disjunction ("or"), negation, and implication. These symbols are independent of one another; for example, $a \supset b$ is not the same as $\neg a \vee b$ [cf. IX (2.2a)]. Then "postulates" or fundamental "formulas" are stated, as was done for the propositional calculus in Chapter IX. We find, for example, the formula

$$\vdash :. a \supset b . \wedge . b \supset c : \supset : a \supset c,$$

where a, b, and c are propositions; also the following, which embodies the Law of Contradiction,

(7.1a) $$\vdash . \neg(a \wedge \neg a)$$

The formula which would embody the Law of the Excluded Middle cannot be derived from the system. However, the formula

(7.1b) $$\vdash . \neg\neg(a \vee \neg a),$$

which may be interpreted as the falsity of the falsity of the Law of the Excluded Middle, is derivable in the system! But, although

(7.1c) $$\vdash . a \supset \neg\neg a$$

occurs in the system, $\neg\neg a \supset a$ does not occur, so the formula $a \vee \neg a$ does not follow from the above. Since the formula $\neg\neg a \supset a$ is derivable from $a \vee \neg a$, it follows, by conjunction of this fact with the remarks just made, that the Law of the Excluded Middle is equivalent to the formula $\neg\neg a \supset a$; see Brouwer [g; 252, footnote 4], who attributes this observation to P. Bernays.

Nevertheless, long strings of negations do not occur. For the formula

(7.1d) $$\vdash . a \supset b . \supset . \neg b \supset \neg a$$

occurs in the system, and if in this formula we replace b by $\neg\neg a$ and make use of (7.1c), we obtain

(7.1e) $$\vdash . \neg\neg\neg a \supset \neg a$$

(as observed in Kolmogoroff [a; 63]; the same proof, in ordinary language rather than in the symbolic form given here, occurs in Brouwer [g; 253], however). Conversely, if we replace a by $\neg a$ in (7.1c), we obtain

(7.1f) $$\vdash . \neg a \supset \neg\neg\neg a$$

and from (7.1e), (7.1f) the equivalence of $\neg a$ and $\neg\neg\neg a$ is obtained.

It follows from the above that every string of negations $\neg\neg \cdots \neg a$ can be reduced to either $\neg\neg a$ or $\neg a$.

It is interesting to observe that Gödel, employing certain results of Glivenko, showed that, if the "logistic" symbols

$$a . b, \ a \vee b, \ \sim a, \ a \supset b$$

are replaced by the following symbols, respectively,

$$a \wedge b, \ \neg(\neg a \wedge \neg b), \ \neg a, \ \neg(a \wedge \neg b),$$

then every formula of the "logistic" calculus is transformed into a valid formula in the intuitionistic logic! (See Gödel [b].)

Moreover, Gödel also showed that a system of axioms for arithmetic (of integers) and number theory due to Herbrand can be interpreted and proved valid in the intuitionist framework, thereby providing an intuitionistic consistency proof for classical arithmetic and number theory. (See Gödel, *loc. cit.*, and Kleene [Kle; § 81].)

7.2 The Kolmogoroff interpretation

The intuitionistic logic assumes a much more natural character, if instead of considering the above a *propositional* calculus, we conceive of $a, b, c, \cdots$, as denoting *problems*, and the calculus itself as a mode for solving problems—a "solution calculus." We can, indeed, following Kolmogoroff [a], approach the matter entirely independently, without any intuitionist presuppositions, purely as a question of formulating methods of solution of mathematical problems (such as, for example, problems in geometric construction).

From the latter point of view, we consider $a, b, c, \cdots$ as denoting problems to be solved, and interpret $a \supset b$ to mean the problem of deriving the solution of b from that of a (or "carrying back" the solution of b to that of a); more explicitly, $a \supset b$ means the problem "assuming that a solution of a is given, find a solution of b." By $a \wedge b$ we would indicate the problem of solving both a and b, and by $a \vee b$ the problem of solving at least one of the problems a and b. By $\neg a$ is indicated "assuming that a solution of a is given, obtain a contradiction." The symbol $\vdash$ is used as an indication of a problem to be solved for all values of the variables (problems) involved; thus,

$$\vdash . a \supset a \wedge a \tag{7.2a}$$

means "for all a, show how to obtain a solution of $a \wedge a$ from a solution of a." Certain basic problems, such as (7.2a), are postulated—that is, considered solved—and solutions of others derived therefrom. For example, if $\vdash . p \wedge q$ is solved, then $\vdash p$ may be considered solved.

The development of the "solution calculus" turns out to coincide, symbolically, with the intuitionist propositional calculus. For instance, we will not expect to find $\vdash . a \vee \neg a$—the Law of the Excluded Middle—in the "solution calculus," since it would be tantamount to having established, for *all* problems a, either a method for solving a or a method for showing that the assumption of a method of solution for a leads to contradiction. As Brouwer asserts, the assumption of the universal validity of the Law of the Excluded Middle would be equivalent to assuming that every problem is solvable. What may be considered the intuitionist form of the Law of the Excluded Middle, (7.1b), is the symbolic formulation of a theorem of Brouwer which asserts that there cannot be given a provably unsolvable problem. It should be noted, incidentally, that the intuitionist logic does not *give permission* to assert the existence of a "middle third"; formula (7.1b) is, in a sense, as far as the logic goes in assertions regarding the general validity of the formula $a \vee \neg a$.

8 General remarks

Enough has been given above, no doubt, to demonstrate that Intuitionism is not a purely negative philosophy such as that of Kronecker. In the hands of Brouwer and his followers a sizable literature of intuitionistic mathematics has been built up. It is interesting to observe that Brouwer, after his early work on the foundations of mathematics during the first decade of the present century, published during the second decade a series of basic and very important articles in the then budding field of topology. But in the third decade he resumed constructive work in the intuitionistic foundations of mathematics, not only laying down the principles of his general theory, but also supplying new proofs of classical theorems which met the requirements of the intuitionist program. For example, in [i] he gave a new proof of the Jordan Curve Theorem† (which he had earlier (1910) proved in his topological papers but not from an intuitionist standpoint); also new proofs of the fundamental theorem of algebra were given both by him and by others. Mention has already been made above of his present researches into partial orderings.‡

8.1 Summary

In a positive way, then, intuitionism builds on the "intuitively given" sequence of natural numbers, using only constructive methods. In the theory of sets this necessitates conceiving of a set not as a ready-made "collection," but as a *law* by means of which the elements of the set can be constructed in a step-by-step fashion. In proofs of existence, the entity whose existence is to be proved must be shown to be *constructible*; it is not sufficient to show that the assumption of its non-existence leads to contradiction.

In attempts to set up a logic for intuitionist mathematics, the latter fact is recognized by not identifying the double negation with affirmation, or,

† If C is a circle in a plane S, then C divides S into two connected parts, called domains, of which C is the common boundary. The Jordan Curve Theorem states that every topological transform $T(C)$ in S does likewise (cf. VII 4).

‡ For a review of the effects of intuitionistic thought on various mathematical theories, the reader is referred to Heyting [He; 22–29].

what is equivalent, by suitably limiting the Law of the Excluded Middle. A corollary of this is that it is not affirmed that every mathematical problem is solvable. It develops that intuitionist mathematics produces its own type of logic, mathematical logic thus becoming a branch of mathematics, or an "applied mathematics." In the view of Brouwer, the chief protagonist of intuitionist mathematics, language or symbolism is not basic to mathematics. While it is recognized that mathematics had its origins in experience, its modern abstract formulation is the product of the pure intellect, and has intuitive, not merely formal, content. In particular, the symbolic logic discussed in 7.1 is not a mathematical tool; it serves, however, as a means of communicating the logic engendered by intuitionism.

8.2 Some examples

The number N defined in Section 2 furnishes a good example of the effect of the intuitionist philosophy on the real number continuum. Of this number it cannot be asserted that either it is equal to zero or it is different from zero. For to show the latter it is necessary to *exhibit* a sequence of seven 7's in π, and to show the former it is necessary to prove that the assumption of the existence of such a sequence leads to contradiction.

The Brouwer continuum is of such a character that a number is fixed only by successive rational approximations or a "closing down" process, such that, for any given number $1/n$, the value of the number is determinate within an error $< 1/n$. Of two numbers N_1 and N_2, it is possible that they are equal (as defined in Section 2), that they are definitely unequal (*entfernt*) as in the case of 3 and π, for example, or such that neither of these assertions can be made, as in the case of 0 and N above.

Or consider the number X which is equal to

$$(-1)^k/(10)^k,$$

where k is the number of the first decimal place in the decimal development of π where a sequence of digits 0123456789 commences; or, if no such number k exists, $X = 0$. The number X is certainly well-defined; but we do not know whether it is negative, positive, or zero! And the number $1 + X$, although certainly positive, is not expressible in the decimal system (since we do not know whether to start it with $1.000 \cdots$ or with $0.999 \cdots$)!

As might be expected, in the light of these limitations and the nature of intuitionist set theory (as set forth above), the building up of analysis—calculus, theory of functions, etc.—in the intuitionist mathematics takes a form quite restricted in comparison with that of the usual analysis. The fundamental theorem of Bolzano, which states that, if $f(x)$ is a continuous single-valued real function over an interval $[a, b]$ such that $f(a) < 0$ and $f(b) > 0$, then there exists a value $\bar{x}$ of x such that $a < \bar{x} < b$ and $f(\bar{x}) = 0$, is lacking, as is also the related theorem on the attainment of its maximum value by a function continuous over an interval. Those portions of analysis which are constructible by actual computational methods are in general to be found in intuitionist mathematics.

Geometry of the analytic variety, subject to the limitations imposed on analysis, is possible in intuitionist mathematics. So, too, are geometries which are built on an axiomatic foundation, so long as the latter is realizable analytically.

In spite of whatever objections may be raised against intuitionist mathematics, it is generally conceded that its methods do not lead to contradiction. Such non-constructively conceived notions as "the set of all sets," "the set of all ordinal numbers," are obviously unattainable by intuitionist methods. But whether such a "state of health" has been achieved by "cutting off the leg to heal the toe" is a matter of opinion, evidently!

Its influence on other foundation philosophies will be touched upon in the sequel.

For further information on intuitionism (besides the already cited Heyting [He_1]), especially as regards more recent developments, the reader is urged to read Chapter 15 of Beth [Be].

ADDITIONAL BIBLIOGRAPHY

Brouwer [h]
Curry [Cu_1]
Dresden [a]
Kleene [Kle]
Menger [a, b]
Weyl [We_1; II]

PROBLEMS

1. If S is a non-empty, simply ordered set of rational numbers which has no first element, give a constructive proof that S contains a $^*\omega$-sequence. (Compare V, Problem 6.)

2. Is the proof (cf. IV, Problem 2) that the set of all finite subsets of the set of natural numbers is denumerable, a constructive proof?

3. Discuss and compare the intuitionistic implication (the "$\supset$" of 7.1) with material implication (IX 2.2.1).

4. Give instances in which the Intuitionist would accept the Law of the Excluded Middle as a basis for an existence proof.

5. Is the Intuitionist being "inconsistent" when he rejects the Law of the Excluded Middle as general basis for an existence proof, and on the other hand proves non-existence by contradiction?

6. The mathematical induction principle (or equivalent thereof) is usually stated as a postulate in the classical theory of the natural numbers (as in the Peano axioms, for instance). Why, then, is the *proof* of it which is indicated in Section 2 valid for the Intuitionist?

7. Although Intuitionism does not accept set theory of the classical type (Chapters III to V), its philosophy does suggest questions concerning the nature of constructive proof therein. Which of the proofs given in the set theory of Chapters III to V would you consider as constructive?

8. Compare the Intuitionist conception of the natural numbers with that described in IV 4.1.2.

9. The Intuitionist (cf. Heyting [He_1; 17]) regards the following as an example in which $\neg\neg p$ does not imply p: Let $r = 0.333\ldots$ unless a sequence "0123456789" occurs in the decimal development of π, in which case if the "9" of the first such sequence occurs in the kth decimal place of π, then r has only "0's" after its kth decimal place. Now let p be the proposition "r is rational" and suppose "$\neg r$." Then $r = 0.33\ldots 3$ (k decimals) would be impossible, and no sequence "0123456789" appears in π, so that $r = 1/3$—again rational. Hence he would assert "$\neg\neg p$." However, we cannot assert "p," since this would mean (for the Intuitionist) that integers p and q can be calculated such that $r = p/q$. Contrast this with the classical mode of argument.

10. What do you consider the significance of the interpretations of classical systems within intuitionist logic as given by Gödel (cited in Section 7.1)?

XI

Formal Systems; Mathematical Logic

In contrast to the intuitionist tenet that language and symbolism are not basic to mathematics, was Hilbert's belief that the axiomatic method offered the most hopeful remedy for the "ills" of mathematics. This was a natural evolution from his early work on axioms for geometry. But, to insure a mathematics free of contradiction, the axiomatic method as used in his *Foundations of Geometry*, or as described above in Chapters I and II, could not suffice. Freedom from contradiction is guaranteed only by consistency proofs; and the proof of consistency by interpretation (II 1, 2) cannot be satisfactory in general, inasmuch as it usually only shifts the question to another domain of mathematics. In the case of the *Foundations of Geometry*, consistency was assured only if real number arithmetic conceals no inconsistencies; and a proof of this is still lacking! A different approach is therefore demanded.

Partly influenced, no doubt, by the work of Peano and his school, as well as by the Russell-Whitehead work, Hilbert decided upon a *union* of the axiomatic and logistic methods. A reduction of mathematics to logic, even if successful, would still leave open the question of the consistency of that "logic." But perhaps by analyzing mathematical concepts and processes, both logical and otherwise, and representing them by an appropriate symbolism, as in a symbolic logic like that discussed in Chapter IX, we may be able to demonstrate that the formula for a contradiction can never be obtained from the fundamental formulas and the rules laid down for manipulating the symbols. For a given branch of mathematics, this would mean combining a system like that of the predicate calculus with axioms using the technical terms. In short, the predicates involved in the formulas would include symbols representing the technical terms and relations.

1 Hilbert's "proof theory"

Hilbert called a formal system thus inadequately described a "proof theory." As described by Hilbert in a preliminary paper (cf. Hilbert [b]), the fundamental idea of any proof theory would be as follows: "Everything which constitutes mathematics today is rigorously formalized, so that it becomes a stack of formulas. These distinguish themselves from the ordinary formulas of mathematics only by the fact that besides the ordinary signs or symbols there enter also the symbols of logic, especially implication ($\rightarrow$) and negation ($\bar{\ }$). Certain formulas which serve as foundation stones for the formal edifice of mathematics are called *axioms*. A proof is a figure, which must lie clearly before us as such; it consists of conclusions by means of the conclusion scheme

$$\frac{\begin{matrix} S \\ S \rightarrow T \end{matrix}}{T} \qquad (\textit{modus ponens})$$

where the premise, S, is either an axiom (or axioms) or the end formula of a proof figure which occurred earlier in the development.† A formula is called *provable* if it is either an axiom or the end formula of a proof.

"To the ordinary, thus formalized mathematics, is added a, in a certain sense, new mathematics, a metamathematics .... In this metamathematics one works with the proofs of ordinary mathematics, these latter themselves forming the object of investigation." (Cf. IX 5.3.)

If F stands for any formula in such a system, then a contradiction is obtained if the formula

(1a) $$F \,.\, \bar{F}$$

(where . is the logical "and," and $\bar{F}$ is Hilbert's form of the negation of F) is provable. If one can show that no such formula as represented in (1a) is provable, then one has shown the "consistency" of the system. In particular, if the symbols and axioms are given mathematical meanings, then by following the proof methods allowed in the theory no contradiction can be obtained in the corresponding mathematical system.

In this connection it is interesting to note Hilbert's further comments (Hilbert [b; 154–155]): "The infinite comes into the proof theory as soon as we introduce the ideas 'all' and 'there exists.' The former is equivalent to an infinite 'and,' the latter to an infinite 'or.' From the Law of the Excluded Middle for finite sets we conclude that either all elements of a given finite set have a certain significant property P, or there exists one that does not have property P.... In ordinary mathematics we argue likewise in regard to infinite sets.

"But if we thoughtlessly apply to infinite totalities a procedure which is permissible in the finite case, then we open gate and door to errors. It is

† Hilbert may have been oversimplifying here, since in the "Grundlagen" (2.1) *modus ponens* is not the only rule of inference required.

the same source of error that we see in analysis; in the latter field the carrying over, to infinite sums and products, of theorems which are valid for the corresponding finite sums and products is permissible only if special convergence conditions, etc., are satisfied. Similarly we cannot treat the infinite logical sums and products in the same manner as the finite, unless our proof theory reveals such treatment to be justified." These remarks, among others, were later (1928) cited by Brouwer [h], as evidence of the influence which Intuitionism had had on Formalism.

2 Actual development of the proof theory

As Hilbert's program developed, there first appeared (in 1928) the Hilbert-Ackermann *Grundzüge der theoretischen Logik* [H-A], cited frequently in Chapter IX. In essence preparatory to the larger work cited below, it is (as its title indicates) a formal treatise on logic in the tradition of Peano and Russell.

2.1 The *Grundlagen der Mathematik*

What may be considered the "Principia Mathematica" of Formalism is the Hilbert-Bernays *Grundlagen der Mathematik*, published in two volumes; volume I appeared in 1934 and volume II in 1939 [H-B]. Originally planned as a detailed exposition of the proof theory, with proofs of freedom from contradiction, its scope had to be enlarged as a result of unforeseen difficulties. (The first forty-four pages of volume I contain some very interesting and elementary discussions of the axiomatic method as applied to geometry of the plane and elementary number theory, and the arguments for its formalization.)

Since it would not be practicable to give here a detailed summary of this work, we shall be content with some remarks regarding methods, the ends actually achieved, and the work of later writers.

2.2 Metamathematical proofs

A question which has probably already occurred to the reader is: What is considered a *proof* in the metamathematics? For plainly the proofs embodied in the formal mathematics, which are to be justified in the metamathematical study, must not be duplicated *in toto* in the latter! Otherwise we would be traveling in a vicious circle. More generally, we may ask, "What *methods of procedure* are used in the metamathematics?" In seeking the answers to these questions, we are at first struck by the intuitionist-like character of the methods used—not that metamathematics and intuitionist mathematics turn out to be the same—but the existence of an entity, for instance, is considered established only if it is constructible by finite methods. For example, existence of a number having a given property is established only by giving a method for constructing such a number. And all proofs have to be of a finite character that is "clearly perceptible" (*anschaulich überblickbar*). Also, the method of definition by induction based on the natural numbers, which Intuitionism regards as basic, is generalized to a more powerful tool called "definition by recursion."

2.3 Recursive definition

To give an indication of this method, we need the notion of "number-theoretic function" by which is meant a function of non-negative integers, each of whose values is a non-negative integer. Such a function, for example, would be a function $f(x)$ which is 0 when x is even and 1 when x is odd; or a function $f(x, y)$ which is 2 when $|x - y|$ is even and 3 when $|x - y|$ is odd. A function $f(x, y_1, \cdots, y_n)$ is called *recursively defined* by the number-theoretic functions $g(y_1, \cdots, y_n)$ and $h(z, x, y_1, \cdots, y_n)$ if for all non-negative integers $y_1, \cdots, y_n$ the following relations hold:

$$f(0, y_1, \cdots, y_n) = g(y_1, \cdots, y_n),$$
$$f(x + 1, y_1, \cdots, y_n) = h(f(x, y_1, \cdots, y_n), x, y_1, \cdots, y_n).$$

If $h(z_1, \cdots, z_m)$ and g_i, $i = 1, \cdots, m$ are given functions, then the function $h(g_1, \cdots, g_m)$ is said to be formed from h and the g_i's by substitution.

A number-theoretic function f is called *recursive* (see Gödel [d] for the definition as stated here; usually now called "primitive recursive"†) if there exists a finite sequence of number-theoretic functions $f_1, f_2, \cdots, f_k$ such that f_k is f, and every f_i, $i = 1, \cdots, k$ is either (1) recursively defined from two of the preceding f_i's, (2) results from preceding f_i's by substitution, or (3) is a constant or a "successor function" $x + 1$.

For example, let $S(x)$ be the successor function; i.e., $S(x)$ is the integer $x + 1$ directly following x in the series of integers. And let $g_i^n(x_1, \cdots, x_n) = x_i$, $i = 1, 2, \cdots, n$, be the identity (or selector) function; i.e., it selects the ith integer from the *n-tuple* $(x_1, \cdots, x_n)$. Then we may define a function $A(x, y)$ as follows:

$$A(0, y) = g_1^1(y),$$
$$h(z, x, y) = S(g_1^3(z, x, y)),$$
$$A(x + 1, y) = h(A(x, y), x, y).$$

† Regarding "general recursiveness," and its relations to "effective calculability" (Church) and the Turing [a, b] computing machine, see Kleene [Kle; Part III], Davis [D], and Beth [Be; 297–316].

As the function $A(x, y)$ is recursively defined from the functions g_i^n and h, it is a recursive function by virtue of the sequence S, g_i^n, h, A. (The reader will recognize $A(x, y)$ as the sum $x + y$; compare the above with VI 3.1.1.)

Such recursive functions are of fundamental importance in both the Hilbert-Bernays theory and modern investigations of "formal systems."† Number-theoretic functions that are recognized in the metamathematics to be recursively definable are deemed to be constructively (or effectively) definable, and hence admissible, in the formal system. We shall see an example of this in Section 3.

2.4 Degree of success of the program

For certain elementary systems the proofs of consistency were successfully carried out, thus incidentally exemplifying what Hilbert would like to have done for a "complete" mathematics. For instance, the first application of the proof theory in the *Grundlagen* [H-B; § 6] to an infinite system is a proof that a certain elementary system of axioms concerning integers is consistent. This system embodies the calculus of propositional functions and the Peano axioms, but not the mathematical induction principle. In the proofs, the case which involves no bound variables (IX 4.1) is first treated, and then extension is made to cases involving bound variables. In connection with the latter it is shown how to reduce a formula to numerical formulas which can be shown true or false (cf. Section 3 below).

It is subsequently shown how to extend such reduction methods to the system obtained by adding the mathematical induction axiom to the other Peano axioms in the system. Unfortunately this does not give a consistency proof for the entire arithmetic of whole numbers. For it will be recalled that in developing arithmetic from the Peano axioms it is necessary to define $+$ and $\times$ (VI 3.1.1, 3.1.2); and in the proof theory this is done by means of recursive definition. Hence it becomes necessary to determine whether the addition of such recursive definitions introduces the possibility of contradiction. That contradiction is not introduced so long as the system embodies only the predicate calculus for free variables was established.

2.4.1 Example. As an example of a system shown to be consistent in the *Grundlagen* we give the following (System "D," vol. I, p. 357). Instead

† See also Grzegorczyk [a].

of $S(a)$, the symbol a' is used. In addition to the axioms and rules of the predicate calculus, one states

$$a = a,$$

$$a = b \rightarrow (A(a) \rightarrow A(b)),$$

$$a' \neq 0,$$

$$a' = b' \rightarrow a = b,$$

$$a + 0 = a,$$

$$a + b' = (a + b)',$$

$$A(0) \,\&\, (x)(A(x) \rightarrow A(x')) \rightarrow A(a).$$

The first two of these axioms establish the role of "="; $A(x)$ is any formula involving the free variable x, and a and b are individual arbitrary instances of x. The next two axioms are counterparts of the Peano axioms (1) and (3) of VI 3.1. The fifth and sixth axioms serve to define "+," and the last axiom is the mathematical induction axiom—number (4) of VI 3.1 (this is really an infinite number of axioms, covering all choices of "A"—sometimes referred to as an "axiom schema").

If to the above are added the following axioms

$$a \cdot 0 = 0,$$

$$a \cdot b' = a \cdot b + a,$$

then a formalization† of the arithmetic of integers results (this is Hilbert-Bernays' System "Z"). No consistency proof for this system was obtained. And a decision procedure such as will be described in the next section, if found for this System "Z," would provide a method for solving the famous Fermat problem. Also, such unsolved problems as "Is every even number the sum of two prime numbers?" and "Is it true that there exist arbitrarily large prime numbers differing by 2?" would become solvable. For these problems may be formulated in the symbols of the above system.

However, the consistency of System "Z" is demonstrable by methods developed by Gentzen (see Gentzen [a, b]). Unfortunately, Gentzen's proof utilizes transfinite induction and, consequently, departs from the finite methods to which the proof theory is restricted. (Cf. [H-B II; 360ff].) As pointed out in X 7.1, if we use the Gödel transformation (cf. X 1.6.1), the question of consistency for the arithmetic of integers is reducible to that of the consistency of the Heyting calculus of intuitionist

† It will be noted that this formalization does not embody the concept of "set," or even of a "sequence," however.

arithmetic; and for the consistency of the latter the Kolmogoroff (X 7.2) "problem-calculus" affords a proof by interpretation (as explained in II 2).

3 Gödel's incompleteness theorem

Whether the complete Hilbert program could be carried out was rendered very doubtful by results due to Gödel (which appeared in 1931, before the actual publication of the *Grundlagen*) and other authors.†

These results may be roughly characterized as a demonstration that, in any system broad enough to contain all the formulas of a formalized elementary number theory, there exist theorems (formulas) that neither can be proved nor disproved within the system. (Cf. II Problem 23.) In particular, then, as Gödel pointed out, a system such as that of Principia Mathematica (see Chapter IX), if consistent, must contain "undecidable" formulas, as must also certain axiom systems for set theory (such as the Zermelo-Fraenkel-von Neumann system) when formalized by the addition of the axioms and rules of conclusion of the predicate calculus. The manner in which this was accomplished is described below.‡

Consider a formal§ system whose primitive symbols consist of individual variables $x, y, z, \cdots$, functional symbols $F, G, \cdots$, and further symbols S, 0, $=$, $\sim$, $\supset$, (,). In interpreting formulas constructed from these symbols we shall regard the individual variables as ranging over the natural numbers, while the functional symbols stand for properties and relations of numbers. S will denote the successor function (so that Sx is the number $x + 1$), while 0, $=$, and the logical connectives have their usual meaning. Thus the formula‖

$$(x)F(x) . \supset . (\exists y)(G(y) \vee (y = SS0))$$

would be read: If every number x has the property F, then there exists a number y such that either y has the property G, or y is 2. In the formal system, incidentally, we shall denote $SS0$ by $\bar{2}$, $SSS0$ by $\bar{3}$, etc.

† A very elementary exposition of Gödel's results and their implications may be found in the little book of Nagel and Newman [N-N].

‡ The material below and in Section 4 is based on lectures given by Professor Leon Henkin at the Mathematical Seminar, University of Southern California, November, 1949. Attention is also called to an elementary exposition given by Rosser [b], and to Weyl [We_1; 219ff].

§ Concerning formal systems in general, see Curry [a].

‖ Symbols like $\vee$, $\wedge$ (for the logical product, analogous to the $\cdot$ of P.M.), $\exists$ may be regarded, in this system, as abbreviating formulas composed only of primitive symbols. For example, $(A \vee B)$ would be taken as short for $((\sim A) \supset B)$, whatever the formulas A and B, while $(\exists x)A$ stands for $\sim((x) \sim A)$.

After setting up formal axioms and rules of inference for obtaining formal proofs of those formulas which express valid propositions about numbers, we proceed to "arithmetize"† the formal system as follows. We first assign to each primitive symbol a *coordinate* in the following manner (in the groupings below, symbols in the first line are within the system, and the numbers in the second line are the coordinates of the symbols directly above):

For the basic logical symbols we make the correspondences:

$$\begin{matrix} 0 & S & \sim & \supset & (&) & = \\ 2 & 3 & 4 & 5 & 6 & 7 & 8 \end{matrix}$$

For the individual variables (which can be denumerably infinite in number):

$$\begin{matrix} x & y & z & \ldots \\ 11 & 11^2 & 11^3 & \ldots \end{matrix}$$

For functions of the first order (functions of individual variables):

$$\begin{matrix} F_1 & G_1 & H_1 & \ldots \\ 13 & 13^2 & 13^3 & \ldots \end{matrix}$$

And so on, using powers of the $(n + 5)$th prime number for function symbols of the nth order.

To a formula in the system will correspond, then, a sequence of numbers, namely, those corresponding to the successive symbols of the formula. Thus, to

(3a) $$(x)F_1(x)$$

will correspond the sequence

$$6 \quad 11 \quad 7 \quad 13 \quad 6 \quad 11 \quad 7$$

We can make a unique number correspond to the formula by taking the product of the successive prime numbers (in their natural order) with powers equal to the numbers of the symbols (in the order in which they occur). Thus to formula (3a) will correspond the number

$$2^6 \cdot 3^{11} \cdot 5^7 \cdot 7^{13} \cdot 11^6 \cdot 13^{11} \cdot 17^7$$

We shall call this the *Gödel number* of the formula.

Notice that the Gödel number can be computed as soon as a formula is given; and that, conversely, when a number is given we can determine (by factoring it into primes) whether or not it is a Gödel number, and if so to which formula it belongs.

† Note the similarity to the process of assigning coordinates to points in geometry.

Finally, to any sequence of formulas we may make correspond a unique number

(3b) $$2^{n(1)} \cdot 3^{n(2)} \cdot \ldots \cdot p_k^{n(k)}$$

where $n(1), n(2), \cdots, n(k)$ are the successive Gödel numbers of the formulas of the sequence. We call (3b) the *sequence number* of the sequence of formulas. In particular, to every formal proof (i.e., proof within the system) corresponds a sequence number, since a formal proof of a formula F consists of a finite sequence of formulas each of which is an axiom, or derived from axioms and preceding formulas by the allowable rules of inference of the system, the last formula of the sequence being F itself.

By means of this arithmetization, statements about the formal system can be "translated" into sentences about numbers. In general, corresponding to each class, relation, or operation on formulas there is an associated class, relation, or operation on (Gödel) numbers.

Now it happens that many of the interesting classes and functions arising in this way are primitive recursive (2.3). Since all primitive recursive notions may be defined by formulas of our system, we see that in a sense the formal system can "talk about itself"! The meaning and some of the consequences of this fact will become evident as we proceed.

Outside the system (in the metamathematics, that is) let $B(x, y)$ denote that y is the sequence number of a formal proof in which the last formula has Gödel number x. And let $f(x, y)$ be the Gödel number of the formula obtained by replacing, in the formula having Gödel number x, each occurrence of the particular individual variable z by $\bar{y}$. Gödel shows that both B and f are recursive.

Now form the relation $B(f(x, x), y)$. This is also recursive; denote it by $G(x, y)$. Being recursively definable, it has a counterpart in the system which we denote by $\Gamma(x, y)$. Then by virtue of the rules for constructing formulas, the system contains the formula

(3c) $$\sim(\exists y)\Gamma(z, y).$$

Let the Gödel number of this formula be i.

A new formula is obtained by replacing z in (3c) by $\bar{\imath}$:

(3d) $$\sim(\exists y)\Gamma(\bar{\imath}, y).$$

Let the Gödel number of (3d) be j. Note that

(3e) $$j = f(i, i).$$

We now prove:

3.1 Theorem. *If a formal system, S, such as that described above, is consistent, then the formula* (3d) *is not provable in S.*

Proof. Suppose that it is provable in S, and let k be the sequence number of such a proof. Then $B(j, k)$ holds and, by virtue of (3e), $G(i, k)$ holds. But then $\Gamma(\bar{\imath}, \bar{k})$ is provable in S, for it may be shown that, whenever particular numbers are in some recursive relationship, the formula of the system which expresses this fact is formally provable. It then follows easily that $(\exists y)\Gamma(\bar{\imath}, y)$ is provable in S. Since our starting assumption was that (3d) is provable in S, it follows that S is not consistent. However, S was assumed consistent, so we conclude that (3d) is not provable in S.

To show that the formula

(3f) $$(\exists y)\Gamma(\bar{\imath}, y)$$

is also not provable, Gödel used the assumption of "ω-consistency";†

Definition. A formal system is ω-inconsistent if it contains a formula $\varphi(x)$ such that both

$$(\exists x)\varphi(x)$$

and

$$\sim\varphi(0), \ \sim\varphi(\bar{1}), \ \sim\varphi(\bar{2}), \cdots$$

are provable in the system. If no such formula exists, then the system is called *ω-consistent*. Evidently ω-consistency implies ordinary consistency (since inconsistency implies all formulas provable), but the converse does not necessarily hold.‡

3.2 Theorem. *If the formal system S is ω-consistent, then the formula* (3f) *is not provable in S.*

Proof. Suppose that it is provable. Then $B(j, k)$ cannot hold for any k, since otherwise (3d) is provable in S and S is inconsistent. Thus, for no k does $B(j, k)$ hold. That is, $B(f(i, i), k)$ holds for no k, implying that $G(i, k)$ holds for no k. Thus, for each natural number k, $\sim\Gamma(\bar{\imath}, \bar{k})$ is provable in S. In particular, then, all formulas

$$\sim\Gamma(\bar{\imath}, 0), \qquad \sim\Gamma(\bar{\imath}, \bar{1}), \qquad \sim\Gamma(\bar{\imath}, \bar{2}), \cdots$$

are provable in S. But, since (3f) is also provable, this implies that S is ω-inconsistent. We must conclude, then, that (3f) is not provable in S.

From Theorems 3.1 and 3.2 it follows that S contains a formula (3d) that is not provable, nor, if S is ω-consistent, is its negation provable in S. From this we get

3.3 Theorem. *If the system S is ω-consistent, then it cannot be complete.*

† See, however, Rosser [a], [b].
‡ Cf. Gödel [c], p. 190.

4 Consistency of a formal system

It is natural to ask if such a formal system S could be made complete by adding to its basic axioms either (3d) or its negation. The answer is no, because, so long as only formulas are added whose Gödel numbers constitute a primitive recursive set of numbers, the resulting system will remain incomplete. In such systems, then, consistency seems to imply incompleteness; that the converse holds is obvious, since an inconsistent system is always complete. As for systems with axioms whose Gödel numbers do *not* form a recursive set—they suffer from the embarrassing drawback that there is no effective test for deciding whether or not a given formula is an axiom!

Regarding consistency, Gödel showed that in a formal system such as S above, it is a recursive notion and hence expressible in the system. This can be done as follows: Since $B(x, y)$ is recursive, it has a counterpart $\beta(x, y)$ in the system, and we can let "Wid" denote

$$(x) \,.\, \sim(\exists y_1)(\exists y_2) \,.\, \beta(x, y_1) \wedge \beta(N(x), y_2)$$

[Here we have indicated by $N(x)$ the counterpart in S of the Gödel number of $\sim A$, where A is the formula with Gödel number x.] Gödel then showed that, if the system is consistent, the consistency is not provable within the system:

4.1 Theorem. *If S is a consistent system, then* Wid *cannot be proved in S.*

Proof. By paralleling within the system our proof of 3.1, we can demonstrate that

$$\text{Wid} \supset \sim(\exists y)\beta(\bar{j}, y). \tag{4.1a}$$

And, since $j = f(i, i)$, and hence $B(j, y) = B(f(i, i), y) = G(i, y)$, (4.1a) gives

$$\text{Wid} \supset \sim(\exists y)\Gamma(\bar{\imath}, y).$$

Now, if Wid were provable, we would then have at once that

$$\sim(\exists y)\Gamma(\bar{\imath}, y)$$

was also a formal theorem. But this contradicts Theorem 3.1.

5 Formal systems in general

Although it appears an impossibility to develop mathematics in a complete, consistent formal system, just as it would be impossible to axiomatize the whole of mathematics in a single system using the type of axiomatics described in Chapters I and II, we can reduce special parts of mathematics and logic to formal systems just as we axiomatize special parts of algebra and geometry. And this has been done with fruitful results. In some of these systems, analogues of the Gödel theorems have been shown to hold. In others, however, positive results have been obtained; decision problems are solvable (see below) and completeness demonstrable. We have already seen an instance of the latter in the case of the propositional calculus [IX 3.9]. Other instances will be found in Tarski's "A decision method for elementary algebra and geometry" [T_1] (cf. also references to other cases therein) and Tarski-Mostowski-Robinson [T-M-R].

5.1 Decision procedures

Given a collection, *C*, of formulas in a theory *T*, a *decision problem* for *C* in *T* is the problem of finding a method—an effective procedure—by which, given any formula, we can decide in a finite number of steps whether it is in *C*. If such a method exists, it may be called a *decision method* or *decision procedure* for *C*. To quote Tarski [T_1], "a decision method must be like a recipe, which tells one what to do at each step so that no intelligence is required to follow it; and the method can be applied by anyone so long as he is able to read and follow directions." The same author cites as an example the collection, *A*, of all true formulas (sentences) of the form "*p* and *q* are relatively prime" in the elementary theory of integers. For any particular sentence, such as "215 and 349 are relatively prime," Euclid's algorithm provides a method for deciding whether it is in *A* or not.

The important case, and the one in which we are most interested, is that where *C* consists of the set of all provable formulas of a formalized system. The decision problem then becomes the problem of finding a decision procedure by which it can be decided of any particular formula whether it is provable by the methods admissible in *T*. If such a procedure exists, it is called simply a *decision procedure* for *T*; and the problem of determining whether a decision procedure exists for *T* is called the *decision problem* for *T*. A theory for which a decision procedure exists is called *decidable*; and if not, *undecidable*.

Now we may justifiably ask: "What is an 'effective procedure' anyway?" To say that it must involve only a finite number of steps, constructively pursued (an "algorithm"), is still rather vague. It is precisely this charge that some have leveled at the intuitionist requirement of "constructive proof"—that it is not precisely defined. Consequently, in attacking a decision problem for a theory *T*, we may expect better chances of success if we can formulate, in some manner, a specific type of procedure *P* and then investigate whether it "works" for *T*.

Continuing the same line of thought, we may achieve greater generality by so formulating P as to make it meaningful not just for one theory, but to a whole class of theories; perhaps, indeed, to any formalized theory. Thus, if a theory T is formalized by using the first-order predicate calculus, symbols for the technical terms, operations, etc., of T, and we "arithmetize" it in the manner devised by Gödel (Section 3), then we may be able to formulate P by recursive methods as exemplified in the "primitive recursion" employed in Section 2.3.

5.2 Precise formulation of the notion of decision procedure

The carrying out of such a program was achieved in several ways, all *a priori* different. Chief among these were *general recursiveness* (Herbrand-Gödel), a generalization of primitive recursion; *λ-definability* (Church-Kleene); and *computability* (Post-Turing). The notion of λ-definability was proposed by Church [d] as a precise definition of what one might mean by the intuitive notion of "effectively calculable." In dealing with an "arithmetized" theory, the decision problem can be transformed into a problem concerning whether a given number-theoretic function (2.3) is effectively calculable or not (cf. Church, *loc. cit.*); hence, Church's proposal. And, as the word implies, computability is based on the idea that if a number-theoretic function is to be effectively calculable, it should be possible to construct, at least theoretically, a machine that would carry out the calculation.

That these notions have perhaps caught the intuitive notion of effective calculability is supported by the fact that all have been proved equivalent (see Kleene [Kle; 320], for instance; also citations therein). In addition, other evidence such as that known effectively calculable functions all turn out to be general recursive, and that search for methods that could yield functions not general recursive seem only to give functions that can hardly be deemed effectively calculable, bolster the belief that in general recursiveness we have a justifiable definition of "effective calculability." And at the same time, we are able to give formal expression to the notion of "effective definability."

To develop these notions in detail would take us beyond the scope of this book, and we refer the reader to more technical works such as that of Kleene [Kle] and Davis [D]. It is interesting to speculate, incidentally, on what Poincaré's attitude toward these developments might have been, considering his ideas regarding definition in mathematics (VIII 7) and, on the other hand, his rejection of the "logistic" investigation of the foundations of mathematics (*loc. cit.*). The attitude of Intuitionism would probably be general rejection of any attempt to confine the notion of constructive definition to a precise "confining" formulation; for a fuller discussion, however, see Kleene [Kle; §§ 80–82].

5.3 Existence of undecidable theories

In 1936, Church [d] showed, by using the notion of λ-definability cited above, the absence of decision procedures for a large class of formal systems. In particular, he showed that there is no decision procedure for Peano arithmetic. On the other hand, for the system D of Section 2.4.1 there does exist a decision procedure (cf. Hilbert-Bernays [H-B; 359–367]); although there does not exist one for system Z. Church's result was extended by Rosser [a] to showing that the Peano arithmetic is *essentially undecidable* in the sense that every consistent extension of it is undecidable. The arithmetic of every finite field is decidable. Also, the arithmetic of the field of real numbers is decidable, and so is that of the field of complex numbers.

Since most theories turn out to be undecidable, the study of decision problems has turned out to be a rather "negative" occupation. Thus, the first order predicate calculus is itself undecidable. In contrast to the real number arithmetic, the field of rational numbers is undecidable. The general theory of rings is undecidable. The theory of groups (and semi-groups) is undecidable (although that of commutative groups is decidable). And various topics in set theory and topology are also known to be undecidable.

For further information regarding decision problems, the reader is referred to such works as Kleene [Kle] and Tarski-Mostowski-Robinson [T-M-R].

6 General significance of formal systems

We have seen how, from the type of axiomatic geometry of the Greeks, there evolved a more rigorous axiomatic method as exemplified in the late nineteenth century and early twentieth century geometries (e.g., Pieri, Hilbert) and in the subsequent axiomatization of a major part of mathematics (groups, fields, real number system, etc.). This type of work still goes on, inasmuch as the so-called "working mathematician" (meaning one who is not a "logician"!) usually employs the kind of axiomatic tools that we described in Chapters I and II, assuming as "universal" not only the logical processes employed, but also the set theory. He may know that the latter conceals contradictions, but since his uses of set theory usually involve only the "safe" portions of the theory, he justifies his methods by observing that they "work." Besides, he would not have time to set up a complete formalization.

Of the earlier investigations in foundations, it seemed that only the Peano axioms for the natural numbers achieved any notable recognition by the general mathematical public, for although the type of attitude Poincaré evinced toward work in the logical foundations may not have been so pronounced among the rest of the mathematical community, it certainly seems to have been rather widely shared. The Peano axioms, stripped of their formalism and couched in the natural language common to the axiomatics of the day, formed a convenient starting point for the teaching of analysis during the early part of the twentieth century (in which the concept of the real number system was developed as in VI 3, or some variant thereof such as the use of Cauchy sequences instead of Dedekind cuts). But of the other work of Peano and his school, as well as of Frege, Russell, and others, the general mathematical public seemed unaware.

With the development of modern mathematical logic, however, the importance of logic and, ultimately, of formal systems in general, became apparent. Realization that fundamental questions in foundations could hardly be resolved without reduction to formalism was probably the major factor in this. This was already evident early in the century, when the study of consistency was the motivating factor in logical studies. On the other hand, devices such as the Frege-Russell definition of number, designed to found mathematics in "logic," met cold reception such as that of Poincaré, together with the query, "Why bother to do it?" A quite satisfactory foundation for a theory could be achieved by using axioms as, for instance, Hilbert did in his geometry; and although the proofs of consistency, completeness, and the like by using models may not have been absolute, they were good enough for the working mathematician; and, we may admit, still seem to be. Axiomatic foundations for set theory, such as those of Zermelo-Fraenkel and von Neumann, seemed unnecessary to most mathematicians, since what they wanted was only a portion of the set theory that we have discussed in Chapters III–V. And to a great extent this is still the case—indeed, so far as transfinite numbers are concerned, often the need is only for the distinction between the countable and the uncountable!

But appreciation of the accomplishments of the mathematical logicians seems to be rapidly growing. Earlier results, such as the well-ordering theorem, the subsequent "discovery" of the Choice Axiom, its equivalence to comparability (V 5), and the like, were originally achieved without resort to logical formalism. But deeper questions, such as the nature of models (cf. the "ideal" models of II 1.2 and II 2), the nature of definition, completeness, decidability, and the like, have required a finer instrument than the language of ordinary mathematical discourse and its ambiguities, and compelled resort to formalized axiomatics. And these have justified their introduction by results whose power and generality were undreamed of at the turn of the century. In addition, such matters as the development of machine computation and the theory of automata seem to have given a stimulus for the admission, into the main stream of mathematical evolution, of what is now called "Mathematical Logic."

ADDITIONAL BIBLIOGRAPHY

Beth [Be]
Carnap [Car_2]
Church [Ch_1]
Curry [a] [Cu_1]
Goodstein [Go]
Journal of Symbolic Logic
Kleene[Kle]
MacLane [a]
Rosenbloom [Ros]
Rosser [Ross]
Stoll [Sto]
Woodger [Wo_1]
Tarski-Mostowski-Robinson [T-M-R]

PROBLEMS

1. Contrast the attitudes of (1) Russell, (2) Brouwer, and (3) Hilbert regarding the role of logic in mathematics.

2. If it develops that we can never devise a mathematics wholly free of contradiction, what will the effect be on (1) fields (such as physics) which use mathematics as a tool; (2) the convictions of those who have relied on mathematics as a secure haven from the uncertainties of the natural world?

3. What conclusions have you reached in regard to the respective conceptions of Brouwer and Hilbert regarding what constitutes mathematics? What are their chief differences?

4. Compare (*a*) impossibility of trisecting an angle of 60° with ruler and compass alone, and (*b*) existence of unsolvable problems in a formalized arithmetic.

5. Contrast the axiomatic method as described in Chapters I and II with the formalized axiomatics such as those described in the present chapter. Note, in particular, that in the former the proof procedures are tacit, while in the latter they are given explicitly in the metamathematics.

6. Note that the result of Section 4 implies that there exists no metamathematical proof of consistency of a formalized arithmetic which can be interpreted within the system. Contrast this circumstance with the Gentzen consistency proof cited in 2.4.1.

7. In view of Gödel's incompleteness theorem, do you believe it would be better to develop arithmetic in non-axiomatic form, as, for instance, Intuitionism does (although not necessarily adopting the limitations imposed by the Intuitionist tenet of constructivity)? Or would you prefer, as a compromise, development of arithmetic by the type of axiomatics described in Chapters I and II?

8. Discuss the following opinion: The axiomatic method as described in Chapters I and II is more suitable for the creation and development of new mathematics than systems formalized in the predicate logic. Its methods are not metamathematically limited, and consequently the possibility of discovering new methods quite open.

9. Discuss the following: Formalized axiom systems are the best instrument for the analysis of fundamental questions such as provability, independence of basic principles (e.g., the Choice Axiom in set theory), completeness, etc.

10. Set theories have been presented in formalized axiomatic form (e.g., Gödel [G], von Neumann [b]). What advantages would you think these might have over the non-formalized presentations (such as that given for the Zermelo set theory in Chapter VIII); what disadvantages?

XII

The Cultural Setting of Mathematics

In the preceding chapters we have tried to give an introduction to the fundamental ideas and methods of mathematics. In Chapters I and II we described the axiomatic method as it is commonly used in modern mathematics, and in Chapters III to V we discussed the basic notions of set theory which we encountered in the actual use of axioms and which are fundamental in modern mathematics. In Chapters VI and VII we applied the axiomatic method and set theory to the definition of basic notions of analysis and algebra. The rigorous formalistic use of axioms, which we found emerging in the works of earlier mathematical logicians (Chapters VIII and IX) and reaching maturity in modern symbolic logic (Chapter XI), is rarely used in actual mathematical practice. To proceed purely formally, setting forth the basic axioms as formulas, and the allowable methods of proof as prescribed rules for deriving new formulas from old (cf. XI 1), would probably be impractical as a mode of either teaching or research, generally speaking. Purely formal methods, although indispensable for the investigation of the foundations (as in questions concerning completeness, consistency, and decision problems, for example), would probably hamper research in higher mathematics rather than facilitate it.

The situation may be compared to that existing between a "natural language" and a formal language; although a commonly spoken language, such as English, is full of ambiguities and illogicalities, its replacement by a formal language, if such were possible, would probably slow up social intercourse, and the formal language undoubtedly would eventually be contaminated by the same kinds of ambiguities and logical defects found in the natural language.

Regarding the effect of the controversies over the theory of sets and the contradictions to which it is liable, it is probably safe to say that the modern worker in the foundations is more interested in the problems to which they have given rise than in the choice or validity of an underlying philosophy; one can be either a follower of Russell or an intuitionist and be interested in the relations between the Russell logic and the intuitionist logic (cf. X 7.1), or in the conditions sufficient for the solvability of a given problem. (We have not, indeed, touched upon all the varieties of foundation philosophies to which past speculation has given birth; for further discussion on philosophic lines the reader is referred to the general literature in the subject. See, for example, Benacerraf and Putnam [B–P], Beth [Be], Black [Bl], Fraenkel [F_2], Kattsoff [K], Maziarz [M], and Weyl [We_1; P. I], and their extensive bibliographies.)

In contrast to the somewhat technical material given in the preceding chapters, we now give some attention to the general cultural setting of mathematics. After all, mathematics was born and nurtured in a cultural environment. Without the perspective which the cultural background affords, a proper appreciation of the content and state of present-day mathematics is hardly possible.

1 The cultural background

To avoid misunderstanding, we must make clear our use of the term "culture" above. We use it in the general anthropological sense, as exemplified in the terms "Chinese culture," "ancient Greek culture." In this sense, a culture is the collection of customs, beliefs, rituals, tools, traditions, etc., of a group of people such as an Indian tribe, or the people of a given region such as the United States. It is not the use of the term as in "a cultured person" that we have in mind.

As human beings we are born into cultures, and it is these that, acting on and interacting with our receptive nervous systems, determine the habits, beliefs, etc., that largely make up our "personalities." Most of what we take for granted is culturally determined. A culture, it should be noticed, is more than the habits, beliefs, skills, etc., of any single one of the persons living under its influence; usually it existed before he was born and will continue to exist after him, although it may undergo change during his lifetime. He possesses some of the "cultural elements," but not all of them. If he is born and brought up in a middle western town of the United States, he may belong to one of the branches of the Christian religion (usually depending on whether his parents did), talks English of the variety spoken in that region, perhaps learns a trade such as tool-making, gets married, has a family, and dies. His best friend's experiences may parallel these, except that the friend's religion may be different and he may become a musician instead of a tool-maker.

Now the tool-making skill, the religions, the musical techniques, etc., were already present in the middle western culture before the above hypothetical persons were born, and will go on after them. As individuals these persons may contribute to the culture new ways of making tools, new techniques of piano playing or musical composition, but these, though "new," are the result of the interaction of the cultural elements concerned with the individual's own capacities. Thus the musical composition will probably be written in the diatonic scale, and contain elements common to the fashions in composition of the time. The effect of the culture on the modes of dress, eating habits, recreation, etc., are too obvious to dwell upon.

2 The position of mathematics in the culture

So far as mathematics is concerned, both the tool-maker and his friend the musician will probably learn to count before entering primary school, study arithmetic in the primary school, and possibly elementary algebra and plane geometry in the secondary school. Probably neither will go any further in mathematics, and the musician will undoubtedly promptly forget all the algebra and geometry he has learned. Both will retain enough knowledge of arithmetic to enable them to do the ordinary computations required in borrowing money, using recipes, etc., and the tool-maker may remember such elementary geometry as is useful to him in his trade. As carriers of the mathematical tradition in their culture, however, they play only a minor role, *viz.*, that of every parent who teaches his child to count and helps him with his "home-work" during the school years. Their conception of mathematics might be considered a modern version of that of the ancient Babylonian and Egyptian computers, to whom mathematics was a body of rules for handling quantitative problems arising in the workaday world.

The principal mathematical element in the culture, embodying the living and growing mass of modern mathematics, will be chiefly possessed by the professional mathematicians. True, certain professions, such as engineering, physics, and chemistry, which employ a great deal of mathematics, carry a sizable amount of the mathematical tradition, and in some of these, as in the case of physics and engineering research, some individuals contribute to the growth of the mathematical element in the culture. But, in the main, the mathematical element of our culture is dependent for its existence and growth on the class of those individuals known as "mathematicians."

Now, if we compare mathematics with such cultural elements as language, religion, and dress, we notice a striking difference; for, although the latter elements are generally peculiar, in their various manifestations, to regional groups, mathematics seems to have a universality that knows no boundary lines. During the present century it has been customary to hold international mathematical congresses every few years. In attendance at these will be found mathematicians representing all manners of language, dress, and religion, as well as all shades of political opinion. But a Russian algebraist, a German algebraist, and an American algebraist all share the same algebra; and a Japanese topologist and an American topologist share the same topology. What differences exist between them generally have to do with details concerning new theories, rather than with the known parts of the subject which they share. This universality of mathematics seems to be one of its most distinguishing characteristics among the various cultural elements.

However, on closer study we shall find that (1) this universality did not always exist, (2) it is not so complete as it seems, and (3) its existence is easily explained in cultural terms.

3 The historical position of mathematics

If we look at the position of mathematics in the various cultures about the time of Christ, we find a radically different situation from that of the present. In the Greek culture, geometry and its attendant logical and dialectical devices were the principal elements of mathematics; they were paralleled by a numerical and algebraic system whose basic symbols were derived from the Sumerian-Babylonian cultures and have been virtually forgotten today. Although the Romans were influenced by the Greek mathematics, they had their own system for writing numerals—a system much more cumbersome than the Greek types for numerical manipulations, yet surviving to this day on monuments, title pages, and other odd places. On the other hand, in the Chinese culture, no geometry of the Greek type was known. Mathematics in China consisted principally of numerical computations and solution of algebraic equations. And, although there were evidently contacts between the Eastern and Western civilizations from early Christian times onward, there are few instances of diffusion of mathematical ideas from one of these culture areas to another until relatively modern times. No geometry of the systematic logical type cultivated by the Greeks gained a foothold in China, where such geometric elements as existed seem to have been of the non-logical, "rule" type characteristic of the pre-Greek Egyptian geometry. Chinese mathematics evidently emigrated to Japan along with other elements of Chinese culture and there, despite the probable introduction during the seventeenth century of Euclid texts in the Chinese language, mathematics continued along the same traditional lines. In the seventeenth and eighteenth centuries we find mathematicians in Japan solving equations of degrees as high as 3000 and 4000! The major resistance to absorption of Western mathematical ideas was not broken down in Japan until the cultural innovations following the Restoration of 1868, when European mathematics began to flow in along with other cultural elements.

On the other hand, conditions in the Mediterranean area during the flourishing of the Greek culture and for centuries thereafter were such that no systematic development of algebra was made. And it was left to the Arabs to preserve and transmit, via Africa, Spain, and Italy, the Hindu-Arabic mathematics (and the system of enumeration which is universal today), along with Greek geometry.

It is frequently very difficult to determine why an element of a culture C_1, which might be very useful in a culture C_2, is not always accepted and integrated with C_2. Thus it was pointed out by Sarton (see Schaaf [S; 72–73]), for example, that although the metric system would have been highly useful to the industrial elements in English culture, it was not accepted by the latter. Incidentally, Sarton used this example to support a contention that mathematics is not always "economically" determined, but dependent to a large extent upon forces "interior" to us. So far as economics is only a single aspect of culture, Sarton was of course correct; but we can question the propriety of setting the economic aspects of English culture, by implication, "external" and other aspects "internal." Evidently there were present in the English culture other elements (perhaps national pride and resistance to innovations from without, for instance) which were stronger in their resistance to adopting the metric system than the attraction of the metric system for the industrial elements of the culture.

These observations (which could be augmented by many others if space permitted) teach us that mathematics is not something which is by its nature universal, absolute, or foreordained; it is subject to laws of development and influence from other cultural elements much as are arts and sciences in general. Of course, we do not expect to find mathematics developing, under cultural influences, exactly like other cultural elements, any more than we expect to find different individuals reacting the same way to their environments. This is due partly to the nature of the subject and partly to its position in the complex of human activities, just as the development of the individual personality is due partly to inherent qualities and partly to the position of the individual in his society and culture.

4 The present-day position of mathematics

In contrast to the historical situation depicted above, mathematics occupies today a seemingly unique position. Although mathematicians speak various "natural languages" around the world, their mathematical language is practically universal. Thus an algebraist can pick up an algebraic article in a journal published in any part of the world, and, after piercing the natural language barrier, find the same algebraic notions and, as a rule, algebraic symbols with which he himself works. The same generally holds in any branch of mathematics (although in a new branch there usually is a period when the terminology is a bit chaotic).

However, variations in symbols and terminology do occur, not only between mathematicians in different countries but also between groups in the same locality. Not infrequently, differences of opinion have arisen regarding the direction further research should take and the important problems to be solved. Such differences lead, temporarily, at least, to different "schools" which, while sharing essentially the same philosophy about foundations, do differ on subject matter and choice of method.

For example, it is well known that usually there are many different ways of solving a given problem. Indeed, once a problem is solved, many new solutions are likely to be found, perhaps of a simpler or more elementary character than the original solution. Of the possible solutions, some will be preferred by one group, others by another group. Perhaps some of the solutions are geometric, for instance; then the geometric-minded will usually prefer them. If some are of an algebraic nature, with no geometric elements entering in, they will almost certainly be preferred by mathematicians who have specialized in algebra. An instance of the latter kind was especially noticeable in the relatively new subject of topology a number of years ago when some, called "set-theorists," preferred to use set theory as their chief tool while others preferred algebraic methods. (Today, mixed methods prevail in topology.)

It is natural also to expect that differences will be found between groups living in widely separated (geographically) areas. Indeed, some mathematics still take on a distinctly national character. French mathematics was long known for its preference for function theory, English for interest in applied mathematics, German for foundations, and Italian for geometry. Today we frequently hear comment among mathematicians regarding the abstract character of mathematics in the United States—a trend which was perhaps somewhat retarded by the call for mathematicians to go into war work in 1941 and subsequent years.

Nevertheless, despite these variations (which are generally due to cultural influences), mathematics today can be considered to be distinguished by a universality foreign to most other human activities.

5 What is mathematics from the cultural point of view?

The question, "What is mathematics?" has probably received as much attention as any question of fundamental character—and as many answers. In view of the discussion of the last four chapters, it would be unlikely that the reader would expect that anyone could give a nice, compact answer that would find general acceptance.

Here we must qualify the question, however. As it stands, no indication is given as to the type of answer expected. It is not the dictionary type of definition that is ordinarily desired in asking the question. More likely, it is expected that one will give the sort of answer that philosophy has sought. And both philosophers and mathematicians have given answers, but none has found general acceptance. The trouble seems to lie chiefly in the assumption that mathematics is by nature something absolute, unchanging with time and place, and therefore capable of being identified once the genius with the eye sharp enough to perceive and characterize it appears on the human scene. And, since mathematics is nothing of the sort (although the layman will probably go on for centuries hence believing that it is), only failure can ensue from the attempt so to characterize it.

Is there any sense, then, in which the question can be answered? Probably only in that it is one of those questions that have to be answered by pointing; we point out what the thing to be defined is, at a certain time and place. We do not thereby assume any responsibility for the appearance of the same thing at another time or place. Thus we can give a pretty fair answer as to what mathematics was in Greece in the year 100 B.C. And much of this Greek mathematics has evolved to become part of what we would call mathematics today—although we should not take this to mean that mathematics has a timeless character as so many have asserted, for it is not eternal; much of mathematics once accepted has since been rejected—is no longer mathematics, that is—or has completely changed in character. Although from a superficial point of view the Greek geometry seems still to be a part of mathematics, in a strict sense it is not. In the Greek culture, geometry was thought to be either an idealistic description of real space dictated by natural phenomena, or a doctrine imposed by a philosophy of absolutes; whereas in our culture the analogue of what the Greeks called geometry is only one of several co-existing geometries, each of which embodies only a special mathematical concept. In its modern axiomatic form (as described in Chapters I and II), it is something quite different from what the Greeks considered it. In this sense, then, the Greek geometry cannot be said any longer to constitute mathematics, any more than does the progenitor of man in the evolutionary scale any longer exist except through his descendant, modern man. Like comments hold for the notion of number.

When the question, "What is mathematics?" is considered a request for a criterion that will enable us to distinguish the principal mathematical element of a culture from other elements, then a satisfactory answer may become possible. We have to bear in mind that during the course of human history, as man came to recognize differences between certain activities, he found it convenient to assign them *names* in order to distinguish them from one another; and then ultimately, in a philosophic mood, he came to believe that *the names themselves had acquired a content and being of their own*; and so he set out to find just what this content was. And the result, as in the case of other abstractions, such as "mind," "life," "beauty," has often been a philosophic muddle. "In the beginning was the word" simply does not hold concerning mathematics—for first there had to be something to be called "mathematics." In primitive cultures having only such rudimentary forms of mathematics as counting, no word for "mathematics" is found.

Why do we differentiate certain elements of the Greek and Chinese cultures from other elements of those cultures and call them "mathematics"? More generally, what enables us to go into any culture and pick out its "mathematical" elements? Obviously, we use what we call "mathematics" *in our own culture* as a criterion; thus, so far as ancient cultures are concerned, we lump together everything numerical, geometrical, and algebraic under the name "mathematics." Conceivably, however, what *we* call "mathematics" in an ancient culture might have been (and probably was) called in that culture by names that would correspond, in our language, to "astrology" or "theology." Inevitably *we are influenced and guided by what* we *call mathematics*, in our determination of the mathematical elements of other cultures. And consequently we may as well narrow down the question to: "What is the cultural element called 'mathematics' in the world of today?" (Because of its essential universality, there is no need to say "in the United States" or "in France"—"in the civilized world" will do.)

One way to answer the question, impractical, to be sure, is to gather into one library copies of all the "mathematical" material in the world and, having done so, answer, "This is mathematics today." We said "impractical," for not only would it be obviously impossible to make such a collection, but also selection of the items to be included would in many instances reflect tastes peculiar to the selector. Recall Kronecker's reported remark concerning Lindemann's proof of the transcendence of π. It is still not uncommon to hear one mathematician characterize the work of another by a contemptuous, "It's not mathematics!" The position of the selector would not be unlike that of the zoologist who is asked to determine whether certain tissues are "animal" or have "life." The "border-line" cases alone would render the task of selection an impossibility. The best that could be done would be to place some items in a department labeled with a question mark, or with the notation "not generally accepted as mathematics, but accepted by some."

There is an interesting historical item in this connection, namely, the first extant history of mathematics—that of J. E. Montucla, published in 1758. Montucla divided "mathematics" into two parts, the one "comprising those things that are pure and abstract, the other those that one calls compound, or more ordinarily physico-mathematics." The first part dealt with what we would today probably call "pure mathematics," and the second with other concepts that could be formulated in mathematical terms. Thus Montucla treated such subjects as mechanics, optics, astronomy, judiciary astrology, navigation, geography, and music. And, for good measure, he flung in some material on applications to the construction of observatories, ships, etc. He cites evidence to show that his inclusion of such topics had a precedent in early Greek histories of mathematics which included music and gnomonics (although otherwise devoted exclusively to geometry). The "division of labor" consequent to the growth of mathematics since Montucla's time has forced a change in our ideas of what constitutes mathematics!

Compare Montucla's history with a history of mathematics written during recent years, and notice the change. Although modern histories of mathematics still contain references to astronomy (because so many early "mathematicians" were what we would today call "astronomers"), these are only incidental and no special section is devoted to astronomy as a rule. Vestigial remains of older cultures are still found in our colleges, in the existence of "departments of mathematics and astronomy," despite the fact that today we have a special class of workers called "astronomers" to whom "mathematics" is a "tool," not a profession.

Actually, the variability of the subject matter covered by the term "mathematics" is so pronounced that it can be quickly discerned by comparing a history written about 50 to 75 years ago with one written during more recent times. Take, for instance, the histories of Ball [B] and Cajori [Caj], which were written shortly before 1900. In Ball's first edition (1888) there is no mention of "logic"; in the fourth edition (1908) there is a remark concerning George Boole to the effect that he "was one of the creators of symbolic or mathematical logic." The index of this book contains no citation to any other reference to logic. Cajori's first edition (1893) contains four remarks of a similarly incidental nature concerning logic. The second edition (1919), however, contains four pages (407–410) of material concerning "mathematical logic" (not cited in the index, for some reason). Compare these books with Bell's *Development of Mathematics* [B_3], published in 1940. Here at least 25 pages are devoted to "mathematical logic," reflecting the developments of the present century sketched in our preceding chapters. Obviously, mathematical logic was coming to be considered part of mathematics, although still referred to by some as an "application" of "mathematics." Another reflection of this development is to be found not only in recent writings but also in the fact that, whereas all the logic formerly taught in our colleges was done under the auspices of "philosophy" (pursuant to the Aristotelian tradition), today in contrast mathematics departments commonly contain men whose major work is devoted to logic and who teach courses embodying the developments in logic since the Renaissance.

We are compelled to conclude that the "absolute" type of definition traditionally sought for "mathematics" probably cannot be given. Cultural changes show no respect for abstract definitions. Susceptibility to definition may be a sign of senility.†

6 What we call "mathematics" today

From what we have said above, it is perhaps clear that we cannot give a definitive answer even to the modified question, "What is mathematics in the world of today?" The best we can do is to "approximate." In the culturological approach, the first move should be to observe what those men whom we call "mathematicians" do. And, since mathematicians also eat, drink, play, etc., like other people, we had best qualify this by saying that we should observe what those men, whom we call mathematicians, do *as mathematicians*. As mathematicians they seem to be dealing with, or studying, abstract forms or structures and relations between them. As we understand cardinal number, for instance, it is a structural property of a set; if we say a set has *four* elements, we have said something about the structure of the set. If we say it is *linearly ordered*, we have assigned another structural property. Similarly, topological or other geometrical statements are statements about the form or structure of some set. In algebra we assign relations between elements or subsets of a set. And so on. The nice thing about this characterization of what mathematicians do today is that it includes mathematical logic, which may be said to be concerned with the forms of proof or of abstract reasoning.

Of course, we immediately ask how this is to be differentiated from art, for example, because art also is concerned with form or structure. However, while mathematics *abstracts* form, art usually *individualizes* it. The mathematician conceives the abstract number 4, but the artist only paints four birds, four flowers, etc. Even "abstract art," so-called, is a process of individualization. Many mathematicians, incidentally, consider mathematics an art (see, for example, the articles of J. W. N. Sullivan and J. B.

† Regarding the cultural nature of mathematics, see L. A. White [a]; the same paper is reprinted, in somewhat altered form, in White [Wh], to which we also refer the reader for a general discussion of the meaning of "culture." See also Wilder [b].

Shaw reprinted by Schaaf [S]). From our point of view, when we abstract the forms which we find in painting, pottery, music, etc., we do mathematics. However, no definition would be sufficient to rule out border-line cases, even when applied only to the present. It is impossible to say with definiteness or assurance just where mathematics stops, and art, or physics, or . . . begins.

The abstract structures with which mathematics deals can be applied to individual special cases—just as when we apply "4" to marbles and say "4 marbles," or apply euclidean geometry to engineering. In this we merely reverse the operation of abstraction. There is nothing mysterious, as some have tried to maintain, about the *applicability* of mathematics. What we get by abstraction from something can be returned. Much so-called "applied mathematics," as a matter of fact, is really (pure) "mathematics," in that, although it has been suggested by some particular structure, it is in itself a study of abstract structure. A great deal of the work in modern physics and related subjects is really mathematics in so far as it is concerned with abstract structures.†

Other "border-line" cases arise in applications of the axiomatic method. Is such a setup as that of Woodger [Wo], in which genetics is placed in an axiomatic form, mathematics or biology? It would seem as though it could be called either; as a study in abstract structure, perhaps we should call it mathematics; but, as soon as the undefined terms "org," "cell," "genet," etc., used therein are given their biological meanings, it should perhaps be called biology (and applied mathematics).

It would seem, then, that present-day mathematics is properly characterized as a living, growing element of culture embodying concepts about abstract structures and relations between these structures. And, as such, its content is variable and subject to cultural forces much as is any other cultural element. Even the symbols which are used for the expression and development of mathematics have variable meanings; a symbol which represented one thing to the mathematicians of the nineteenth century may represent something quite different, because of the evolution of mathematical thought, to mathematicians of the twentieth century. And, correspondingly, the mathematical element of our culture during the nineteenth century was something quite different from that of today, even

† Perhaps we should point out that we have not really fallen here into the error of setting up a definition into which we are now trying to herd various activities that are usually called by other names. The "physicists" (and others) to whom we refer are usually quite willing to agree that what they are doing, in the sense referred to, is "mathematics." We shall not attempt to define what "physicists" do; but we can agree either that some "mathematics" is also what we call "physics," or that physicists sometimes are "mathematicians," no matter how they are classified on the payroll.

though many of the symbols remain the same (consider the varied meanings of dy/dx since its introduction by Leibniz). The concepts held by mathematicians of the nineteenth century may still be called "mathematical," although many of them may not actually form part of present-day mathematics; properly speaking, such concepts are mathematical only in the sense that they did form part of mathematics at a certain time and place.

7 The process of mathematical change and growth

One of the most persistent illusions of the layman about mathematics is that mathematics is a fixed, immutable body of "truth" epitomized in such statements as "2 and 2 are 4." Often quoted is that favorite nineteenth century orator (whose address at Gettysburg eclipsed Lincoln's famous address of the same occasion, in the opinion of their audience): "In the pure mathematics we contemplate absolute truths, which existed in the divine mind before the morning stars sang together, and which will continue to exist there, when the last of their radiant host shall have fallen from heaven"—a pronunciamento which would appear to make of the mathematician an inspired prophet. In the same address, Everett quoted "an ancient sage" (probably Plato: "God geometrizes continually") as saying that "God is a geometer," an opinion which was later repeated by the astronomer Jeans.

Even professional mathematicians share the illusion. Thus G. H. Hardy wrote [a; 4]: "It seems to me that no philosophy can possibly be sympathetic to a mathematician which does not admit, in one manner or another, the immutable and unconditional validity of mathematical truth. Mathematical theorems are true or false; their truth or falsity is absolute and independent of our knowledge of them." New mathematics, according to Hardy [Ha; 63–64], "which we describe grandiloquently as our 'creations,' are simply our notes of our observations"; in his view it is already a part of some "mathematical reality" "outside us" and we simply "discover or observe it." As Bell observed [a], "The impregnable strength of this creed is that it can be neither proved nor disproved." And, he might have added, its inevitable weakness lies in its utter inability to *explain* anything about mathematics.

The cultural viewpoint, by way of contrast, does explain much about mathematics. Evidently this opinion was shared by D. J. Struik, who, in the Introduction to his history [St], apologized that lack of space forced "insufficient reference to the general cultural and sociological atmosphere in which the mathematics of a period matured—or was stifled. Mathematics has been influenced by agriculture, commerce, and manufacture, by warfare, engineering, and philosophy, by physics and by astronomy. The influence of hydrodynamics on function theory, of Kantianism and of surveying on geometry, of electromagnetism on differential equations, of Cartesianism on mechanics and of scholasticism on the calculus could only be indicated in a few sentences—or perhaps a few words—yet an understanding of the course and content of mathematics can only be reached if all these *determining factors* are taken into consideration." [Italics mine. RLW.]† A similar viewpoint seemed to be in the mind of D. Jackson when he wrote [b; 411]: "On the plane of social rather than individual psychology, there is a fascinating subject of inquiry in the relation between mathematical advances and the general consciousness of the age that produces them."

Like any other cultural element, mathematics grows by *evolution* and *diffusion.* Given a suitable juxtaposition of ideas, either in the mind of an individual or in the minds of a group, certain syntheses take place and new concepts come forth. Thus we can trace the evolution of the concept of an abstract group in the writings of nineteenth century mathematicians in the same manner as we can trace the evolution of the theory of biological evolution in the works of Darwin's predecessors and his own, and contemporary, writings. Every research mathematician is familiar with the numerous instances of the simultaneous announcement of new theorems and theories by two or more mathematicians. This is an entirely similar phenomenon to that which occurs when a new invention is about to be made; when the cultural conditions are right, it will appear without fail. As Spengler stated [S; v. 2, 507], "a task that historic necessity has set *will* be accomplished with the individual or against him." Mathematics does not grow because a Newton, a Riemann, or a Gauss happened to be born at a certain time; great mathematicians appeared because the cultural conditions—and this includes the mathematical materials—were conducive to developing them. There were just as great potential analysts and algebraists living during the Greek era as during the age which produced Weierstrass and Kronecker, but among other things the necessary analytic and algebraic elements were lacking in Greek culture. This is not to belittle the greatness of great men, but rather to mourn for those who were and are denied the opportunity to develop their talents. What great man, if honest with himself, has not observed the passing chimney-sweep without remarking, "There, but for the 'concatenation of events,' go I."

† In an earlier paper Struik speaks of "sociological" determination of mathematics. However, the quotation from [St] given above (with permission of Dover Publications, Inc.) seems to indicate that what he actually had in mind was "culturological" determination. It is interesting to note that, in a book edited by W. L. Schaaf [S], the essay referred to is contrasted with the remarks of Sarton cited above; actually, however, Sarton and Struik appear to be substantially in agreement if their writings are suitably interpreted in terms of cultural influences.

We can, of course, agree that mathematics could not advance without "creative genius" being available; but we can still insist that the genius cannot operate in an intellectual vacuum, and that without proper cultural stimuli his "genius" will never become known. As the anthropologist Ralph Linton remarked [Li; 319]†: "The mathematical genius can only carry on from the point which mathematical knowledge within his culture has already reached. Thus if Einstein had been born into a primitive tribe which was unable to count beyond three, lifelong application to mathematics probably would not have carried him beyond the development of a decimal system based on fingers and toes."

How do "favorable" conditions for the development of mathematics come about? So far as the general sociological and cultural conditions are concerned, we can say that, in the first place, only in a culture which affords specialization of occupations can mathematics develop to any great extent. Much of early mathematics developed through the needs of a priesthood or for religious purposes (the earliest mathematical work in the United States worthy of note seems to have been in the hands of men trained by the Jesuits to calculate the correct annual date for Easter‡). But the process of counting developed much earlier, being found in all primitive cultures (a fact that could be considered an argument for the intuitionist insistence upon the counting process as the proper basis of all mathematics; cf. X 1). As the needs of agriculture, religion, navigation, etc., make demands, new mathematical tools are created; witness the development of the rudiments of geometry in ancient Egypt. In commenting on the arithmetical and geometrical rules that Sumerian clerks applied ("true prototypes of the quantitative laws of modern science"), the British anthropologist Childe remarks [Chi; 102]§: "Obviously we need not bother to ask the names of the laws' discoverers. They are too patently social products called forth by the needs of a society affected by the urban revolution and discovered with the aid of the spiritual equipment produced by the revolution." The point hardly needs belaboring, it would seem, in the face of the abundant historical evidence.

In the second place, an adequate symbolism must be available. This is necessary for two reasons: (1) All human activity, as distinguished from such general animal activities as eating and sleeping, is based on symbolism, and for a pursuit of such an abstract nature as mathematics, symbolism becomes a *sine qua non*. (2) Symbols are the vehicle for the communication and diffusion of mathematical ideas. No mathematician needs to be

† Quoted with permission of Appleton-Century-Crofts, Inc.

‡ See the Smith-Ginsburg [S-G; 3–4] history; this excellent little work, incidentally, shows a remarkable awareness of the cultural nature of mathematics.

§ Quoted with permission of Penguin Books, Ltd.

convinced of (1); indeed, much mathematical work goes into the construction of suitable symbolic apparatus. And, so far as (2) is concerned, it is in large measure responsible for the present-day universal character of mathematics.

Thirdly—and associated with the above—suitable means should be available for the diffusion of mathematical concepts. Given proper cultural conditions in various societies, if mathematical ideas are to diffuse from one society to another, good means of communication must have been established. Greek and Chinese mathematics developed along such different lines chiefly because of the lack of diffusion of ideas from one culture to another at that time. In modern times, with good means of communication, together with the establishment of mathematical journals, visits of scholars to foreign countries, international mathematical gatherings, etc., mathematics achieves its current practically universal character. Depending only upon the use of symbols, and not necessitating the possession of intricate tools or expensive apparatus, it can achieve a universality not so easily attained in other fields. Even in the face of such barriers to diffusion as are sometimes raised by the conflicts of international politics, mathematics has not, in any appreciable manner, developed different culture areas in modern times.

This is not to imply that mathematics has not been affected by such cultural forces as are comprised by political and social events. The influx to the United States of mathematicians fleeing from the Nazi philosophy and persecution of 1930–1945 had a great influence on mathematics. The resultant contacts between leaders in their fields has led to syntheses of ideas that might not yet, if ever, have occurred under the former conditions of cultural separation. Unfortunately, on the other side of the ledger must be noted those losses, not only in the lives of brilliant mathematicians, but also in the breaking up of various groups who were active in mathematical research (Göttingen, Warsaw, etc.).

8 Differences in the kind and quality of mathematics

It needs to be emphasized that from the cultural point of view such differences in the development of mathematical ideas as occurred between the Greek and Chinese, as well as those slight differences between nations today, are due not to the "nature" or intrinsic qualities of the peoples involved, but to the cultures to which they are subject. From the cultural point of view the peoples themselves may be regarded as constants; it is the cultures into which they are born and to whose evolution they contribute that differ. The mathematicians in a given culture possess a subculture—the mathematical subculture—which is affected not only by diffusion from other mathematical cultures, but also by the larger culture in which it is embedded. If the larger culture makes demands such as those recently made in the United States by scientific, military, and industrial developments, the subculture cannot avoid being affected. Thus we find mathematicians today working on high-speed computers, and various applications (which will themselves inevitably give rise to new concepts), who otherwise might have followed a different course of mathematical development.

The direction that mathematics takes in a culture is guided by the cultural needs or attributes—religious, philosophical, agricultural, navigational, industrial, as well as mathematical—of the culture. The different directions taken by Greek and Chinese mathematics were determined by the primitive cultural conditions prevailing in those cultures during their prehistoric periods. And, having received an initial push in a certain direction, the mathematical subcultures no doubt proceeded to a certain extent on their own momentum, operating mainly under evolutionary forces within (in the case of Chinese mathematics) or under both evolutionary and diffusionary forces (from other parts of the general culture; especially philosophy in the case of Greek mathematics).

Today, mathematics grows under the influence of a complexity of evolutionary and diffusionary forces. Many authors have emphasized the *freedom* which a mathematician has in the choice of direction of his research. Without going so far as Spengler's "freedom to do the necessary . . .," we still have to admit that only those mathematical constructs can evolve and be destined to live that are so related to the evolutionary development of mathematics as to prove fruitful in their consequences in some way. The arbitrary setting up of postulational systems, for example, having no regard for their relation to existing systems, may not be regarded by the mathematics profession as creating mathematics.

The *importance* of a theory or a problem in mathematics is similarly determined, i.e., by its relation to the evolutionary development of mathematics. However, what is not mathematics today (because of its lack of relation to the existing mathematics) can one day become mathematics.

We have mentioned some of the instruments of diffusion above—journals, international and national gatherings, emigration of mathematicians under political or other pressure—but little about the nature of evolutionary forces. These forces present a difficulty, in that they are not so easily detected and pointed out as are the diffusionary forces. It seems reasonable, however, to expect that, once a new branch of mathematics, offering a fertile field for research, is present in the minds of a sufficient number of mathematicians, then that branch will grow along certain lines, irrespective (except for minor details) of the idiosyncrasies of the particular individuals possessing it.† Much overlapping of results ensues (cf. Merton [a], [b]), and the activity in the subject reaches a maximum intensity, which is followed by a waning of interest, due perhaps to the most important and interesting problems having been solved, or perhaps to the branch having been absorbed in a more encompassing new branch of mathematics. Ultimately, only scattered workers may be left in the original branch, the majority of the younger men in the field having become occupied with newer theories.

It will be noticed how similar this process is to that of *styles* or *fashions*. Various fashions arise in mathematics and pursue a course not unlike that of fashion in other parts of the culture. A mathematician of the early part of this century would probably be appalled by the mathematics of today. It was not uncommon for a mathematician who participated in the rapid development of classical analysis to express his disdain for the subject matter and methods of modern mathematics. As in other fields of human affairs, the elders deplore the new fashions favored by youth. But the youth are no more to be blamed for the changes in fashion than are the women whose husbands complain about the changes in styles of clothes. Kroeber has shown that in fashions definite cycles occur, the changes being not in any way subject to the whim of fashion designers (who, like the Sumerian clerks, are "patently social products"). It is possible that, two thousand years hence, similar cycles may be discerned in the history of mathematics. (Although we are accustomed to think of mathematics as one of the most ancient of man's creations, the rapid development of mathematics under both evolutionary and diffusionary forces is a distinctly modern development.) Perhaps the abstract form assumed by modern mathematics is to be superseded by a form more like classical analysis, especially in its relations to the needs of other fields of knowledge. There are already signs that such may be the case, in the applications of mathematical logic, postulational methods, modern algebra, topology, etc., especially in the social sciences. Where classical analysis found application and, indeed, origin in physics, chemistry, and other natural sciences, the new mathematics may find its applications and many of its concepts in the social sciences.‡ Such a development would only be in agreement

† Compare: "Mathematical science, like all other living things, has its own natural laws of growth."—C. N. Moore, *Bull. Amer. Math. Soc.*, vol. 37 (1931), p. 240. Also Weyl [b; 538].

‡ Those who think that a social "study" cannot be a "science" or an "exact science" until it is capable of being put into the same type of mathematical form as modern physics seem to be guilty of putting the cart before the horse. Much of classical mathematics, especially analysis, is due to the needs of physics or to physical influences, and to try to squeeze a modern social science into the same framework

with our view of mathematics as a subculture of our general culture, developing not only under its own evolutionary momentum but also subject to forces, sometimes concealed, that affect it from without.†

9 Mathematical existence

What is the effect of the cultural point of view on the question of mathematical existence? For example, what light does it throw on Kronecker's rejection of infinite sets (VIII 2.1.2)? So far as such an attitude is based on the desire to avoid possible contradiction, or even on an esthetic impulse to found mathematics in a manner free of certain principles, or the like, it may be held to have justification—or at least to provide room for argument. But when such attitudes are based, as they frequently are, on a philosophy which conceives of mathematics as an ideal entity or as having an absolute nature which we endeavor to *discover*, much as the physicist or the explorer seeks natural data, they appear to be unwarranted. For mathematics, like other cultural entities, is what it is as a result of collective human effort directed along evolutionary and diffusionary lines. And what it becomes will not be determined by the discovery of "mathematical truth" now hidden from us, but by what mankind, via cultural paths, makes it.

Consequently, if for sufficient reason, such as the discovery of new contradictions or of more pleasing alternative ideas, uncountably infinite sets are no longer used in mathematics, it will not be because they do not "exist"—they will continue to "exist" as much as they ever did. They would merely cease to be considered "mathematical." Similar remarks may be made concerning the Choice Axiom, well-ordering, continuum hypothesis, etc. It is even conceivable that future political conditions might lead to such a pronounced separation of cultures that one culture, C_1, will reject much of classical mathematics, such as the Cantor set theory—as it would if it adopted wholesale the intuitionist philosophy—

---

is unjustified. This is not to say that some classical analysis, such as is used in statistics, for instance, may not find significant application in some aspects of social science. But the type of mathematics needed by the social sciences, if any, will be developed either as a result of direct need or of the simultaneous acting of hidden cultural forces. (Examples will continue to occur, of course, of mathematical materials which evolved along purely mathematical lines and of "no practical value," which later find application in another field; this is not to gainsay our main point, however.)

† Sarton remarks (see Schaaf's reprint [S; 71]): "On the basis of my historical experience, I fully believe that mathematics of the twenty-fifth century will be as different from that of today as the latter is from that of the sixteenth century." (Quoted with permission of G. Sarton and Harper and Brothers.)

while another culture, C_2, proceeds along the lines of mathematical thought as it exists in the United States today. There would then be no question as to which mathematics is the "true" mathematics—that of C_1 or of C_2—each having been culturally determined and neither having any preferred claim to "truth."

In short, mathematics is what we make it; not by each of us acting without due regard for what constitutes mathematics in our culture, but by seeking to build up new theories in the light of the old, and to solve outstanding problems generally recognized as valuable for the progress of mathematics as we know it. Until we make it, it fails to "exist." And, having been made, it may at some future time even fail to be "mathematics" any longer.†‡

† For an idea of the large amount and variety of new mathematics published annually, the reader is advised to consult the files of *Mathematical Reviews* in any mathematics library. This journal publishes abstracts of articles embodying the results of current mathematical research—articles which have appeared in journals throughout the world.

‡ For a description of an actual "case history" of the cultural development of a mathematical concept as well as further general commentary, see Wilder [e]. (A more extended study of the manner in which mathematics develops will be found in the author's forthcoming book "Evolution of Mathematical Concepts.")

Bibliography

BOOKS AND MONOGRAPHS

Albert, A. A.

[Al] *Modern Higher Algebra*, Chicago, Univ. of Chicago Pr., 1937.

Ball, W. W. R.

[B] *A Short Account of the History of Mathematics*, London, Macmillan, 1888; 4th ed., 1908.

Baumgartner, L.

[Ba] *Gruppentheorie*, Sammlung Göschen, No. 837, Berlin, de Gruyter, 1921.

Bell, E. T.

[B_1] *The Queen of the Sciences*, Baltimore, Williams and Wilkins, 1931.

[B_2] *The Search for Truth*, Baltimore, Williams and Wilkins, 1934.

[B_3] *The Development of Mathematics*, New York, McGraw-Hill, 1940.

[B_4] *Mathematics, Queen and Servant of Science*, New York, McGraw-Hill, 1951.

Benacerraf, P., and Putnam, H.

[B-P] *Philosophy of Mathematics*, Englewood Cliffs, N. J., Prentice-Hall, 1964.

Bernays, P.

[Ber] *Axiomatic Set Theory*, Amsterdam, North-Holland Publ. Co., 1958.

Beth, E. W.

[Be] *The Foundations of Mathematics*, Amsterdam, North-Holland Publ. Co., 1959.

Birkhoff, G.

[Bi] *Lattice Theory*, Amer. Math. Soc. Coll. Pub., v. 25, rev., New York, 1948.

Birkhoff, G., and MacLane, S.

[B-M] *A Survey of Modern Algebra*, New York, Macmillan, 1944, 1953.

Black, M.

[Bl] *The Nature of Mathematics*, New York, Harcourt, Brace, 1934.

Borel, E.

[Bo] *Leçons sur les théorie des fonctions*, 3d ed., Paris, Gauthier-Villars, 1928.

Bourbaki, N.

[B_1] *Théorie des ensembles* [Actualités scientifiques et industrielles, No. 846], Paris, Hermann, 1939.

Cajori, F.

[Caj] *A History of Mathematics*, New York, Macmillan, 1893; 2d ed., 1919.

Campbell, A. D.

[Ca] *Advanced Analytic Geometry*, New York, Wiley, 1938.

Cantor, G.

[C] *Gesammelte Abhandlungen*, herausg. von Ernst Zermelo, Berlin, Springer, 1932.

[C_1] *Contributions to the Founding of the Theory of Transfinite Numbers*, transl. by P. E. B. Jourdain, Chicago, Open Court, 1915.

Carnap, R.

[Car] *Abriss der Logistik*, Vienna, Springer, 1929.

[Car_1] *The Logical Syntax of Language*, New York, Harcourt, Brace, 1937.

[Car_2] *Foundations of Mathematics and Logic*, Int'l Enc. of Unified Sc., v. 1, No. 3, Chicago, Univ. of Chicago Pr., 1939.

Childe, G.

[Chi] *What Happened in History*, New York, Penguin Books, 1946. (Orig. published in England, 1942, by Penguin Books Ltd.)

Church, A.

[Ch] *Introduction to Mathematical Logic*, Part I, Annals of Math. Studies No. 13, Princeton, Princeton Univ. Pr., 1944.

[Ch_1] *Ibid.*, rev. ed., 1956.

Chwistek, L.

[Chw] *The Limits of Science. Outline of Logic and Methodology of the Exact Sciences*, transl. by H. C. Brodie and A. P. Coleman, New York, Harcourt, Brace, 1948.

Courant, R., and Robbins, H.

[C-R] *What Is Mathematics?*, London and New York, Oxford Univ. Pr., 1941.

Couturat, L.

[Co_1] *L'Algèbre de la logique*, Scientia, Phys.-Math. Série, No. 24, Paris, Gauthier-Villars, 1905.

[Co_2] *De l'infinie mathématique*, Paris, Felix Alcan, 1896.

Curry, H. B.

[Cu] *A Theory of Formal Deducibility*, Notre Dame Math. Lectures, No. 6, Notre Dame, Indiana, 1950.

[Cu_1] *Foundations of Mathematical Logic*, New York, McGraw-Hill, 1963.

Davis, M.

[D] *Computability and Unsolvability*, New York, McGraw-Hill, 1958.

Dedekind, R.

[D_1] *Stetigkeit und irrationale Zahlen*, Braunschweig, Vieweg, 1872.

[D_2] *Was sind und was sollen die Zahlen?*, Braunschweig, 1888 (also appears in [D_3]).

[D_3] *Gesammelte mathematische Werke*, Braunschweig, Vieweg, 1932.

Eaton, R. M.

[E] *General Logic*, New York, Scribner, 1931.

Eves, H. and Newsom, C. V.

[E-N] *An Introduction to the Foundations and Fundamental Concepts of Mathematics*, New York, Rinehart & Co., 1958.

Fraenkel, A.

[F_1] *Einleitung in die Mengenlehre*, 3d ed., Berlin, Springer, 1928. (Republished by Dover Pubs., New York, 1946.)

[F_2] *Zehn Vorlesungen über die Grundlegung der Mengenlehre*, Leipzig, Teubner, 1927.

[F_3] *Abstract Set Theory*, Amsterdam, North-Holland Publ. Co., 1961.

Fraenkel, A., and Bar-Hillel, Y.

[F-B] *Foundations of Set Theory*, Amsterdam, North-Holland Publ. Co., 1958.

Frege, G.

[Fr] *The Foundations of Arithmetic*, Oxford, Basil Blackwell, 1950.

[Fr_1] *Begriffschrift*, Halle, Louis Nebert, 1879.

Gödel, K.

[G] *The Consistency of the Axiom of Choice and the Generalized Continuum-Hypothesis with the Axioms of Set Theory*, Princeton, Princeton Univ. Pr., 1940.

Goodstein, R. L.

[Go] *Mathematical Logic*, Leicester, University Press, 1957.

Grelling, K.

[Gr] *Mengenlehre*, Leipzig, Teubner, 1924.

[Gr_1] *Die Paradoxien der Mengenlehre*, Stuttgart, Math. Büchlein, 1925.

Hall, D. W. and Spencer, G. L.

[H-S] *Elementary Topology*, New York, Wiley, 1955.

Hall, M.

[H] *The Theory of Groups*, New York, Macmillan, 1959.

Halmos, P. R.

[Hal] *Naive Set Theory*, Princeton, N. J., D. Van Nostrand Co., Inc., 1960.

Hardy, G. H.

[Ha] *A Mathematician's Apology*, Cambridge, England, The University Pr., 1941.

Heath, T. L.

[Hea] *The Thirteen Books of Euclid's Elements*, Cambridge, England, The University Pr., 1908.

Henkin, L., Suppes, P., and Tarski, A.

[HST] *The Axiomatic Method*, Amsterdam, North-Holland Publ. Co., 1959.

Heyting, A.

[He] *Mathematische Grundlagenforschung. Intuitionismus. Beweistheorie.* Ergebnisse der Mathematik und ihrer Grenzgebiete, vol. 3, No. 4, Berlin, Springer, 1934.

[He_1] *Intuitionism. An Introduction*, Amsterdam, North-Holland Publ. Co., 1956.

Hilbert, D.

[H_1] *Grundlagen der Geometrie*, Leipzig, Teubner, 1899. (Published in *Festschrift zur Feier der Enthüllung des Gauss-Weber-Denkmals in Göttingen.*)

[H_2] *Grundlagen der Geometrie*, 7th ed., Leipzig, Teubner, 1930.

[H_3] *The Foundations of Geometry*, Chicago, Open Court, 1902. (English translation of [H_1], incorporating material from French edition, by E. J. Townsend.)

Hilbert, D., and Ackermann, W.

[H-A] *Principles of Mathematical Logic*, New York, Chelsea Publ. Co., 1950.

Hilbert, D., and Bernays, P.

[H-B] *Grundlagen der Mathematik*, vol. I (1934), vol. II (1939), Berlin, Springer. (Reprinted in lithoprint by Edwards Bros., Ann Arbor, 1944.)

Hobson, E. W.

[Ho] *The Theory of Functions of a Real Variable and the Theory of Fourier's Series*, vol. 1, 2d ed., Cambridge, England, The University Pr., 1921; 3d ed., 1927.

Hohn, F. E.

[Hoh] *Applied Boolean Algebra*, New York, Macmillan, 1960.

Huntington, E. V.

[Hu] *The Continuum and Other Types of Serial Order*, 2d ed., Cambridge, Mass., Harvard Univ. Pr., 1917. [Originally published under the title *The Continuum as a Type of Order*, Annals of Math., vol. 6 (1905), pp. 151–184.]

Kamke, E.

[Ka] *Mengenlehre*, Sammlung Göschen, Berlin and Leipzig, de Gruyter, 1928.

[Ka_1] *Theory of Sets* (translation of 2d ed. of [Ka] by F. Bagemihl), New York, Dover Pubs., 1950.

Kattsoff, L. O.
[K] *A Philosophy of Mathematics*, Ames, Iowa State College Pr., 1948.
Keene, G. B.
[Kee] *Abstract Sets and Finite Ordinals*, New York, Pergamon Press, 1961.
Kelley, J. L.
[Ke] *General Topology*, New York, Van Nostrand, 1955.
Kershner, R. B., and Wilcox, L. R.
[K-W] *The Anatomy of Mathematics*, New York, Ronald, 1950.
Kleene, S. C.
[Kle] *Introduction to Metamathematics*, New York, D. Van Nostrand, 1952.
Klein, F.
[Kl] *Elementary Mathematics from an Advanced Standpoint*, transl. by E. R. Hedrick and C. A. Noble, Part I, New York, Macmillan, 1932.
Kline, M.
[Kli] *Mathematics in Western Culture*, New York, Oxford Univ. Pr., 1953.
[Kli_1] *Mathematics and the Physical World*, New York, T. Y. Crowell, 1959.
König, D.
[Kö] *Theorie der endlichen und unendlichen Graphen*, Leipzig, Akad. Verlagsgesellschaft, 1936.
Landau, E. G. H.
[La_1] *Foundations of Analysis; the Arithmetic of Whole, Rational and Complex Numbers*, transl. by F. Steinhardt, New York, Chelsea Publ. Co., 1951.
[La_2] *Grundlagen der Analysis*, Leipzig, Akad. Verlagsges. M. B. H., 1930.
Lewis, C. I.
[Le] *A Survey of Symbolic Logic*, Berkeley, Univ. of Calif. Pr., 1918.
Lewis, C. I., and Langford, C. H.
[L-L] *Symbolic Logic*, New York, The Century Co., 1932.
Linton, R.
[Li] *The Study of Man*, New York, Appleton-Century-Crofts, Inc., 1936.
Lusin, N.
[L] *Leçons sur les ensembles analytiques*, Paris, Gauthier-Villars, 1930.
McCoy, Neal H.
[Mc] *Rings and Ideals*, Carus Mathematical Monographs, No. 8, Math. Assoc. of Amer., Buffalo, 1948.
Matthewson, L. C.
[Ma] *Elementary Theory of Finite Groups*, Boston, Houghton Mifflin, 1930.
Maziarz, E. A.
[M] *The Philosophy of Mathematics*, New York, Phil'l Lib., 1950.
Miller, G. A., Blichfeldt, H. F., and Dickson, L. E.
[MBD] *Theory and Applications of Finite Groups*, New York, Stechert, 1938.
Moore, E. H.
[Mo_1] *Introduction to a Form of General Analysis*, New Haven, Yale Univ. Pr., 1910.
Nagel, E., and Newman, J. R.
[N-N] *Gödel's Proof*, New York, New York Univ. Pr., 1958.
Newsom, C. V.
[Ne] *An Introduction to Mathematics*, Univ. of New Mexico Bull. No. 295, Albuquerque, Univ. of New Mex. Pr., 1936.
Northrop, E. P., and other members of the College Mathematics Staff, University of Chicago.
[N] *Fundamental Mathematics*, vol. 1, 3d ed., Chicago, Univ. of Chicago Pr., 1938.

Pasch, M.

[Pa] *Vorlesungen über neuere Geometrie*, Leipzig, Teubner, 1882.

Pasch, M., and Dehn, M.

[P-D] *Vorlesungen über neuere Geometrie*, Berlin, Springer, 1926.

Peano, G.

[P_1] *I Principii di Geometria*, Logicamenta Esposti, Turin, Bocca, 1889.

[P_2] *Arithmetices Principia*, Turin, Bocca, 1889.

[P_3] *Formulaire de Mathématiques*, vols. I–V, Turin, Bocca, 1894–1908.

Pierpont, J.

[Pi] *Lectures on the Theory of Functions of Real Variables*, New York, Ginn, 1905, vol. I.

Poincaré, H.

[Po] *The Foundations of Science*, transl. by G. B. Halsted, Lancaster, Pa., Science Pr., 1946.

Quine, W. V.

[Q] *Mathematical Logic*, rev'd ed., Cambridge, Harvard Univ. Pr., 1951.

[Q_1] *From a Logical Point of View*, Cambridge, Harvard Univ. Pr., 1953.

Ramsey, F. P.

[Ra] *Foundations of Mathematics*, London, Paul, Trench, Trubner, 1931.

Richardson, M.

[R] *Fundamentals of Mathematics*, New York, Macmillan, 1941.

Robinson, Gilbert de B.

[Ro] *The Foundations of Geometry*, Toronto, Univ. of Toronto Pr., 1940.

Rosenbloom, P. C.

[Ros] *The Elements of Mathematical Logic*, New York, Dover Pubs., 1950.

Rosser, J. B.

[Ross] *Logic for Mathematicians*, New York, McGraw-Hill, 1953.

Russell, B.

[R_1] *Introduction to Mathematical Philosophy*, London and New York, Macmillan, 1919; 2d ed., 1920.

[R_2] *Principles of Mathematics*, New York, Norton, 1903; 2d ed., 1937.

Schaaf, W. L., editor.

[S] *Mathematics Our Great Heritage: Essays on the Nature and Cultural Significance of Mathematics*, New York, Harper, 1948.

Sierpinski, W.

[S_1] *Leçons sur les nombres transfinis*, Paris, Gauthier-Villars, 1928.

[S_2] *Hypothèse du continu*, Warsaw-Lwow, Monografje mat., vol. IV, 1934.

[S_3] *Cardinal and Ordinal Numbers*, Warsaw, 1958.

[S_4] *General Topology*, Toronto, Univ. of Toronto Pr., 1952.

Smith, D. E., and Ginsburg, J.

[S-G] *A History of Mathematics in America before 1900*, Carus Mathematical Monographs, No. 5, Math. Assoc. of Amer., Buffalo, 1934.

Spengler, O.

[S] *The Decline of the West*, transl. by C. F. Atkinson, New York, Knopf; vol. 1, 1926; vol. 2, 1928.

Stoll, R. R.

[Sto] *Set Theory and Logic*, San Francisco, W. H. Freeman, 1963.

Struik, D. J.

[St] *A Concise History of Mathematics*, 2 vols., New York, Dover Pubs., 1948.

Suppes, P.

[Su] *Axiomatic Set Theory*, New York, D. Van Nostrand, 1960.

Tarski, A.

[T] *Introduction to Logic*, New York, Oxford Univ. Pr., 1941.

[T_1] *A Decision Method for Elementary Algebra and Geometry*, Project RAND, publication R-109, Santa Monica, RAND Corp., 1948.

[T_2] *Logic, Semantics, Metamathematics*, Papers from 1923 to 1938, transl. by J. H. Woodger, Oxford, England, The Clarendon Press, 1956.

Tarski, A., Mostowski, A., and Robinson, R. M.

[T-M-R] *Undecidable Theories*, Amsterdam, North-Holland Publ. Co., 1953.

Veblen, O., and Whitehead, J. H. C.

[V-W] *The Foundations of Differential Geometry*, Cambridge, England, The University Pr., 1932.

Veblen, O., and Young, J. W.

[V-Y] *Projective Geometry*, New York, Ginn, 1910, vol. I.

[V-Y_2] *Projective Geometry*, New York, Ginn, 1918, vol. II.

van der Waerden, B. L.

[Wa] *Moderne Algebra*, 2d ed., 2 vols., Berlin, Springer, 1937 and 1940; photostat reprint, New York, Ungar, 1943.

Weyl, H.

[We] *Das Kontinuum*, Leipzig, von Veit, 1918.

[We_1] *Philosophy of Mathematics and Natural Science*, Princeton, Princeton Univ. Pr., rev. ed., 1949.

White, L. A.

[Wh] *The Science of Culture*, New York, Farrar, Straus, 1949.

Whitehead, A. N., and Russell, B.

[P.M.] *Principia Mathematica*, Cambridge, England, The University Pr., vol. 1, 2d ed., 1925; vol. 2, 2d ed., 1927.

Wilder, R. L.

[Wi] *Evolution of Mathematical Concepts* (forthcoming).

Wittgenstein, L.

[Wt] *Tractatus Logico-philosophicus*, New York, Harcourt, Brace, 1922.

Woodger, J. H.

[Wo] *The Axiomatic Method in Biology*, Cambridge, England, The University Pr., 1937.

[Wo_1] *The Technique of Theory Construction*, Int'l Enc. of Unified Sc., vol. 2, No. 5, Chicago, Univ. of Chicago Pr., 1939.

Young, J. W.

[Y] *Lectures on Fundamental Concepts of Algebra and Geometry*, New York, Macmillan, 1916.

Young, J. W. A., editor.

[Yo] *Monographs on Modern Mathematics*, New York, Longmans, Green, rev. ed., 1915.

PAPERS, ETC.

Ackermann, W.

[a] *Mengentheoretische Begründung der Logik*, Math. Ann., vol. 115 (1937–38), pp. 1–22.

Bachmann, E.

[Encyk] *Aufbau des Zahlensystems*, Encyklopädie der Math. Wiss., vol. I 1, Heft 2, 1939, pp. 1–28.

Bell, E. T.

[a] *Confessions of a mathematician*, Sc. Monthly, vol. 54 (1942), p. 81 (review of G. H. Hardy's "A Mathematician's Apology").

Bernays, P.

[a] *A system of axiomatic set theory*, Journ. Symb. Logic, vol. 2 (1937), pp. 65–77; vol. 6 (1941), pp. 1–17; vol. 7 (1942), pp. 65–89, 133–145; vol. 8 (1943), pp. 89–106; vol. 13 (1948), pp. 65–79.

[b] *Axiomatische Untersuchung des Aussagen-Kalkuls der "Principia Mathematica,"* Math. Zeit., vol. 25 (1926), pp. 305–320.

Bernstein, F.

[a] *Über die Reihe der transfiniten Ordnungszahlen*, Math. Ann., vol. 60 (1905), pp. 187–193.

[b] *The continuum problem*, Proc. Nat. Acad. Sci., vol. 24 (1938), pp. 101–104.

Beth, E. W.

[a] *On Padoa's Method in the theory of definition*, Kon. Akad. van Wetenschappen, Proceedings, vol. 56 (1953), pp. 330–339.

Birkhoff, G. D.

[a] *A set of postulates for plane geometry, based on scale and protractor*, Annals of Math., vol. 33 (1932), pp. 329–345.

Borel, É.

[a] *Quelques remarques sue les principes de la théorie des ensembles*, Math. Ann., vol. 60 (1905), pp. 194–195.

Brouwer, L. E. J.

[a] *Zur Begründung der intuitionistischen Mathematik, I.*, Math. Ann., vol. 93 (1925), pp. 244–257.

[b] *Zur Begründung der intuitionistischen Mathematik, II*, ibid., vol. 95 (1926), pp. 453–472.

[c] *Zur Begründung der intuitionistischen Mathematik, III*, ibid., vol. 96 (1926), pp. 451–488.

[d] *Remarques sur la notion d'ordre*, Comptes Rendus Acad. Sci., Paris, vol. 230 (1950), pp. 263–265.

[e] *Sur la possibilité d'ordonner le continu*, ibid., pp. 349–350.

[f] *Intuitionism and formalism*, Bull. Amer. Math. Soc., vol. 20 (1913–1914), pp. 81–96 (transl. by A. Dresden).

[g] *Intuitionistische Zerlegung Mathematischer Grundbegriffe*, Jahresbericht d. Deut. Math. Ver., vol. 33 (1925), pp. 251–256.

[h] *Intuitionistische Betrachtungen über den Formalismus*, Proc. Akad. Wet. Amsterdam, vol. 31 (1928), pp. 374–379.

[i] *Intuitionistischer Beweis der Jordanschen Kurvensatzes*, ibid., vol. 28 (1925), pp. 503–508.

[j] *Points and spaces*, Canadian Jour. Math., vol. 6 (1953), pp. 1–17.

Cavaillès, J.

[a] *Sur la deuxième définition des ensembles finis donnée par Dedekind*, Fund. Math, vol. 19 (1932), pp. 143–148.

Church, A.

[a] *A bibliography of symbolic logic*, Journ. Symb. Logic, vol. I (1936), pp. 121–216.

[b] Additions and corrections to *A bibliography of symbolic logic*, ibid., vol. 3 (1938), pp. 178–212.

[c] *A note on the Entscheidungsproblem*, Jour. Symb. Logic, vol. 1 (1936), pp. 40–41, 101–102.

[d] *An unsolvable problem of elementary number theory*, Amer. Journ. Math., vol. 58 (1936), pp. 345–363.

[e] *A formulation of the simple theory of types*, Journ. Symb. Logic, vol. 5 (1940), pp. 56–68.

Chwistek, L.

[a] *The theory of constructive types*, Annales de la Société Polonaise de Mathématique, vol. 2 (1924), pp. 9–48; vol. 3 (1925), pp. 92–141.

Curry, H. B.

[a] *Some aspects of the problem of mathematical rigor*, Bull. Amer. Math. Soc., vol. 47 (1941), pp. 221–241.

Dresden, A.

[a] *Brouwer's contributions to the foundations of mathematics*, Bull. Amer. Math. Soc., vol. 30 (1924), pp. 31–40.

Dushnik, B., and Miller, E. W.

[a] *Partially ordered sets*, Amer. Journ. Math., vol. 63 (1941), pp. 600–610.

Emch, A. F.

[a] *Implication and deducibility*, Journ. Symb. Logic, vol. 1 (1936), pp. 26–35.

Faber, G.

[a] *Über die Abzählbarkeit der rationalen Zahlen*, Math. Ann., vol. 60 (1905), pp. 196–203.

Frink, O.

[a] *New algebras of logic*, Amer. Math. Mo., vol. 45 (1938), pp. 210–219.

Furstenberg, H.

[a] *On the infinitude of primes*, Amer. Math. Mo., vol. 62 (1955), p. 353.

Gentzen, G.

[a] *Die Widerspruchsfreiheit der reinen Zahlentheorie*, Math. Ann., vol. 112 (1936), pp. 493–565.

[b] *Neue Fassung des Widerspruchs-freiheitsbeweises für die reine Zahlentheorie*, Forschungen zur Logik und zur Grundlegung der exakten Wissenschaften, Neue Folge, Heft 4, pp. 19–44, Leipzig, 1938.

Gödel, K.

[a] *What is Cantor's continuum problem?*, Amer. Math. Mo., vol. 54 (1947), pp. 515–525. [Rev. and expanded in Benacerraf and Putnam [B-P]; 258–273.]

[b] *Zur intuitionistischen Arithmetik und Zahlentheorie*, Erg. Math. Kolloqu., 1933, Heft 4, pp. 35–38.

[c] *Die Vollständigkeit der Axiome des logischen Funktionen-kalküls*, Monats. für Math. u. Phys., vol. 37 (1930), pp. 349–360.

[d] *Über formal unentscheidbare Sätze der Principia Mathematica und verwandter Systeme I*, Monats. für Math. u. Phys., vol. 38 (1931), pp. 173–198.

Golomb, S.

[a] *A connected topology for the integers*, Amer. Math. Mo., vol. 66 (1959), pp. 663–665.

Hall, M. S., Jr.

[a] *The word problem for semi-groups with two generators*, Journ. Symb. Logic, vol. 14 (1949), pp. 115–118.

Halldén, S.

[a] *A note concerning the paradoxes of strict implication and Lewis's system S1*, Journ. Symb. Logic, vol. 13 (1948), pp. 138–139.

Hamel, G.

[a] *Eine Basis aller Zahlen und die unstetigen Lösungen der Funktionengleichung:* $f(x + y) = f(x) + f(y)$, Math. Ann., vol. 60 (1905), pp. 459–462.

Harary, F.

[a] *A very independent axiom system*, Amer. Math. Mo., vol. 68 (1961), pp. 159–162.

[b] *A measure of axiomatic independence*, Mind, vol. 72 (1963), pp. 143–144.

Hardy, G. H.

[a] *Mathematical proof*, Mind, vol. 30 (1929), pp. 1–25.

Hartogs, F.

[a] *Über das Problem der Wohlordnung*, Math. Ann., vol. 76 (1915), pp. 438–443.

Helmer, O.

[a] *On the theory of axiom-systems*, Analysis, vol. 3 (1935), pp. 1–11.

Henkin, L.

[a] *Fragments of the propositional calculus*, Journ. Symb. Logic, vol. 14 (1949), pp. 42–48.

[b] *The completeness of the first-order functional calculus*, ibid., pp. 159–166.

[c] *On mathematical induction*, Amer. Math. Mo., vol. 67 (1960), pp. 323–338.

Henle, P.

[a] *The independence of the postulates of logic*, Bull. Amer. Math. Soc., vol. 38 (1932), pp. 409–414.

Heyting, A.

[a] *De telbaarheidspraedicaten van Prof. Brouwer* (The countability-predicates of Prof. Brouwer), Nieuw Archief voor Wiskunde (2), vol. 16, no. 2 (1929), pp. 47–58.

Hilbert, D.

[a] *Über das Unendliche*, Math. Ann., vol. 95 (1925), pp. 161–190; also published in French translation in Acta Math., vol. 48 (1928), pp. 91–122.

[b] *Die logischen Grundlagen der Mathematik*, Math. Ann., vol. 88 (1922), pp. 151–165.

Huntington, E. V.

[a] *A complete set of postulates for the theory of absolute continuous magnitude*, Trans. Amer. Math. Soc., vol. 3 (1902), pp. 264–279.

[b] *A new set of postulates for betweenness, with proof of complete independence*, ibid., vol. 26 (1924), pp. 257–282.

[c] *Simplified definition of a group*, Bull. Amer. Math. Soc., vol. 8 (1962), pp. 296–300.

Jackson, D.

[a] *A comment on "differentials,"* Amer. Math. Mo., vol. 49 (1942), p. 389.

[b] *The human significance of mathematics*, Amer. Math. Mo., vol. 35 (1928), pp. 406–411.

Jones, F. B.

[a] *Connected and disconnected plane sets and the functional equation* $f(x) + f(y) = f(x + y)$, Bull. Amer. Math. Soc., vol. 48 (1942), pp. 115–120.

Keyser, C. J.

[a] *Charles Sanders Peirce as a pioneer*, Galois Lectures, Scripta Mathematica Library, no. 5, pp. 87–112; Scripta Mathematica, Yeshiva College, N. Y., 1941.

Kleene, S. C.

[a] *General recursive functions of natural numbers*, Math. Ann., vol. 112 (1936), pp. 727–742.

[b] *A note on recursive functions*, Bull. Amer. Math. Soc., vol. 42 (1936), pp. 544–546.

[c] *On the interpretation of intuitionistic number theory*, Journ. Symb. Logic, vol. 10 (1945), pp. 109–124.

[d] *Recursive predicates and quantifiers*, Trans. Amer. Math. Soc., vol. 53 (1943), pp. 41–73.

Klein, F.

[a] *A comparative review of recent researches in geometry*, Bull. N. Y. Math. Soc., vol. 2 (1892–3), pp. 215–249; transl. by M. W. Haskell.

Kneser, H.

[a] *Eine direkte Ableitung des Zornschen Lemmas aus dem Auswahlaxiom*, Math. Zeit., vol. 53 (1950), pp. 110–113.

Kolmogoroff, A.

[a] *Zur Deutung der intuitionistischen Logik*, Math. Zeit., vol. 35 (1932), pp. 58–65.

Kuratowski, C.

[a] *Sur la notion de l'ordre dans la Théorie des Ensembles*, Fund. Math., vol. 2 (1921), pp. 161–171.

Lewis, C. I.

[a] *Emch's calculus and strict implication*, Journ. Symb. Logic, vol. 1 (1936), pp. 77–86.

Lindenbaum, A.

[a] *Remarques sur une question de la méthode axiomatique*, Fund. Math., vol. 15 (1930), pp. 313–321.

Lusin, N.

[a] *Sur les ensembles analytiques nuls*, Fund. Math., vol. 25 (1935), pp. 109–131.

MacLane, S.

[a] *Symbolic logic*, Amer. Math. Mo., vol. 46 (1939), pp. 289–296.

McKinsey, J. C. C.

[a] *On the independence of undefined ideas*, Bull. Amer. Math. Soc., vol. 41 (1935), pp. 291–297.

Menger, K.

[a] *Der Intuitionismus*, Blätter für Deutsche Phil., vol. 4 (1930), pp. 311–325.

[b] *The new logic* (transl. by H. B. Gottlieb and J. K. Senior), Phil. of Sci., vol. 4 (1937), pp. 299–336.

Merton, R. R.

[a] *Priorities in scientific discovery: A chapter in the sociology of science*, Amer. Soc. Review, vol. 22 (1957), pp. 635–659.

[b] *Singletons and multiples in scientific discovery: A chapter in the sociology of science*, Proc. Amer. Phil. Soc., vol. 105 (1961), pp. 470–486.

Miller, E. W.

[a] *A note on Souslin's problem*, Amer. Journ. Math., vol. 65 (1943), pp. 673–678.

Miller, G. A.

[a] See collected works, Urbana, Univ. of Ill. Pr., vol. I, pp. 38–78, reprints from vols. 2 and 3 of Amer. Math. Mo.

Moore, E. H.

[a] *On the projective axioms of geometry*, Trans. Amer. Math. Soc., vol. 3 (1902), pp. 142–158.

Moore, R. L.

[a] *On the foundations of plane analysis situs*, Trans. Amer. Math. Soc., vol. 17 (1916), pp. 131–164.

[b] *A note concerning Veblen's axioms for geometry*, ibid., vol. 13 (1912), pp. 74–76.

Nagel, E.

[a] *The formation of modern conceptions of formal logic in the development of geometry*, Osiris, vol. 7 (1939), pp. 142–222.

von Neumann, J.

[a] *Eine Axiomatisierung der Mengenlehre*, Journ. für die reine u. angewandte Math., vol. 154 (1925), pp. 219–240. [Also see Berichtigung, ibid., vol. 155 (1926), p. 128.]

[b] *Die Axiomatisierung der Mengenlehre*, Math. Zeit., vol. 27 (1928), pp. 669–752.

Nicod, J.

[a] *A reduction in the number of the primitive propositions of logic*, Proc. Camb. Phil. Soc., vol. 19 (1916), pp. 32–42.

Niven, I.

[a] *Note on a paper by L. S. Johnston*, Amer. Math. Mo., vol. 55 (1948), p. 358.

Padoa, A.

[a] *Essai d'une théorie algébrique des nombres entiers, précédé d'une introduction logique à une théorie déductive quelconque*, Bibliothèque du Congrès International de Phil., vol. 3 (1900).

[b] *Un nouveau système irréductible de postulats pour l'algèbre*, C. R. Deuxieme Congres International des Mathematiciens, Paris, 1902, pp. 249–256.

[c] *Le problème No. 2 de M. David Hilbert*, L'Enseignement Math., vol. 5 (1903), pp. 85–91.

Peano, G.

[a] *Sui fondamenti della Geometria*, Rivista di Matematica, vol. 4 (1894), pp. 51–90.

Post, E. L.

[a] *Introduction to a general theory of elementary propositions*, Amer. Jour. Math., vol. 43 (1921), pp. 165–185.

[b] *Finite combinatory processes—formulation* I, Journ. Symb. Logic, vol. 1 (1936), pp. 103–105.

[c] *Computability and λ definability*, Journ. Symb. Logic, vol. 2 (1937), pp. 153–163.

Quine, W. V.

[a] *Element and number*, Journ. Symb. Logic, vol. 6 (1941), pp. 135–149.

[b] *New foundations for mathematical logic*, Amer. Math. Mo., vol. 44 (1937), pp. 70–80.

[c] *Paradox*, Sci. Amer. 206 (1962), pp. 84–96.

Rosenthal, A.

[a] *Über das dritte Hilbertsche Verknüpfungsaxiom*, Math. Ann., vol. 69 (1910), pp. 223–226.

[b] *Vereinfachungen des Hilbertschen Systems der Kongruenzaxiome*, Math. Ann., vol. 71 (1912), pp. 257–274.

[c] *Eine Bemerkung zu der Arbeit von Fräulin S. Weinlös: "Sur l'indépendance des axiomes de coïncidence et de la parallétité . . .,"* Fund. Math., vol. 13 (1929), pp. 304–306.

Rosser, B.

[a] *Extensions of some theorems of Gödel and Church*, Journ. Symb. Logic, vol. 1 (1936), pp. 87–91.

[b] *An informal exposition of proofs of Gödel's theorems and Church's theorem*, ibid., vol. 4 (1939), pp. 53–60.

[c] *The Burali-Forti paradox*, ibid., vol. 7 (1942), pp. 1–17.

[d] *On the many-valued logics*, Amer. Journ. Physics, vol. 9 (1941), pp. 207–212.

[e] Review of F. Bernstein [b], Journ. Symb. Logic, vol. 3 (1939), p. 86.

Russell, B.

[a] *Mathematical logic as based on the theory of types*, Amer. Journ. Math., vol. 30 (1908), pp. 222–262.

[b] *Whitehead and Principia Mathematica*, Mind, vol. 57 (1948), pp. 137–138.

Sheffer, H. M.

[a] *A set of five independent postulates for Boolean algebras*, Trans. Amer. Math. Soc., vol. 14 (1913), pp. 481–488.

Sierpinski, W.

[a] *Sur une hypothèse de M. Lusin*, Fund. Math., vol. 25 (1935), pp. 132–135.

[b] *L'hypothèse généralisée du continu et l'axiome du choix*, ibid., vol. 34 (1947), pp. 1–5.

[c] *Sur un problème de la théorie générale des ensembles équivalent au problème de Souslin*, ibid., vol. 35 (1948), pp. 165–174.

Skolem, T.

[a] *Einige Bemerkungen zur axiomatischen Begründung der Mengenlehre*, 5th Skand. Mat. Kong., Helsingfors, 1923, pp. 217–232.

[b] *Über die Grundlagendiskussionen der Mathematik*, 7th Skand. Mat. Kong., Oslo, 19–22 August 1929, Oslo, 1930, pp. 3–21.

Stone, M. H.

[a] *The theory of representations for Boolean algebras*, Trans. Amer. Math. Soc., vol. 40 (1936), pp. 37–111.

Szele, T.

[a] *On Zorn's lemma*, Publicationes Mathematicae, vol. 1 (1950), pp. 254–256; *erratum*, ibid., p. 257.

Tarski, A.

[a] *Sur les ensembles finis*, Fund. Math., vol. 6 (1924), pp. 45–95.

[b] *A general theorem concerning primitive notions of euclidean geometry*, Indag. Math., vol. 18 (1956), pp. 468–474.

Turing, A. M.

[a] *On computable numbers, with an application to the Entscheidungsproblem*, Proc. Lond. Math. Soc., ser. 2, vol. 42 (1937), pp. 230–265.

[b] *A correction* (to [a]), ibid., ser. 2, vol. 43 (1937), pp. 544–546.

Veblen, O.

[a] *A system of axioms for geometry*, Trans. Amer. Math. Soc., vol. 5 (1904), pp. 343–384.

Wang, H.

[a] *A new theory of element and number*, Journ. Symb. Logic, vol. 13 (1948), pp. 129–137.

[b] *A formal system of logic*, ibid., vol. 15 (1950), pp. 25–32.

Weinlös, S.

[a] *Sur l'indépendance des axiomes de coïncidence et de parallélitité dans un système des axiomes de la géométrie euclidienne à trois dimensions*, Fund. Math., vol. 11 (1928), pp. 206–221.

[b] *Remarques à propos de la note de M. Rosenthal: "Eine Bemerkung zu der Arbeit von Frl. Weinlös . . ., etc.,"* Fund. Math., vol. 15 (1930), pp. 310–312.

Wiener, N.

[a] *A simplification of the logic of relations*, Proc. Camb. Phil. Soc., vol. 17 (1912–14), pp. 387–390.

Weyl, H.

[a] *Mathematics and logic*, Amer. Math. Mo., vol. 53 (1946), pp. 2–13.

[b] *A half-century of mathematics*, ibid., vol. 58 (1951), pp. 523–553.

White, L. A.

[a] *The locus of mathematical reality; an anthropological footnote*, Phil. of Sci., vol. 14 (1947), pp. 289–303.

Wilder, R. L.

[a] *Concerning R. L. Moore's axioms Σ_1 for plane analysis situs*, Bull. Amer. Math. Soc., vol. 34 (1928), pp. 752–760.

[b] *The cultural basis of mathematics*, Proc. Int'l Cong. of Math'ns, Cambridge, Mass., 1950, vol. 1, pp. 258–271.

[c] *Axiomatics and the development of creative talent*, in Henkin, L., Suppes, P., and Tarski, A. [HST].

[d] *Topology: its nature and significance*, The Math. Teacher, vol. 55 (1962), pp. 462–475.

[e] *The origin and growth of mathematical concepts*, Bull. Amer. Math. Soc., vol. 59 (1953), pp. 423–448.

[f] *Mathematics: A cultural phenomenon*, in "Essays in the Science of Culture," ed. by G. E. Dole and R. L. Carneiro, New York, T. Y. Crowell, 1960.

Zermelo, E.

[a] *Beweis, dass jede Menge wohlgeordnet werden kann*, Math. Ann., vol. 59 (1904), pp. 514–516.

[b] *Neuer Beweis für die Möglichkeit einer Wohlordnung*, ibid., vol. 65 (1908), pp. 107–128.

[c] *Untersuchungen über die Grundlagen der Mengenlehre*, ibid., pp. 261–281.

Additional Bibliography
Books

Addison, J. W., Henkin, L. and Tarski, A., eds.,

[Ad] **The Theory of Models,** Amsterdam, North-Holland Publ. Co., 1965.

Brouwer, L. E. J.

[Br] **Collected Works,** 2 vols., ed. H. Freudenthal, Amsterdam, North-Holland Publ. Co., and N.Y., American Elsevier Publ. Co., 1976.

Christie, Dan E.

[Chr] **Basic Topology, A Developmental Course for Beginners,** N.Y., Macmillan Publ. Co., 1976.

Cohen, P. J.

[Co] **Set Theory and the Continuum Hypothesis,** N.Y., W. A. Benjamin, Inc., 1966.

Conference Board of the Mathematical Sciences

[Con] **The Role of Axiomatics and Problem Solving in Mathematics,** N.Y., Ginn and Co., 1966.

Dauben, J W.

[Da] **Georg Cantor: His Mathematics and Philosophy of the Infinite,** Cambridge, Mass., Harvard Univ. Pr., 1980.

Lakatos, I.

[La] **Proofs and Refutations,** Ed. J. Worral and E. Zahar, Cambridge Univ. Pr., 1976

Lyndon, R. C.

[Ly] **Notes on Logic,** N.Y., van Nostrand, 1966

Munkres, J. R.

[Mu] **Topology: A First Course,** Englewood Cliffs, N.J., Prentice-Hall, 1975

Robinson, A.

[Rob] **Non-Standard Analysis,** Amsterdam, North-Holland Publ. Co., 1966.

Rosser, J. B.

[Ross] **Simplified Independence Proofs, Boolean Valued Models of Set Theory,** N.Y., Academic Pr., 1969.

Thompson, E. S. W.

[Th] **Sociocultural Systems: An Introduction to the Structure of Contemporary Models,** Dubuque, Iowa, Wm. C. Brown Co., 1977.

Monk, J. D.

[a] **On the Foundation of Set Theory,** Amer. Math., Mo., vol. 77, (1970), pp. 703-711.

Shoenfield, J. R.

[a] **Open Sentences and the Induction Axiom,** Jour. of Symb. Logic, vol. 23 (1958), 7-12.

Index of Symbols

ε, 59
$\subset$, 59
$\cup$, $\bigcup$, 62
$\cap$, $\bigcap$, 62
$\{\}$, $\{|\}$, 61
$\emptyset$, 60
$\bar{\bar{A}}$, 104
c, 86ff
f, 106
$\aleph_0$, 86
$\aleph_1, \aleph_2, \cdots$, 131
A^S, 100
2^S, $2^{\bar{\bar{S}}}$, 106
$\approx$, 48
$(C, <)$, 47
$\bar{A}$, 116, 127
ω, $^*\omega$, 116
ω_1, 131
S/x, x/S, 121
Λ, 151
$[A, B]$, 118
$x \circ y$, 167
x^{-1}, 167
$(X, +, \times)$, 174
$\mathfrak{M}(I)$, 25
$\Rightarrow$, 52, 132
$\mu(\Sigma)$, 49
A, 83
E, 88
E^1, E^2, 89
F, 81
I, 89
N, 64
R, 84
R^+, 145
$\bar{R}^1$, 112
$\sim$, 29, 223
$\vee$, 62, 223, 257
$.$, 229
$\wedge$, 257
$\vdash$, 222
$\supset$, 60, 223, 257
$\exists$, 234
$\forall$, 234
$\hat{x}(\varphi x)$, 239
$\hat{x}\hat{y}(\varphi xy)$, 240
$\neg$, 257
$-\!\!3$, 231
$\Diamond$, 231
$/$, 230
$\rightarrow$, 108, 265

Index of Topics and Technical Terms

Aleph-null, 86
Alephs, 130
Algebra, VII
  of logic (Boole), 207-208
Algebraic numbers, 83
  denumerability of, 83
  Kronecker's acceptance of, 202
Algorithm, 94, 232
Analysis, mathematical, 158
  beginnings of, 199ff
  intuitionist, 261ff
Analysis situs, 184
Antinomy, *see* Contradiction
Arabs, transmission of mathematics to Spain by, 285
Arithmetic, Frege's, 208
  of real numbers, 155
Art and mathematics, 290ff
Association for Symbolic Logic, 58
Associative law, of groups, 167
  of real number system, 156
Ausdehnungslehre, Grassmann's, 6
Axiomatic method, I, II
  advantages of, 43
  disadvantages of, 45
  economy of effort achieved by, 14, 38
  evolution of, 3
Axiom systems, 11, 19
  categoricalness of, 36
  complete independence of, 30
  completeness of, 33
  consistency of, 23, 43
  logical basis of, 38
  monotransformable, 42
  satisfiable, 26
Axioms, I, II
  and postulates, 3ff
  conditional, 25
  definition of, 19
Axioms, I, II
  denial of, 29
  for geometry, 4
  for propositional calculus, 224ff
  for predicate calculus, 236
  Hausdorff, 190
  independence of, 29
  negation symbol for, 29
  selection of, 18ff
  source of, 18
  vacuous satisfaction of, 25

Bernstein equivalence theorem, 109, 254
Binary fractions, 99
Binary number system, 99
Binary relation, 47, 166, 216; *see also* Relations
Bolzano theorem, intuitionist attitude toward, 262
Boolean algebra, 206-208
Borel theorem, 89

Calculus, 199ff
  of logic, 206, IX, XI
  of propositions, 225
  functional, 235
  of propositional functions, 235
Cantor axiom, 88f, 150
Cardinal number of the continuum, c, 86ff
  proof that set of all irrational numbers has, 89ff
  proof that set of all points in plane has, 89
  proof that set of all transcendental numbers has, 90
Cardinal numbers, 101ff
  addition of, 146f
  Frege-Russell definition of, 102
  intuitionist concept of, 252

Cardinal numbers, 101ff
norms for, 102ff
on basis of Zermelo system, 216
ordering of, 105
representative sets for, 105
to have the same, 86
transfinite, 107
Cartesian product, 166
Categoricalness, 36
Characterization, 38
Chinese mathematics, 267ff
Choice Axiom, 72ff, 79, 108, 215, 238f
comment on, 74ff, 122
independence of in set theory, 133, 238f, 244
Choice Principle, 73
Class, as defined in P.M., 238; *see also* Set, Sets
Class decomposition (corresponding to an equivalence relation), 49
Cofinal, 118
Collection, 14
empty, 15, 60, 63
Comparability, 132
Completeness, desirability of, 37
of an axiom system, 32, 52 (Problem 21)
of an axiom system relative to its undefined terms, 42
of Hilbert's geometry, 37
of predicate calculus, 236ff
of propositional calculus, 233
Commutative laws of real number system, 156
Complex number system, foundation of, on Peano axioms, 162
Computability, 276
Concept underlying an axiom system, 18ff
Consistency, of a formal system, 265, 268, 274
of an axiom system, 19, 23
of arithmetic, 269
ω-, 273
proof of, 26
relative, 28
Constructive procedures, 91ff
Continuum, *see* Linear continuum
Continuum hypothesis, 132, 238f
generalized, 132
independence of, in set theory, 133, 238f, 244
Continuum problem, 132
Contradiction, 56, 210
Burali-Forti, 56, 130
Grelling, 77, 241
in a formal system, 265
lack of in intuitionist mathematics, 262
Richard, 113, 241
Russell, 54, 77
set of all sets, 113 (Problem 28)
set of all singletons, 113 (Problem 29)
Contradiction, Law of, 26, 39
intuitionist attitude toward, 257
Coset, 170
right, 170f
Countable, 87
effectively, 95
intuitionist definition of, 255
Cut, 118
Dedekind, 118f, 151
half-, 160
of the real number system, 147

Decimals, finite, 146
density of, in real number system, 146
Decision problem, 275
Decision procedure, 275
Definition, 10, 54
by induction, 125
effective, or constructive, 75, 94ff, 203
nature of, in axiomatic systems, 40ff
non-predicative, 211
primitive recursive, 267
recursive, 267
Density, 146
Denumerable, 87
effectively, 95
intuitionist definition of, 254
Derivative, 199ff
Derived set (or derivative) of a point set, 130f
Diagonal procedure, 85, 91ff
general, 97
use in constructing irrational numbers, 91
use in constructing transcendental numbers, 97
Differential, 199
Diffusion of mathematical ideas, 293ff
Distance function, 183
Distributive laws of real number system, 157
Division ($\div$), 179

Domain of integrity, 178
Double negative, 227
  relation to Law of Excluded Middle, 258

Effective calculability, 267f, 276
Effective procedures, *see* Constructive procedures
Equivalence relation, 48
  between species, 254
Equivalence of Choice Axiom, Well-Ordering Theorem and Comparability, 135ff
Euclid's algorithm, 275
Euclid's *Elements*, 4, 142
Euclid's fifth postulate ("parallel postulate"), 5
Euler's polyhedral formula, 186
Evolution in mathematics, 293 ff
Excluded Middle, Law of, 26, 39, 226, 243, 246, 258
  equivalence to Law of Contradiction in P.M., 229
  intuitionist attitude toward, 249, 257, 259
  use of relative to infinite sets, 265
Exist, to (as used in an axiom), 18
Existence, mathematical, 202ff
  cultural view of, 298

Factor group, 170
Falsehood, as used in logic, 222
Fashions in mathematics, 297
Fermat's "last theorem," 269
Field, 178
  non-commutative, 178
  ordered, 181
Finite, Brouwer definition of, 250
  Dedekind, 66
  ordinary, 64ff
  species, 253
First element of a simply ordered set, 117f
Fluxions, 200
Formal systems, XI
Formalism, XI
Formula, logical, 225, 235
  undecidable, 270
Frege-Russell definition of number, 102, 209
Frege-Russell thesis, 219
  degree of general acceptance of, 243
Function, 34
  analytic, 200
  domain and range of, 34
  inverse, 34
  number-theoretic, 267
  recursive, 267
  with no derivative, 200
Fundamental Theorem of Algebra, 203
  intuitionist proofs of, 260

General recursiveness, 276
Geometry, affine, 187
  analytic, 201
  elementary school, 3, 9f
  equiform, 187
  Euclid's, 4
  Greek, 287
  Hilbert's, 7, 30, 36, 38
  in the sense of Klein, 183
  intuitionist, 262
  non-euclidean, 5
  on the basis of P.M., 242
  Pasch's, 7
  Pieri's, 8
  plane, 9
  projective, 187
  Veblen's, 8, 41
Gödel incompleteness theorem, 270
Gödel number, 271ff
Greek mathematics, 284ff, 287
Group, 167
  affine, 168
  of integers mod m, 170
  of non-singular square matrices, 168
  of nth roots of unity, 167
  of rigid motions, 187
  of rotations of a circle, 168
  order of, 167
  projective, 187
  symmetric, of degree n, 169
Groups, VII
  abelian (= commutative), 167
  cancellation law for, 171
  factor, 170
  isomorphic, 172
  order of, 167
  semi-, 173
  sub-, 169
  substitution (= permutation), 168, 173
Grundlagen der Mathematik, 266

Hamel basis for real numbers, 125ff
Hilbert-Bernays System "Z," 269
Homeomorphism, 188

Ideal, 177
Identity correspondence, 35
Implication, 17
  intuitionist, 257
  material, 223
  strict, 231
Independence of axioms, 28
  complete, 30
Infinite, Dedekind, 66
  intuitionist definition of, 253
  ordinary, 65
Integers, foundation of, on Peano axioms, 158ff
  mod *m*, 170
  as topological space, 192
Integral domain, 178
Interpretation of an axiom system, 11, 24
  vacuous, 25
Interval, 154
  closed, 154
  open, 154
Intuitionism, X
  Kronecker's, 201-204
Invariant (geometric), 185
  topological, 189
Irrational numbers, order type of, 164 (Problem 17)
Isomorphic, 35
Isomorphism, 35
  as an equivalence relation, 49
Italian school of logic, 8, 209, 220

Japan, mathematics of, in 17th and 18th centuries, 284
Jesuits, influence on early American mathematics by, 294
Jordan Curve Theorem, 260f
  Brouwer proof of, 260

Klein's Erlanger Program, 186f

Lambda-definability, 276
Language, and Intuitionism, 247
  contrast of natural and formal, 281
  universality of mathematical, 285
Last element of a simply ordered set, 117f
Law of the excluded third, *see* Excluded Middle, Law of
Laws of Signs, for a general number system, 175
  for a real number system, 155, 157, 182
  for a system, 175
Limit, 199ff
Limit point, 184
Limits of ordinal numbers, 134, 185
Line, euclidean, order type of, 150
Linear continuum, 151
  axioms for, 151
  Brouwer conception of, 261
Linear order = simple order (q.v.)
Logic and Intuitionism, 248
  Aristotelian, 26, 39, 45, 72
  as a basis for mathematics, 208, IX, XI
  Boole's algebra of, 206-208
  calculus of, 205ff
  intuitionist, 256ff
  Kronecker's viewpoint concerning, 204
  use of, in axiomatic method, 26, 38ff
Logical deduction, 17, 225
Logical product, 229
"Logistic thesis," 208, 219
  Poincaré's views concerning, 212
Löwenheim-Skolem theorem, 237

Magnitudes, 142
Mapping, 100
Mathematical induction principle, 67
  definition by, 68
  Poincaré's views concerning, 212
  proof by, 68
Mathematical Logic, IX, XI
Metalanguage, 243
Metamathematics, 265
Metric, *see* Distance function
Model, 24
  of one axiom system within another, 28
Modus ponens, 225
Monotonic law of real number system, 157

National features of mathematics, 286
Natural numbers, 46
  as basis of intuitionist mathematics, 247, 249
  Kronecker's viewpoint concerning, 202
  natural order of, 64f
  ordering of in ω^2-sequence, 128
  Peano axioms for, 158

Negation, as used in logic, 223
 Double, 227
 Intuitionist theory of, 258
Neighborhood, 184
Newton's *Principia*, 4
Non-euclidean geometries, 5
Normal subgroups, 169

(1-1)-correspondence, 34
Operations, 166
 tables of, 172
  for a field, 179
Or = and/or, 47f, 148f
Order type, 115ff
 addition of, 117
 have the same, 115
 ω, axiomatic definitions of, 117ff, 120
 ω, $^*\omega$, 116
 ω_1, 131, 133
  axiomatic definition of, 134
Ordered *n*-tuple, 114, 116
Ordered pair, 216
Ordinal numbers, 127
 addition of, 117
 classes of, 131
 intuitionist concept of, 256
 of the first and second kind, 135
 ordering of, 129
 product of, 128
 second class of, 133
Ordinal numerals, 116
Osiris, 6f, 7f, 8f

P.M. (= *Principia Mathematica*), IX
Parallel postulate, 5, 10
Partial order, 51 (Problem 16)
Peano axioms for natural number system, 158
 as basis for real numbers, 158ff
Peano school, 209
Pi (π), decimal value of, 251
 existence of, 203
 transcendence of, 91, 203
Point, 185
Point set, 185
 closed, 190
 connected, 192
 open, 190
Postulates, *see* Axioms
Predicate calculus, 234ff
Present-day mathematics, general character of, 285ff
Primitive terms (of an axiom system), 18
Principia Mathematica, IX
Product set, 166
Proof, as used in axiomatic method, 13ff
 formal, 265
 Kronecker's concept of, 203
 metamathematical, 266
 "reductio ad absurdum," 9, 39, 203, 227
Proof theory (Hilbert), 264ff
Proposition, 221
Propositional calculus, 224ff
 intuitionist, 257ff
Propositional function, 222

Quantifiers, 234
Quaternions, 163f

Rational numbers, 81
 denumerability of, 82
 foundation of, on Peano axioms, 159ff
Real numbers, 84
 axiomatic definition of order type of, 150
 concept of, 143
 density property of, 146
 foundation of, in axiomatic form, 181
  on Peano axioms, 158ff
 Kronecker's viewpoint concerning, 202
 operations (addition, multiplication, etc.) with, 154ff
 ordering of, 144ff
 separability of, 145ff
 uncountability of set of all, 84
Recursive definition, 267
Reducibility Axiom, 239ff
Reductio ad absurdum, 9, 203, 227; *see also* Proof
Reflexive, reflexiveness, 48
Relations, 216, 238, 240
 binary, 216
 ternary, *n*-ary, 218 (Problem 3)
Ring, 175ff
 commutative, 176
 of integrity, *see* Integral domain
 quotient, 177
 with a unit, 176
"Russell set," 57

Satisfiable, 26
Section (of a simply ordered set), 121
Separability, 146
  axiom, for linear continuum, 151
  in the ordered real number field, 181fn
Set, 55
  Brouwer's definition of, 249ff
  Cantor's definition of, 55
  finite, *see* Finite
  inductive, 67
  infinite, *see* Infinite
  null, 60
  spanning, 118
  symbol for definition of, 60
  synonyms for, 55
Set of all sets, self-contradictory nature of, 113 (Problem 28)
Sets, difference of, 62
  disjoint, 63
  intersection of, 62
  intuitionist theory of, 248
  non-degenerate, 63
  non-empty (non-vacuous), 63
  operations with, 61
  relations between, 59ff
  symbols for, 59
  theory of, III-V
  union of, 62
  Zermelo theory of, 212ff
Sheffer "stroke" symbol, 230
Simple order, axioms for, 47
Skolem's paradox, 237ff
Social sciences,mathematics and, 297
Souslin's problem, 154
Space, 183
  topological, 184, 190
Span, 118
Species, 251ff
  relations between, 252ff
Spread, 250
  -law, 250
Statement, I-Σ-, 27
  Σ-, 17
Styles and fashions in mathematics, 297
Subset, 59
  proper, 60
Substitution, in logic, 225
Successor, 66
  immediate, 120
Symbolic logic, 58, 205ff, IX, XI
  intuitionist, 257ff
Symbolism, dependence of mathematics on, 294
Symmetric, symmetry of a binary relation, 48
System (X, +, ×), 174

Tautology, 225, 232
Technical terms, 18
Tertium non datur, see Excluded Middle, Law of
Theorems, 10
Topology, 89f, 188ff
  base for, 191
  Brouwer's work in, 260
  order, 185
Transcendental numbers, 90
  construction of, 97
  existence of, 90
  order type of, 164 (Problem 17)
Transfinite induction principle, 123
  definition by, 125
  proof by, 124
Transfinite numbers, 55, 202, 204
Transformation, 183
  affine, 187
  group, 183
  product of, 183
  projective, 187
  similarity, 187
  topological, 188
Transitivity, of a binary relation, 48
Truth, as used in axiom systems, 20
Truth, as used in logic, 222
Truth-tables, 232ff
Types, theory of, 239
  ramified, 241
  simple, 240

Uncountable, 87
Undecidable theories, 277
Undefined terms, 9
  independence and completeness of, 39ff
  logical, 11
  technical, 11, 18
Universal terms, 11, 18

Variable, bound (apparent), 234
  free,.234
  in logic, 222, 234
Vector, 180
  linear independence of, 180

Vector spaces, 179ff
  base for, 180
  dimension of, 181
  over rings, 180f

Well-Ordered, 119
Well-Ordering, 119
  minimal, 140 (Problem 21)
  type, 119
Well-Ordering Theorem, 121, 122, 210
  intuitionist attitude toward, 256
  proof of, 137ff

Zermelo Postulate; this term is frequently employed to denote the Choice Axiom (q.v.)
Zermelo theory of sets, 212ff
Zorn's Lemma, 138, 181f
Zero, as a number, 63

Index of Names

Abel, N. H., 200
Ackermann, W., 217, 235, 236, 237, 242, 266
Albert, A. A., 177, 181f, 193
Archimedes, 4
Aristotle, 3, 4, 11, 205
Augustine, St., 202f

Bachmann, F., 162
Baire, R., 212
Ball, W. W. R., 289
Banach, S., 109f
Bar-Hillel, Y., 217
Baumgartner, L., 173f, 193
Bell, E. T., xii, 7f, 21, 43f, 65f, 193, 202f, 217, 289, 292
Benacerraf, P., 282
Berkeley, G., 200
Bernays, P., 37, 50, 208, 217, 218, 225f, 231, 258, 266, 268, 269, 277
Bernoulli, Johann, 199
Bernstein, F., 56f, 109, 130f, 132f, 147, 254
Beth, E. W., 43, 72f, 139, 247f, 262, 267f, 279, 282
Birkhoff, G., 50, 139, 181f, 182, 193
Birkhoff, G. D., 88f
Black, M., 244, 282
Blichfeldt, H. F., 173f
Bolyai, J., 5
Bolzano, B., 65f, 262
Boole, G., 206-208, 219, 289
Borel, E., 89, 131, 136f, 212
Bourbaki, N., 87
Brouwer, L. E. J., 189, 246ff, 266, 279
Burali-Forti, C., 56, 130, 197

Cajori, F., 289
Campbell, A. D., 193
Cantor, G., 55, 56, 65f, 76, 80, 83, 85, 86, 91, 98, 108, 109f, 116, 130, 132, 150, 160, 201, 202, 204, 205, 211, 219, 220, 221f, 298
Carnap, R., 241, 242, 279
Cauchy, A., 126, 200, 251, 278
Cavaillès, J., 66
Childe, V. G., 294
Church, A., 58, 240, 242, 267f, 276, 277, 279
Chwistek, L., 239, 240
Cohen, P., 133
Courant, R., 128f, 188
Couturat, L., 203
Curry, H. B., 231, 262, 270f, 279

D'Alembert, J. L., 200
Darwin, C. R., 293
Davis, M., 267f, 276
Dedekind, R., 65ff, 118f, 151, 158, 160, 201, 202, 218, 219, 256, 278
Dehn, M., 21
De Morgan, A., 206, 219
Dewey, John, 36f
Dickson, L. E., 173f
Dresden, A., 262
Dushnik, B., 154f

Eaton, R. M., 244
Einstein, A., 294
Emch, A. F., 231
Euclid, 4, 39, 141, 191, 275
Euler, L., 186, 199
Everett, E., 292
Eves, H., 21, 163

Faber, G., 83
Fermat, P., 269
Fraenkel, A., 80f, 111, 139, 213f, 214, 215f, 216, 217, 218, 238, 278, 282
Frege, G., 102, 208, 209, 210, 219, 221,

Frege, G., 243, 253, 278
Furstenberg, H., 192

Galileo, 65f
Gauss, C. F., 5, 184, 200, 293
Gentzen, G., 269, 279
Ginsburg, J., 294f
Glivenko, V., 258
Gödel, K., 28f, 132f, 133, 139, 217, 237 238, 258, 259, 263, 267, 269, 270 271, 272, 273, 274, 275, 276, 279, 280
Golomb, S., 192f
Goodstein, R. L., 279
Grassman, H., 6
Grelling, K., 57, 76, 77, 111, 139, 241
Grzegorczyk, 268f

Hadamard, J., 212
Hall, D. W., 192
Halldén, S., 231
Halmos, P., 139, 215f
Hamel, G., 125, 126, 138, 139
Harary, F., 31f
Hardy, G. H., 292
Hartogs, F., 136f
Hausdorff, F., 190
Heath, T. L., 3f, 5f
Helmer, O., 31
Henkin, L., xii, 159f, 270f
Henle, P., xii, 225f
Herbrand, J., 259, 276
Hermite, C., 91
Heyting, A., 247, 248, 250, 254f, 256, 257, 260f, 262, 263, 269
Hilbert, D., 7, 8, 9f, 21, 30, 36, 37, 38, 50, 88f, 89f, 130f, 132f, 208, 231, 235, 236, 237, 242, 246, 264, 266, 268 269, 270, 277, 278, 279
Hobson, E. W., 76, 111, 130f, 139, 162, 163
Hohn, F. E., 207
Huntington, E. V., 36f, 47, 50, 75f, 139 147, 150f, 163, 167f

Jackson, D., 293
Jeans, J. H., 292
Jevons, W. S., 207, 219
Jones, F. B., 126f
Jourdain, P. E. B., 55f

Kac, M., 190f, 234f
Kamke, E., 76, 111, 139
Kant, I., 247
Kattsoff, L. O., 282
Kelley, J. L., 217
Kerschner, R. B., 21, 193
Kleene, S. C., 247f, 259, 262, 267f, 276, 277, 279
Klein, F., 45, 155f, 162, 163f, 186, 188, 189
Kneser, H., 132
König, D., 109f
Kolmogoroff, A., 257, 258, 259
Kowalewski, S., 202f
Kroeber, A. L., 297
Kronecker, L., 201ff, 203, 205, 210, 211, 219, 244, 246, 247, 248, 260, 288, 293, 298
Kummer, E. E., 201f
Kuratowski, C., 216

Lagrange, J. L., 4, 199
Landau, E. G. H., 159f, 162
Langford, C. H., 231
Lebesgue, H., 212
Lefschetz, S., 181f
Leibniz, G. W., 199, 206, 208, 219, 292
Levi, Beppo, 72f
Lewis, C. I., 217, 231
Lincoln, A., 292
Lindemann, F., 91, 203, 288
Lindenbaum, A., 30f, 133f, 203
Linton, R., 294
Lobachevski, N. I., 5
Löwenheim, L., 237, 238
Lukasiewicz, J., 232
Lusin, N., 131, 132f

MacLane, S., 181f, 193, 244, 279
McCoy, N. H., 193
McKinsey, J. C. C., 43
Matthewson, L. C., 173f
Maziarz, E. A., 282
Menger, K., 212, 217, 262
Meray, C., 160
Merton, R. K., 297
Miller, E. W., 154f
Miller, G. A., 173f
Montucla, J. E., 289
Moore, C. N., 297f
Moore, E. H., 30
Moore, R. L., 8, 30f

Mostowski, A., 275, 277, 279

Nagel, E., 6, 7f, 8, 21, 270f
Neumann, J. von, 217, 218, 278, 280
Newman, J., 270f
Newsom, C. V., xii, 21, 163
Newton I., 4, 199, 200, 293
Nicod, J., 230
Niven, I., 83
Northrup, E. P., 21, 193

Obreanu, P., 83f, 112f

Padoa, A., 40, 41, 43
Pasch, M., 7, 8, 13, 21, 162, 221f
Peano, G., 7, 8, 45, 48, 67, 72f, 76, 144, 158, 161, 162, 181, 209, 219, 220, 221, 249, 264, 266, 268, 269, 277, 278
Peirce, C. S., 207, 208
Pieri, M., 8, 221f
Pierpont, J., 162, 163
Plato, 4, 292
Poincaré, H., 21, 189, 210, 211ff, 247, 278
Poncelet, V., 186
Post, E. L., 232, 276
Proclus, 3f, 5
Putnam, H., 282

Quine, W. V., 230f, 231, 242

Ramsey, F. P., 239, 240, 241
Richard, J., 113, 241
Richardson, M., 21, 28, 50, 111
Riemann, B., 5, 186, 189, 200f, 293
Robbins, H., 128f, 188
Robinson, G. de B., 8
Robinson, R. M., 275, 277, 279
Rosenbloom, P. C., 208, 244, 279
Rosenthal, A., 30f
Rosser, J. B., 132f, 139, 270f, 273f, 277, 279
Russell, B., 45, 56ff, 75, 77, 102, 162, 197, 208, 209, 210, 212, 219, 221, 225f, 229, 239, 240, 242, 243, 244, 253, 264, 266, 278, 279, 281

Sarton, G., 285, 293f, 298f
Schaaf, W. L., 285, 291, 293f, 298f
Schröder, E., 109f, 207, 208
Shaw, J. B., 290-291
Sheffer, H. M., 230
Sierpinski, W., 74f, 75, 76, 87, 91, 93f, 102, 109f, 111, 117, 118f, 128f, 131, 132f, 133, 137f, 139, 150f, 154f, 190
Skolem, T., 217, 237, 238
Smith, D. E., 294f
Souslin, M., 154
Spencer, G. L., 192
Spengler, O., 293, 296
Staudt, K. G. C. von, 186, 221f
Stevin, S., 142
Stoll, R. R., 163, 208, 217, 279
Stone, M. H., 208
Struik, D. J., 200f, 202f, 206f, 217, 292, 293f
Sturm, J. C. F., 97
Sullivan, J. W. N., 290
Suppes, P., 76, 216f, 217
Szele, T., 132

Tarski, A., 41, 42, 50, 66, 77, 133f, 231, 275, 277, 279
Turing, A. M., 267f, 276

Veblen, O., 8, 21, 36f, 41, 50, 52f, 188f, 189, 221f, 242

Waerden, B. L., Van der, 177, 180f, 182, 193
Weierstrass, K., 160, 200-201, 204, 205, 219, 293
Weinlös, S., 30f
Weyl, H., 193, 242, 262, 270f, 282, 297f
White, L. A., 290f
Whitehead, A. N., 209, 219, 221, 229, 239, 242, 264
Whitehead, J. H. C., 188f
Wiener, N., 216
Wilcox, L. R., 21, 193
Wilder, R. L., 44, 102, 103, 189, 290f, 299
Wittgenstein, L., 239
Woodger, J. H., 19f, 279, 291

Young, J. W., x, 8f, 13f, 21, 37f, 50, 76, 77, 78, 89f, 108f, 111, 151f, 162, 163, 182f
Young, J. W. A., 8f

Zeno, 4
Zermelo, E., 72, 121, 122, 133, 136f, 138, 210, 212, 213, 214f, 216f, 217, 218, 238, 256, 278, 280
Zippin, L., xii
Zorn, M., 138, 139, 181f